# Electro-Optics and Dielectrics of Macromolecules and Colloids

# Electro-Optics and Dielectrics of Macromolecules and Colloids

Edited by

Barry R. Jennings
*Brunel University*
*Uxbridge, England*

PLENUM PRESS · NEW YORK AND LONDON

Library of Congress Cataloging in Publication Data

Main entry under title:

Electro-optics and dielectrics of macromolecules and colloids.

Includes index.

1. Colloids—Electric properties—Congresses. 2. Colloids—Optical properties—Congresses. 3. Macromolecules—Electric properties—Congresses. 4. Macromolecules—Optical properties—Congresses. I. Jennings, Barry Randall, 1939-

QD549.E44 547'.7 79-10117

ISBN-13: 978-1-4684-3499-6 e-ISBN-13: 978-1-4684-3497-2

DOI: 10.1007/978-1-4684-3497-2

Proceedings of the Conference on Electro-Optic and Dielectric Studies on Macromolecules and Colloids, held at Brunel University, Uxbridge, United Kingdom, April 12–14, 1978.


Softcover reprint of the hardcover 1st edition 1979
A Division of Plenum Publishing Corporation
227 West 17th Street, New York, N.Y. 10011

'For any man with half an eye,
what stands before him may espy;
But optics sharp it needs, I ween,
to see what is not to be seen.'

JOHN TRUMBULL (1750-1831) McFingal

# Preface

'The time has come', the Walrus said, 'to talk of many things:'
Lewis Carroll (1832-1898) The Walrus and the Carpenter

In 1875, a scientific paper appeared in the Philosophical Magazine under the title 'A new relation between electricity and light'. Its author was John Kerr, a 51 year old ordained minister of the Free Church of Scotland who, after ordination, seems to have applied himself to the teaching and study of the natural philosophy to which he had partially applied his mind during his university career at Glasgow University. In those early days he had worked in what was the first physical laboratory in a British university. Its founder and director at the time was Professor William Thomson (Lord Kelvin).

At the age of 33, John Kerr was appointed a lecturer in mathematics at the Glasgow Free Church Training College. He held this position for some 44 years. During much of this period, he spent his spare hours searching for an effect that would demonstrate the interrelation between electrical and optical phenomena. In this respect, he followed in the train of the renowned Michael Faraday. Using nothing better than a paraffin candle or sunlight reflected from a bright cloud as his light source, together with the potential difference generated across the secondary terminals of a Ruhmkorff induction machine as the origin of his electric field, Kerr demonstrated the phenomenon of 'electric birefringence' in glass and liquids. Thus was born the Kerr electro-optic effect. Briefly stated, it is the ability of an apparently isotropic medium to act as a uniaxial, anisotropic system when an electric field is applied across it. The medium is viewed between crossed polarising devices in a direction perpendicular to the electric field, application of which is accompanied by the transmission of light. Since its discovery, the effect has been used in a host of devices of which the best known is the electro-optic shutter. This has been adapted for the measurement of the velocity of light, for high speed photography and, more recently, as an optical modulator in laser technology.

Some seven years before Kerr's death, Larmor proposed that electric birefringence had its origin in the orientation of anisotropic molecules or elements within the apparently isotropic medium. The theory for this concept was formulated by Langevin. During the next half century, occasional measurements were made both to characterise the phenomenon and to evaluate the relevant physico-chemical parameters of pure liquids and molecular fluids.

During the 1930-40 era, Staudinger and others demonstrated the existence in nature of giant molecules and colloidal particles. Since that time it has slowly but increasingly been realised that these big molecules or particles often have relatively large dipole moments, are generally anisotropic in structure and hence, in solution or suspension, give rise to significant electric birefringence signals. Furthermore, there have been three electronic innovations which have greatly eased the experimental measurement of the effect for such materials. These were the development of photomultiplier tubes for detection, of oscilloscopes for display and of high voltage generators developing bursts or pulses of potential difference. The last mentioned enable the experimenter to study the Kerr effect not only for its amplitude but also in the time domain. The rates of molecular response to the switching of the electric field lead directly to information on the size and geometry of the constituent molecules and particles in a dilute solution or suspension. It is the combination of electrical, optical and geometrical information on a sample from a single, fast experimental method that has resulted in the great escalation of interest in electro-optical studies of macromolecules and colloids during the past twenty-five years.

It would be misleading and very restrictive to give the reader the impression that the Kerr-Effect is the only electro-optical phenomenon. One of the most rewarding and interesting aspects of current studies is the way that younger researchers have been active in predicting, manifesting, measuring and utilising a host of novel electro-optical phenomena for macromolecular characterisation. In response to pulsed applied fields, changes in the scattered intensity, the fluctuations in the scattered light, the absorption, linear dichroism, fluorescence, optical rotation and circular dichroism have been measured. They all have their origins in the structural anisotropy of the molecules of the medium studied. It is not however, a refractive index anisotropy as in the case of birefringence. Rather, each is a manifestation of a different anisotropic optical and hence structural characteristic of the molecules. It is this factor which promises so much for future investigations. Furthermore, recent work in this laboratory has shown that the oscillating electric vector associated with high powered laser beams is itself capable of inducing electro-optical effects in macromolecules. These are sometimes called optico-optical effects. They herald yet further advances in the study of high frequency polarisation processes and, because of the extremely short duration of switched laser pulses, offer the possibility of studying the rotation rates of ever smaller molecules.

To date electro-optical methods have already been used to study a wide variety of material classes. These include synthetic polymers, polyelectrolytes, polypeptides, polynucleotides, nucleic acids, micro-organisms, cells, bacteria, vesicles, nerve axons, proteins, viruses, membranes, lipids, liquid crystals, clay minerals, colloids and surfactants.

It was against this historical background and in the face of future promise, that the Electro-Optics Group of Brunel University staged an International conference in April 1978. Such is the enthusiasm of the ever-expanding community of researchers using electro-optic methods for the study of macromolecules and colloids, that the participants came from all over the world. No single country, not even that of the hosts, dominated the meeting. The papers and discussion covered such a diversity of experimental methods, classes of materials, analytical procedures and future proposals that the material of the meeting well represented the current state-of-the-art of macromolecular and colloidal electro-optical studies. For this reason, it was decided that these proceedings be published. All but two of the works presented at the conference are contained herein. It is hoped that, for those who attended the meeting, this book will both remind them of a memorable occasion and provide stimuli for future developments. The more important function however, is to disseminate results on the materials studied and to introduce the wider community to electro-optic methods and the ways that these can be utilised to evaluate molecular information on macromolecules and colloids of biological and synthetic origin.

In editing these proceedings I should like to thank the following members of the Brunel Electro-Optics Group for helping with details of the conference and for proof reading the present text; K. Ballinger, R. Fairey, E.Y. Hawkins, P. Hammond, P. Kolinsky, V.J. Morris, D. Oakley and P.J. Ridler. Special thanks are extended to A.R. Foweraker who co-organised the conference. Finally, thanks are expressed to Mrs. S. Dunster who, over many months, has laboriously but accurately retyped all the manuscripts and texts.

Brunel University
Uxbridge

B. R. JENNINGS

# Contents

'Knowledge is of two kinds. We know a subject ourselves, or we know where we can find information upon it.'
Samuel Johnson (1709-1784) Letter to Lord Chesterfield

THEORETICAL ADVANCES

## ABSORPTION PHENOMENA



## NUCLEIC ACIDS AND POLYNUCLEOTIDES

## BIOLOGICAL SYSTEMS



## POLYELECTROLYTES AND POLYMERS

## COLLOIDAL SYSTEMS



## LASER AND HIGH FIELD EFFECTS

# *Theoretical Advances*

'It is certainly not the least charm of a theory that it is refutable.'

FREDERICH W. NIETZSCHE (1844-1900)

Beyond good and evil

# SPECIFIC KERR CONSTANTS OF RIGID, ELLIPSOIDAL MACROMOLECULES IN CONDUCTING SOLUTION AT VERY LOW IONIC STRENGTH

Sonja Krause, Bina Zvilichovsky and Mary E. Galvin

Department of Chemistry, Rensselaer Polytechnic Institute, Troy, N.Y., U.S.A.

The theoretical treatment of the Kerr Constant of rigid, conducting, dipolar macromolecules of O'Konski and Krause (1970), has been extended to very low ionic strength solutions. The O'Konski and Krause theoretical treatment postulated surface conductivity directly along the surface of each macromolecule; for charged macromolecules, this surface conductivity was generally assumed to be caused by movement of the counterions of the macromolecule. In the present work, it has been assumed that, at very low ionic strength, the average counterion is at the Debye characteristic distance from the surface of each charged macromolecule and contributes to surface conductivity at that distance, not directly at the surface of the macromolecule.

## INTRODUCTION

When a D.C. electric field is imposed on a solution of macromolecules, the molecules of both solute and solvent tend to orient in the field if they are anisometric and/or electrically anisotropic. The extent of this orientation can be studied by many methods, one of which involves the determination of the birefringence of the solution. If the solution is dilute, one may assume that the birefringence contributed by solute and solvent is additive, i.e., that the birefringence contributed by the solute can be calculated by subtracting the birefringence of pure solvent at the same electric field from the birefringence of the solution.

When the solution of macromolecules conducts an electrical current, these electric birefringence experiments cannot be done using D.C. electric fields because of the inevitable Joule heating of the solution that would occur. Instead, a square pulsed electric field may be employed. Such square pulsed electric fields allow the observation of transient effects associated with the rotation of the macromolecules to their steady-state alignment after the field is turned on, they allow observation of the steady-state alignment of the macromolecules, and they allow observation of the transient effects associated with the rotational diffusion of the macromolecules back to random orientation after the field is turned off. This paper deals only with the steady-state orientation of the macromolecules in an electric field of magnitude E as observed by their birefringence $\Delta n$, defined as the difference between the refractive indices parallel and perpendicular to the electric field direction.

The specific Kerr Constant, $K_{sp}$, of a macromolecule in solution is obtained from the steady-state birefringence at low electric fields:

$$K_{sp} = \frac{\Delta n}{nE^2 C_v} \qquad (1)$$

where n is the refractive index of the solution and $C_v$ is the volume fraction of macromolecules in the solution. At higher electric fields, the birefringence is not proportional to $E^2$ so that eq. 1 does not yield a constant value for $K_{sp}$. In this paper, we shall deal only with electric fields that are low enough to yield constant values for $K_{sp}$.

For a macromolecule which can be represented as an ellipsoid of revolution with semi-major axis "a" and semi-minor axes "b" and whose principal axes for the refractive index and dielectric tensors are also "a" and "b",

$$K_{sp} = \frac{2\pi}{15n^2}(g_a - g_b)(P_a - P_b + Q) \qquad (2)$$

where $(g_a - g_b)$ is an optical anisotropy factor for the molecules which was first calculated by Peterlin and Stuart (1), $P_a$ and $P_b$ are permanent dipole terms, and Q is the induced dipole term for the macromolecule.

## INSULATING SOLUTIONS

The first solutions to be considered theoretically were insulating solutions which contained anisotropic ellipsoids with arbitrarily oriented dipole moments immersed in an isotropic solvent. The equations shown below are a compendium of the ideas of Peterlin and Stuart (1), O'Konski et al. (2), and Holcomb and Tinoco (3). One obtains:

$$P_i = B_i^2(\varepsilon)\,\mu_i^2 / k^2T^2 \tag{3}$$

$$Q = \frac{v\,\varepsilon_o}{kT}\left\{B_a(\varepsilon)(\varepsilon_a - \varepsilon_1) - B_b(\varepsilon)(\varepsilon_b - \varepsilon_1)\right\} \tag{4}$$

where $\mu_i$ = the component of the molecular dipole along the i axis of the molecule

$\varepsilon_o$ = the permittivity of vacuum

$\varepsilon_1$ = dielectric constant of the solvent

$\varepsilon_i$ = principal value of the dielectric constant of the ellipsoid along its i axis

$k$ = the Boltzmann Constant

$T$ = the absolute temperature

$v$ = $\frac{4}{3}\pi a b^2$ (the volume of the ellipsoid)

$$B_i(\varepsilon) = \left[1 + \left\{\frac{\varepsilon_i}{\varepsilon_1} - 1\right\} A_i\right]^{-1} \tag{5}$$

$$A_a = \frac{ab^2}{2}\int_o^{\infty}\frac{d\lambda}{(a^2+\lambda)^{3/2}(b^2+\lambda)} \tag{6a}$$

$$A_b = \frac{ab^2}{2}\int_o^{\infty}\frac{d\lambda}{(b^2+\lambda)^2(a^2+\lambda)^{1/2}} \tag{6b}$$

## CONDUCTING SOLUTIONS

When these equations for insulating systems were used to calculate the expected value of the specific Kerr Constant for a strain of tobacco virus that appeared to have no permanent dipole moment (so that $P_a$ and $P_b$ in equation 2 vanished), the theoretical value obtained was only 2% of the actual experimental value (4). In addition, the $K_{sp}$ of the same strain of tobacco mosaic virus decreased noticeably as the electrolyte concentration was increased at constant pH and furthermore, the specific Kerr Constant also varied with pH, that is, with the magnitude of the charge on the macromolecule.

These results indicated that the conductivity of the solution, as well as its dielectric constant, was probably involved in the polarization of the macromolecule in an electric field; in addition, it seemed that the magnitude of the charge on the macromolecule had also to be implicated. Using such considerations, O'Konski and Krause (5) derived equations for the specific Kerr Constant of a dipolar ellipsoid with an anisotropic dielectric constant and an anisotropic volume conductivity in an isotropic dielectric and conducting solvent. Most macromolecules exhibit no true volume conductivity, but O'Konski (6) had shown earlier that a surface conductivity, which must exist around any macromolecule in conducting solution, is theoretically equivalent to an anisotropic volume conductivity in the macromolecules. These considerations led to the following equations for $P_i$ and for Q if the macromolecules are, again, considered as ellipsoids of revolution:

$$P_i = \frac{B_i^2(\kappa)\,\mu_i^2}{k^2\,T^2} \tag{7}$$

$$Q = \frac{V\,\varepsilon_1\,\varepsilon_o}{kT}\left[\left\{\frac{\kappa_a}{\kappa_1} - \frac{\varepsilon_a}{\varepsilon_1}\right\} B_a^2(\kappa) - \left\{\frac{\kappa_b}{\kappa_1} - \frac{\varepsilon_b}{\varepsilon_1}\right\} B_b^2(\kappa) + \left\{\frac{\kappa_a}{\kappa_1} - 1\right\} B_a(\kappa) - \left\{\frac{\kappa_b}{\kappa_1} - 1\right\} B_b(\kappa)\right] \tag{8}$$

where most symbols have the same significance as in insulating solution and

$\kappa_1$ = the conductivity of the solvent

$\kappa_i$ = principal value of the effective conductivity of the ellipsoid along its i axis

$$B_i(\kappa) = \left[1 + \left\{\frac{\kappa_i}{\kappa_1} - 1\right\} A_i\right]^{-1} \tag{9}$$

$$\kappa_i = \kappa_i^o + \kappa_i' \tag{10}$$

where $\kappa_i^o$ = true volume conductivity along the i axis

$\kappa_i'$ = volume conductivity along the i axis equivalent to the surface conductivity of the ellipsoid.

O'Konski (6) showed how $\kappa_a'$ and $\kappa_b'$ could be calculated in an ellipsoid with equivalent surface conductivity $\lambda$. If $\lambda$ is a constant on the surface of the ellipsoid, $\kappa_a'$ and $\kappa_b'$ vary with position within the ellipsoid (6), but must be considered as constant values in the theoretical treatment of O'Konski and Krause (5), generally calculated, for ellipsoids of revolution as

$$\kappa_a' = \frac{2\lambda}{b} \qquad (11a)$$

$$\kappa_b' = \frac{C\,(a,b)\lambda}{\pi\, ab} \qquad (11b)$$

where C (a,b) is the circumference of an ellipse with semi-axes a and b.

Eqs. 11a and 11b actually give the values of $\kappa_a'$ and $\kappa_b'$ at the center of the ellipse. For rodlike molecules, $\kappa_a'$ and $\kappa_b'$ do not change with position in the rod and

$$\kappa_a' = \frac{2\lambda}{b} \qquad (12a)$$

$$\kappa_b' = \frac{(a+b)\lambda}{ab} \qquad (12b)$$

where a is the half-length of the cylinder

b is the radius of the cylinder.

For a highly charged macromolecule, one may assume as a first approximation that the surface conductivity $\lambda$ arises mainly from the movement of the molecule's counterions on the surface of the macromolecule. This assumption should be reasonable at ionic strengths at which the Debye radius is small, i.e., at ionic strengths so large that the average counterion is close to the surface of the macromolecule. In addition, if the charge density on the macromolecule is large enough (corresponding to a distance of about 7 Å or less between charges on a linear macromolecule in water at room temperature), at least some of the counterions will be condensed on to the macromolecular surface even at low ionic strength (7,8).

Assuming the presence of mobile counterions directly on the surface of the macromolecule,

$$\lambda = n_c uZe/S \qquad (13)$$

where $n_c$ = number of mobile ions on the surface of the macromolecule, generally assumed equal to the number of counterions as a first approximation

$Z$ = charge on each mobile ion on the surface

$e$ = electron charge

$u$ = mobility of mobile ions on surface, generally assumed to equal the mobility in the bulk of the solution as a first approximation

$S$ = surface area of the ellipsoidal macromolecule

For our ellipsoid of revolution,

$$S = 2\pi b^2 + 2\pi \frac{ab}{e} \arcsin e \tag{14a}$$

where

$$e = \left\{1 - \frac{b^2}{a^2}\right\}^{1/2} \tag{14b}$$

This theoretical treatment yielded a value of the specific Kerr Constant of tobacco mosaic virus which had twice the experimental value (5), that is, it had the correct order of magnitude, when it was assumed that all the counterions to each molecule (assumed to be 6000 potassium ions) were moving along the macromolecular surface. If one assumes only partial condensation of the counterions on the macromolecular surface, exact agreement between theory and experiment can be obtained.

## VERY LOW IONIC STRENGTH

For macromolecules with low surface charge density in solutions of very low ionic strength, one cannot assume the presence of a layer of counterions on the surface of the macromolecules. In these cases, the counterions are distributed throughout the solution, more or less in accord with Debye-Hückel Theory, at least when the concentration of polyelectrolyte is low.

In this work, an attempt is made to calculate a first approximation to the specific Kerr Constant for such cases, assuming that the average counterion moves around the macromolecule at a distance from the surface equal to the Debye radius in the solution. All the surface conductivity, $\lambda$, in this model, exists at a distance equal to the Debye length, d, away from the macromolecule.

$$d = \left\{ \frac{\varepsilon_o \, \varepsilon_1 kT}{2 N_o e^2 I} \right\}^{1/2} \tag{15a}$$

where $N_o$ = Avogadro's number

and $$I = \sum \frac{1}{2} Z_i^{\,2} m_i, \tag{15b}$$

the ionic strength of the solution, and

$m_i$ = molality of the ion with charge $Z_i$.

Using the ellipsoid of revolution model for the polyelectrolyte, it was thus assumed that the counterions, on the average, move around the macromolecule on an ellipsoidal surface whose semi-major axis is $a' = a + d$, and whose semi-minor axes are $b' = b + d$. The volume of the ellipsoid whose semi-axes are a and b is designated $v_2$ and the volume of the ellipsoid whose semi-axes are a' and b' is designated as $v_2 + v_3$. Therefore, the volume of solution between the actual physical ellipsoid and the ellipsoidal surface at the Debye radius is $v_3$.

Values of bulk conductivities equivalent to the surface conductivity $\lambda$ are now calculated using a' and b' instead of a and b in eqs. 11 and 12. Derivation of Q for this model (9) is exactly like the derivation of eq. 8 (ref. 5) with the following complication. There is a surface discontinuity for the dielectric constant at the physical boundary of the ellipsoidal macromolecule whose semi-major and semi-minor axes are a and b, and there is a different surface of discontinuity for the conductivity at the Debye radius which is the surface of the ellipsoid with semi-major and semi-minor axes a' and b'. A steady-state charge accumulates at both surfaces of discontinuity while current flows.

The only change in $P_i$, eq. 7, that occurs in this approximation is in the calculation of $B_i(\kappa)$ for which a' and b' must be used in place of a and b. In the calculation of Q, the same change is made in $B_i(\kappa)$ and eq. 16 results.

Since the value of the ionic strength of the solution is needed for calculation of a' and b' (needed to calculate $\kappa_a$, $\kappa_b$, $v_3$, $B_a(\kappa)$ and $B_b(\kappa)$), and since the bulk conductivity of the same solution is needed for calculation of $B_a(\kappa)$, $B_b(\kappa)$, and Q, both quantities must be known for any solution of interest. Ionic strength can be calculated from the composition of the solution, but the conductivity must be measured.

$$Q = \frac{\varepsilon_o \varepsilon_1}{kT} \left[ \left\{ \frac{\kappa_a}{\kappa_1} (v_2 + v_3) - \frac{\varepsilon_a}{\varepsilon_1} v_2 - v_3 \right\} B_a^{\,2}(\kappa) \right.$$

$$+ (v_2 + v_3) \left\{ \frac{\kappa_a - \kappa_1}{\kappa_1} \right\} B_a(\kappa) - \left\{ \frac{\kappa_b}{\kappa_1} (v_2 + v_3) - \frac{\varepsilon_b}{\varepsilon_1} v_2 - v_3 \right\} B_b^{\,2}(\kappa)$$

$$\left. - (v_2 + v_3) \left\{ \frac{\kappa_b - \kappa_1}{\kappa_1} \right\} B_b(\kappa) \right] \tag{16}$$

For the purposes of making predictions using this theoretical treatment, however, we have found that the ratio of $\kappa_o$ to I is fairly constant in solutions in which protons do not carry appreciable current; we have used $\kappa_o / I = 80\ cm^2\ ohm^{-1}\ mole^{-1}$ $(8 \times 10^{-3}\ m^2\ ohm^{-1}\ mole^{-1})$. Table I shows some literature values of $\kappa_o / I$ which can be used in these calculations. In table I, the ionic strength has been calculated in terms of molarity, i.e., mol $dm^{-3}$, instead of molality, as in eq. 15b.

Table I. Experimental Comparison of Ionic Strength and Conductivity

A. In Absence of Macromolecules (Handbook Values)

| Electrolyte | I (mol $dm^{-3}$) | $\kappa_o$ ($\Omega^{-1} m^{-1}$) | $(\kappa_o/I) \times 10^3$ ($m^2 \Omega^{-1} mol^{-1}$) |
|---|---|---|---|
| $KHC_8H_4O_4$ | 0.024 | 0.20 | 8.3 |
| $KH_2PO_4$ | 0.036 | 0.30 | 8.3 |
| $K_2HPO_4$ | 0.063 | 0.52 | 8.3 |
| NaCl | 0.017 | 0.17 | 10.3 |
| NaCl | 0.085 | 0.82 | 9.6 |
| $Na_3C_6H_5O_7$ | 0.102 | 0.74 | 7.2 |
| Citric Acid | 0.023 | 0.12 | 5.2 |
| HCl | 0.012 | 4.51 | 34.0 |

B. In Presence of Macromolecules

| Macromolecule | pH | Electrolyte | $I \times 10^4$ (mol $dm^{-3}$) | $\kappa_o \times 10^3$ ($\Omega^{-1} m^{-1}$) | $(\kappa_o/I) \times 10^3$ ($m^2 \Omega^{-1} mol^{-1}$) |
|---|---|---|---|---|---|
| TMV[1] (ref. 4) | 7.0 | phosphate | 3.3 | 1.92 | 5.8 |
| TMV[1] (ref. 4) | 7.0 | phosphate | 3.3 | 2.68 | 8.2 |
| TMV[1] (ref. 4) | 7.0 | phosphate + KCl | 4.9 | 4.62 | 9.4 |
| TMV[1] (ref. 4) | 7.0 | phosphate + KCl | 8.5 | 10.2 | 12.0 |
| TMV[1] (ref. 4) | 5.6 | phthalate | 12.8 | 7.17 | 5.6 |
| BSA[2] (ref. 10) | 5.2 | KCl | 5. | 18. | 36. |
| BSA[2] (ref. 10) | 5.3 | KCl | 25. | 52.6 | 21. |
| BSA[2] (ref. 10) | 5.0 | $ZnCl_2$ | 75. | 62.6 | 8.3 |
| BSA[2] (ref. 10) | 5.1 | $ZnCl_2$ | 7.5 | 7.55 | 10.3 |
| Paramyosin (ref. 11) | 3.2 | citrate | 3.6 | 30. | 83. |

1. Tobacco Mosaic Virus
2. Bovine Serum Albumin

When predictions are made using this theoretical treatment, appreciably higher Specific Kerr Constants are calculated for macromolecules with no condensed counterions at very low ionic strength than are calculated using the equations of O'Konski and Krause (5). A comparison of the predictions from the two theoretical treatments will be presented in a later publication.

## ACKNOWLEDGEMENTS

We thank the National Science Foundation for supporting this work under Grants No. PCM75-06456 and CHE77-10046 and one of us (SK) would like to thank the National Institutes of Health for support by means of a Research Career Award. One author, (BZ) is on leave from the Israel Fiber Institute, Jerusalem.

## REFERENCES

1 Peterlin A and Stuart H A, Z. Phys. 112 (1939) 129.
2 O'Konski C T, Yoshioka K and Orttung W H, J. Phys. Chem. 63, (1959) 1558.
3 Holcomb D N and Tinoco, Jr., I, J. Phys. Chem. 67 (1963) 2691.
4 O'Konski C T and Haltner A J, J. Am. Chem. Soc. 79 (1957) 5634.
5 O'Konski C T and Krause S, J. Phys. Chem. 74 (1970) 3243.
6 O'Konski C T, J. Phys. Chem. 64 (1960) 605.
7 Manning G S, J. Chem. Phys. 51 (1969) 924.
8 Oosawa F, "Polyelectrolytes", Marcel Dekker, Inc., New York (1971).
9 Krause S, in preparation for publication.
10 Krause S, Ph.D. Thesis, University of California, Berkeley (1957).
11 DeLaney D E, Ph.D. Thesis, Rensselaer Polytechnic Institute (1975).

# TRANSIENT ELECTRIC BIREFRINGENCE OF MACROMOLECULAR SOLUTIONS AT REVERSING FIELDS OF ARBITRARY STRENGTH AND DURATION

G. Koopmans, Joh. de Boer and J. Greve

Physics Laboratory of the Free University
Biophysics Department, De Boelelaan 1081
Amsterdam, The Netherlands

The theory of the transient behaviour of the electric birefringence of solutions of rigid macromolecules is extended to fields of arbitrary strength and duration. This is achieved by expanding the orientation distribution function in Legendre polynominals. The diffusion equation for this function then transforms into a matrix equation which is numerically solved. At low orientational energies the results agree with the curves as calculated by Matsumoto et al (1). The formulas are applied to reversing pulse experiments on solutions of T2LO and T6 bacteriophages.

## INTRODUCTION

Transient electric birefringence is an important tool to obtain information about hydrodynamic and electrical properties of macromolecules in solution. The theory of electric birefringence has been the subject of many investigations (1-5). For orientational energies small as compared to kT, it appeared possible to obtain analytic expressions for the transient behaviour of the electric birefringence. Expressions valid at orientational energies of the order of kT have been obtained by Matsumoto et al. (1). These authors also treat the transient behaviour obtained with reversing fields, as introduced by O'Konski and Haltner (9). We have extended the theory of Matsumoto to a more general case where (a) the orientational energy is arbitrary and (b) no steady state is assumed at field reversal.

## THEORY

In this section we calculate the birefringence of solutions of rotationally symmetric rigid particles. The particles move independently through the solution. The birefringence of such a solution is proportional to the orientation factor $\Phi(t)$, as defined by Peterlin and Stuart. $\Phi$ is directly related to the orientation distribution function f of the symmetry axis of the macromolecules by (6)

$$\Phi = \int_{-1}^{1} f(u) \frac{(3u^2 - 1)}{2} du \, . \, 2\pi \tag{1}$$

Here u is the cosine of the angle which the symmetry axis of the macromolecule makes with the field $\bar{E}$. Throughout this paper we will use the more compact Dirac notation common in quantum mechanics to describe these integrals. We will work in the Hilbert space spanned by the orthonormal basis $|\ell>$ defined by

$$|\ell> = (\ell + \tfrac{1}{2})^{\frac{1}{2}} P_\ell (u) \tag{2}$$

where $P_\ell(u)$ are Legendre polynomials. Eq. 1 is written in this notation as

$$\Phi = <2|f> \cdot (2/5)^{\frac{1}{2}} \, 2\pi \tag{3}$$

In order to calculate $\Phi$ we have to solve the diffusion equation for $|f>$:

$$F\,|f> = \frac{1}{D}\frac{\partial}{\partial t}|f> \tag{4}$$

where the operator F is given by

$$F = (1 - u^2)\frac{\partial^2}{\partial u^2} - 2u\frac{\partial}{\partial u} - (1 - u^2)(\beta + 2\gamma u)\frac{\partial}{\partial u} + 2\,(\beta u + \gamma(3u^2 - 1)) \tag{5}$$

D is the rotational diffusion coefficient of the particle with respect to an axis of rotation perpendicular to its symmetry axis. $\beta$ and $\gamma$ are constants related to the anisotropy in the polarizability $\Delta\alpha$ of the particle and the dipole moment $\vec{\mu}$ of the particle along its symmetry axis.

$$\beta = \mu E/kT$$
$$\gamma = \Delta\alpha\, E^2/2kT \tag{6}$$

The general solution of eq. 4 is given by

$$|f\rangle = \sum_i c_i \exp(\lambda_i D t) |h_i\rangle \tag{7}$$

where $|h_i\rangle$ and $\lambda_i$ are eigenvectors and eigenvalues of F

$$F |h_i\rangle = \lambda_i |h_i\rangle \tag{8}$$

Matsumoto transformed this equation into selfadjoint form by the solution

$$|h_i\rangle = Ex |y_i\rangle \tag{9}$$

with

$$Ex = \exp\left(\frac{\beta u + \gamma u^2}{2}\right) \tag{10}$$

This results in

$$G |y_i\rangle = Ex^{-1} F\, Ex |y_i\rangle = \lambda_i |y_i\rangle \tag{11}$$

G may be expressed in terms of Legendre polynomials by calculating the matrix $\langle i | G | j \rangle$. This matrix may be appropriately truncated and diagonalized by computer. This gives the values of $\lambda_i$ and $\langle i | y_i \rangle$. Using the completeness of the vectors $|y_i\rangle$ it is now possible to find from the boundary condition the values of the coefficients $c_i$ in eq. 5. At $t = 0$ we may write

$$|f\rangle = \frac{1}{\sqrt{2}} |0\rangle$$

$$= \frac{1}{\sqrt{2}} \sum_i |h_i\rangle \langle y_i| Ex^{-1} |0\rangle \tag{12}$$

Therefore it follows from eq. 7 that for $t \neq 0$:

$$|f\rangle = (\tfrac{1}{2})^{\frac{1}{2}} \sum_i Ex |y_i\rangle e^{\lambda_i D t} \langle y_i| Ex^{-1} | 0\rangle \tag{13}$$

In the same way it may be derived that after field reversal at $t = T$ the distribution function is given by

$$t>T: \ |f> = (\tfrac{1}{2})^{\frac{1}{2}} \sum_i \ Ex \ |y_i> \ e^{\lambda_i D(t-T)} \ <y_i| \ Ex^{-1} \sum_j |j>(-1)^j .$$

$$<j|\sum_k \ Ex \ |y_k> \ e^{\lambda_k DT} \ <y_k| \ Ex^{-1} | \ 0> \qquad (14)$$

In order to calculate $\Phi$ by eq. 1 we have to express all operators in the Legendre basis. By completeness of the Legendre polynomials we may write

$$<j| \ Ex \ |y_k> \ = \ \sum_\ell <j| \ Ex|\ell> \ <\ell|y_k> \qquad (15)$$

The calculation of $\Phi$ therefore reduces to simple matrix multiplication after $<i|G|j>$ has been diagonalized.

## APPLICATIONS

(a) An example of a pulse where a high degree of orientation is achieved is shown in fig. 1. The solution consists of T2LO bacteriophages suspended in 25 mM phosphate buffer with pH = 8. In this buffer the phages have their fibers fully extended and their dipole moments are relatively large. The pulse has been fitted with a calculated curve corresponding to $\beta$ = 3.6 and $\gamma$ = 0.

(b) In fig. 2 some results are shown obtained by Greve (7) from reversing pulse experiments on solutions of bacteriophages suspended in a borate buffer of pH = 9. He calculated the diffusion constant from the decay of the birefringence ($D_{decay}$) and from the minimum of the birefringence ($D_{min}$). If this minimum occurs at $t_m$ seconds after field reversal Tinoco (8) derived that for orientational energies small as compared to kT

$$D_{min} \ = \ \ln 3 \ / \ 4 \ t_m \qquad (16)$$

$D_{min}$ appeared to be dependent on field strength whereas $D_{decay}$ remained constant. We simulated the pulse shapes by the formulas developed above and calculated $D_{min}$. In this way we found the full line drawn in fig. 2, which fits the experimental data very well.

(c) In table 1 we list estimates of the range of validity of various formulas that describe transient electric birefringence.

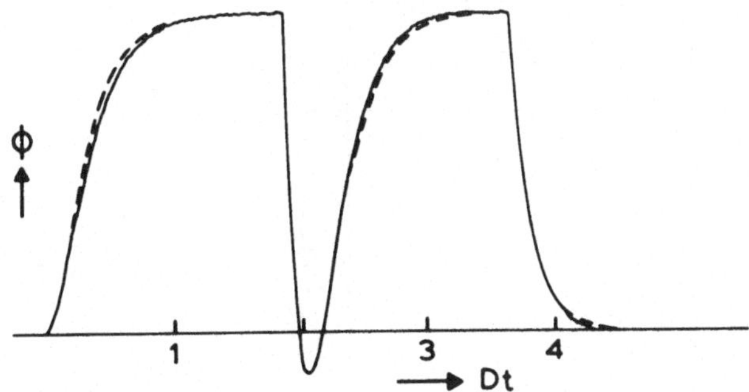


Fig. 1. Transient birefringence of a solution of T2LO phages in 25 mM phosphate buffer pH = 8. The full line is experimentally obtained. The dashed line has been calculated with $\beta$ = 3.7 and $\gamma$ = 0.

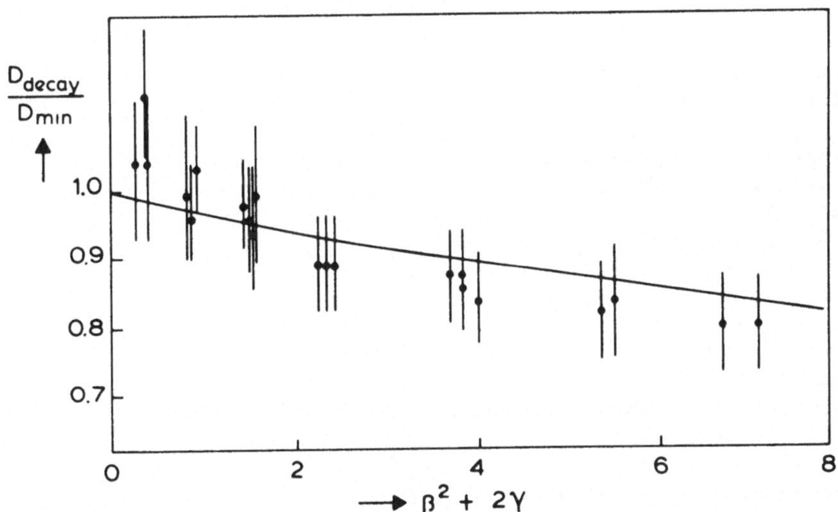


Fig. 2. Plot of the field dependence of $D_{decay}/D_{min}$ for a solution of T2LO phages in 25 mM borate buffer pH = 9. The drawn line has been derived from simulated pulse shapes calculated with $\beta^2/2\gamma$ = 60.

Table 1

| Formulas | Applicable for $(\beta^2+2\gamma)$ smaller than | $\Phi$ (t) |
|---|---|---|
| Benoit (2), Tinoco (8) | 0.4 | $1 + c_i e^{-2Dt} + c_2 e^{-6Dt}$ |
| Matsumoto (1) | 1.5 | $\sum_{i=1}^{4} c_i \exp(\lambda_i Dt)$ |
| present | 13 (N = 6) | $\sum_{i=1}^{N} c_i \exp(\lambda_i Dt)$ |
| | 80 (N = 15) | $\sum_{i=1}^{N} c_i \exp(\lambda_i Dt)$ |

NOMENCLATURE

$D$ rotational diffusion coefficient of the particle
$\vec{E}$ electric field
Ex $\exp\{(\beta u + \gamma u^2)/2\}$
F differential operator from diffusion equation
$|f\rangle$ orientation distribution function
G $Ex^{-1}F\ Ex$
$|h_i\rangle$ $i^{th}$ eigenvector of F
$|i\rangle$ $i^{th}$ normalized Legendre polynomial
k Boltzmann's constant
T temperature
$\beta$ $\mu E/kT$
$\gamma$ $\Delta\alpha\ E^2/2kT$
$\Delta\alpha$ anisotropy of the polarizability tensor of particle
$\Phi$ orientation factor
$\lambda_i$ $i^{th}$ eigenvalue of F
$\vec{\mu}$ dipole moment of the particle

REFERENCES

1 Matsumoto M, Watanabe H, Yoshioka K, J. Phys. Chem., 74 (1970) 2182.
2 Benoit H, Ann. Phys., 6 (1951) 561.
3 O'Konski C T, Yoshioka K, Orttung W H, J. Phys. Chem., 62 (1959) 1558.
4 Nishinari K, Yoshioka K, Koll. Z.Z. Polym., 235 (1969) 1189.

5 Groot G de, Boontje W, Greve J, Boersma H J, Biopol., 16 (1977) 1377.
6 Peterlin A, Stuart H A, Z. Phys., 112 (1939) 129.
7 Greve J, thesis, Vrije Universiteit, Amsterdam.
8 Tinoco I, Yamaoka K, J. Phys. Chem., 63 (1959) 423.
9 O'Konski C T and Haltner A J, J. Am. Chem. Soc., 78 (1956) 3604.

# THE LIMIT OF THE NUMERICAL METHOD OF INVERTING THE LAPLACE TRANSFORMATION AND THE UNIQUENESS OF RELAXATION DISTRIBUTION FUNCTION OBTAINED BY THE METHOD

Mitsuhiro Matsumoto* and Hiroshi Watanabe**

*Department of Chemistry, College of General Education
Tokushima University, Tokushima, Japan

**Department of Chemistry, College of General Education
University of Tokyo, Komaba, Meguro, Tokyo, Japan

It is shown that the high frequency components of the relaxation distribution function (RDF) lose their significance in the process of numerical inversion of the Laplace transformation. Five different methods of approximating the RDF are compared. The methods of approximating the RDF by a polynomial, that is by a continuous function, reproduce all features of the original function, whereas the method of approximating the RDF by line spectra shows a great deal of arbitrariness in the result. This can be understood in terms of the loss of high frequency components in the process of transformation. The effect of approximating the RDF by a set of rectangles or trapezia on the inverting procedure is also discussed.

## INTRODUCTION

As is well known, the relaxation process of any physical quantity y(t) can be related with the relaxation distribution function (RDF) $f(\alpha)$ by the Laplace transformation (1).

$$y(t) = \int_0^\infty f(\alpha) \exp(-\alpha t)\, d\alpha, \tag{1}$$

or equivalently

$$y(t) = \int_0^1 g(x)\, x^{t-1}\, dx, \tag{2}$$

where t is the time, $\alpha$ is the inverse relaxation time, $x = \exp(-\alpha e)$ and e is the time expressed in an arbitrary unit.

Electro-optical measurements in which the electric field is applied as a rectangular pulse produce such decay curves. We therefore commonly have the problem of how to derive the RDF $f(\alpha)$ or $g(x)$ from the experimentally obtained relaxation curve $y(t)$. In the present paper we discuss the general features of this problem. One important point is the fact that high frequency components in the distribution function lose their significance in the process of transformation.

In the second part of the paper we compare the effect of approximating the RDF by several different functions. An artificially composed decay curve

$$y(t) = \frac{1}{2.6}\,(0.2\,e^{-2t} + 0.6\,e^{-4t} + e^{-6t} + 0.6\,e^{-8t} + 0.2e^{-10t}) \tag{3}$$

is used as a "yardstick" for the approximating methods. The results are discussed with relation to the loss of the high frequency components of the RDF in the process of transformation.

## THE CONTRIBUTION OF HIGH FREQUENCY COMPONENTS

When the $g(x)$ in eq. 2 is equal to a parabolic function, $2x(1-x)$, then the decay curve is given by

$$y_0(t) = 2/(1+t)(2+t) \tag{4}$$

If we assume the parabolic $g(x)$ has a sinusoidally vibrating term $\sin(2n+1)\pi x$ in the coefficient, the corresponding decay curve $y_n(t)$ is

$$y_n(t) = A\int_0^1 2x(1-x)\left\{1+\sin(2n+1)\pi x\right\}x^{t-1}\,dx, \tag{5}$$

where A, a normalization constant, is equal to $\left\{0.5+1/(2n+1)\pi\right\}^{-1}$. The differences

$$\Delta y(t) = y_n(t) - y_0(t), \tag{6}$$

for different values of n are shown in fig. 1. Obviously, the difference $\Delta y(t)$ decreases with increasing number of n. For n = 1, the maximum difference is about 0.05, and reduces to a value less than 0.005 for n = 40. The broken line in fig. 1 corresponds to the original decay curve $y_0(t)$. That clearly indicates that the high frequency components in the RDF do not contribute to the $y(t)$. In other words, the $y(t)$ gives only a little information on the high frequency components of RDF. We therefore can derive only a rounded off form of the RDF from a given relaxation curve.

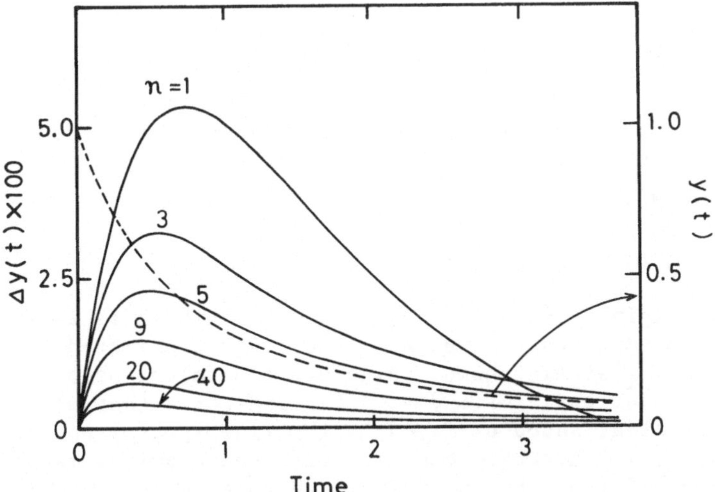


Fig. 1. The contribution of the high frequency components to the decay curve. The $\Delta y(t)$ is the difference defined by eq. 6; $y_0(t)$ is the original decay curve defined by eq. 3.

## A COMPARISON OF DIFFERENT APPROXIMATING METHODS

In the previous work we have examined the method of approximating the RDF by a polynomial with a converging factor (2)

$$\overline{f(\alpha)} = \exp(-B\alpha) \sum A_i (\alpha - \alpha_0)^i \qquad (7)$$

where B and $\alpha_0$ are parameters to be chosen appropriately, and $A_i$'s are coefficients to be determined so as to minimize the difference between the original decay curve y(t) and the recomposed decay curve $\overline{y(t)}$ from $\overline{f(\alpha)}$. As a measure of difference, we use the standard deviation

$$\sigma = \left\{ \sum_{i=1}^{N} \left[ y(t_i) - \overline{y(t_i)} \right]^2 \right\}^{\frac{1}{2}} / N. \qquad (8)$$

For $\overline{y(t)}$, we have

$$y(t) = \exp\left[ -(B+t) \right] \alpha_0 \sum_{i=1}^{M} \frac{i!\ A!}{(B+t)^{i+1}} \qquad (9)$$

In the present work we further examine several different methods of approximating the RDF $f(\alpha)$ and g(x). In order to make the contrast, we firstly examine the method of approximating $f(\alpha)$ by a set of impulse functions as follows:

$$\overline{f(\alpha)} = \sum_{i=1}^{M} A_i \, \delta(\alpha - \alpha_i), \tag{10}$$

where $\alpha_i$'s are constant, and $A_i$'s are coefficients to be determined by the least squares method. This approximates y(t) as

$$\overline{y(t)} = \sum_{i=1}^{M} A_i \exp(-\alpha_i t). \tag{11}$$

In fig. 2 an original $f(\alpha) = \sum A_i \delta(\alpha - \alpha_i)$ which is used to generate the master decay curve eq. 3 and the approximated $\overline{f(\alpha)}$ for three sets of $\{\alpha_i\text{'s}\}$ are shown. Although the values of $\sigma$ are very small, we got a variety of $\overline{f(\alpha)}$'s. This indicates that we cannot surmise the original shape of $f(\alpha)$ by the line spectra approximation. The result makes a remarkable contrast with the previous method. In fig. 3, reproduced $\overline{f(\alpha)}$ by eq. 7 from the same master curve given by eq. 3 is shown by a broken curve. In spite of the discrepancy in the nature between the original RDF (discrete) and the reproduced one (continuous), the reproduced $\overline{f(\alpha)}$ well expresses the overall feature of the original RDF. The reason for the poor reproducibility of eq. 10 can now be understood on account of a large contribution of the high frequency components in the impulse function.

We also examined another approximating continuous function

$$\overline{g(x)} = (x - x_0)(x - x_M) \sum_{i=1}^{M} A_i x^i, \tag{12}$$

which will be normalized as

$$\int_0^1 \{ g(x) / x \} \, dx = 1 \tag{13}$$

where $x_0$ and $x_M$ define the lower and upper limits of the RDF, respectively ($0 \leq x_0 < x_M \leq 1$). We have then

$$\overline{y(t)} = A_i (T_i - S_i T_0 / S_0) + T_0 / S_0, \tag{14}$$

where

$$S_i = (x_M^{i+2} - x_0^{i+2}) / (i+2) - (x_M + x_0)(x_M^{i+1} - x_0^{i+1}) / (i+1) + x_M x_0 (x_M^i - x_0^i) / i, \tag{15}$$

$$T_i = (x_M^{t+i+2} - x_0^{t+i+2}) / (t+i+2) - (x_M - x_0)(x_M^{t+i+1} - x_0^{t+i+1}) / (t+i+1) + x_M x_0 (x_M^{t+i} - x_0^{t+i}) / (t+i), \tag{16}$$

$$S_0 = x_M x_0 \ln(x_M / x_0) - (x_M^2 - x_0^2), \tag{17}$$

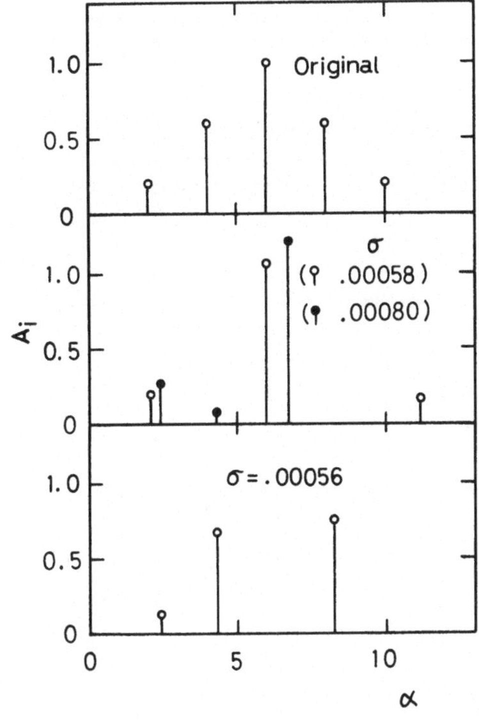


Fig. 2. Reproduction of RDF by the sets of impulse functions as eq. 10.

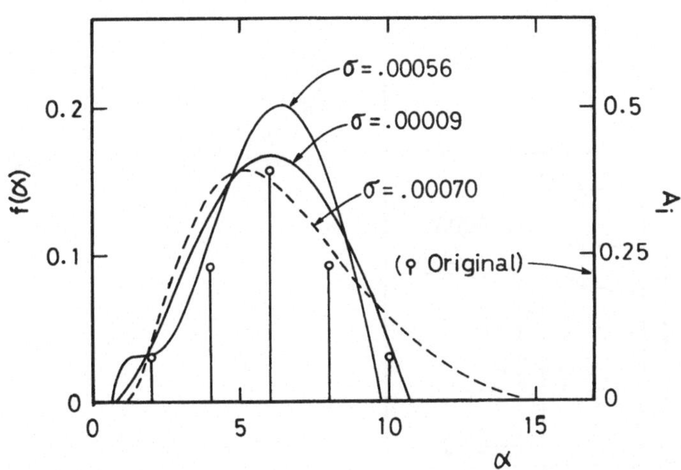


Fig. 3. Reproduction of RDF by the continuous functions. Broken curve is the RDF by eq. 7 and solid curves are the RDF by eq. 12.

Solid curves in fig. 2 represent the results, which also express the reliability of the method. This is important because we thus have a means of checking the uniqueness of the derived RDF from a given experimental decay curve.

We also tried a third type of function as approximating the RDF. In this case the integrals of eq. 1 or 2 are divided into several regions and in each region f $(\alpha)$ or g $(x)$ are replaced by a simple function:

$$\overline{y(t)} = \sum_{i=1}^{M} \int_{\alpha_{i-1}}^{\alpha_i} f_i(\alpha) \exp(-\alpha t)\, d\alpha \tag{18}$$

If we let $f_i(\alpha) = C_i$ (constant), then

$$\overline{y(t)} = \sum_{i=1}^{M} C_i \left\{ \exp(-\alpha_{i-1} t) - \exp(-\alpha_i t) \right\} / t \tag{19}$$

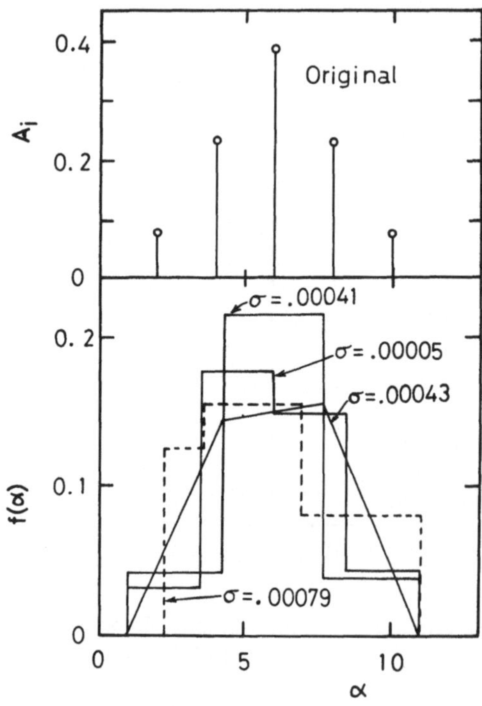


Fig. 4. Reproduction of RDF by the intermediate procedures defined by eq. 19 and eq. 20.

and if we let $f_i(\alpha) = A_i \alpha + B_i$ with boundary conditions $A_{i-1}\alpha_{i-1} + B_{i-1} = A_i \alpha_{i-1} + B_i$ and $A_1 \alpha_0 + B_1 = 0$ and $A_M \alpha_M + B_M = 0$ where $A_i$'s and $B_i$'s are constant, then

$$y(t) = \sum_{i=1}^{M} A_i \left\{ \exp(-\alpha_{i-1} t) - \exp(-\alpha_i t) \right\} / t^2. \quad (20)$$

For the integral (20), if we let $g_i(x) = C_i$ (constant), we have the same equation as eq. 19:

$$y(t) = \sum_{i=1}^{M} C_i (x_{i-1} - x_i) / t. \quad (21)$$

All the constants $A_i$'s, $B_i$'s and $C_i$'s are to be determined so as to minimize the value of $\sigma$ by the least squares method. The RDF thus obtained and the values of $\sigma$ are presented in fig. 4. As the results suggest, the procedure provides us with an intermediate form between the continuous function approximation and the line spectra approximation.

## REFERENCES


1 Matsumoto M, Watanabe H and Yoshioka K, Kolloid-Z, Z Polym., 250 (1972) 298.

2 Tsuji K, Watanabe H and Yoshioka K, Adv. Md. Relaxation Prcoesses, 8 (1976) 49.

# LOW ANGLE APPROXIMATIONS TO THE THEORY OF ALTERNATING ELECTRIC FIELD LIGHT SCATTERING

Thomas J. Herbert

Laboratory for Quantitative Biology, Department of Biology, University of Miami, Coral Gables, Florida 33124, U.S.A.

The intensity of light scattered from a solution of rotationally asymmetric macromolecules placed in a small alternating electric field is calculated for the limit of small scattering angles. These calculations demonstrate that the alternating double frequency component of the scattered light intensity is proportional to $b_0b_2$ where $b_\ell$ is the coefficient of the spherical harmonic $Y_\ell^0$ in an expansion of the scattering amplitude. Similarly, the intensity of the steady component of scattering is proportional to $b_0^2$. Using the Rayleigh-Gans-Debye approximation, the $b_\ell$ are calculated for prolate and oblate spheroids.

## INTRODUCTION

When subject to an alternating electric field, nonspherical isotropic particles in solution tend to align at some angle with respect to the field direction; the preferred direction being dependent upon the magnitude and orientation of the molecular polarizability tensor and the permanent dipole moment. At low frequencies of the applied field, orientation is almost complete but as the frequency is increased the fraction of oriented macromolecules will decrease. As a result, the intensity of light scattered from a solution of macromolecules placed in an alternating electric field will depend upon the frequency of the field, the dielectric properties of the molecule, and the degree of deviation from a spherical shape. Wippler (1) and Jennings (2-4) have shown that the scattering intensity consists of a dc term and an alternating term with frequency equal to twice that of the applied electric field. Both terms contain a frequency dependent dispersion factor which reaches a maximum value at very low frequencies and falls to zero at very high frequencies. The widths of the dispersion factors are a function of the rotational diffusion constant of the macromolecules and the magnitude of the dc and double-frequency components is dependent on the dielectric properties of the macromolecule and the degree of deviation from a sphere.

Analysis of the dispersion term to yield a rotational diffusion constant is straightforward. However, determination of the magnitudes of the polarizability and permanent dipole moment is complicated by a lack of simple functional forms for electric field light scattering intensities from many particle shapes. The computational difficulties involved can be avoided somewhat if an assumption of small scattering angles is made. With this assumption, the scattering intensity can be described as a function of the first two coefficients of an expansion in spherical harmonics of the single particle scattering amplitude. The expansion coefficients are identical to those coefficients which appear in the description of the intensity autocorrelation function of light scattered from a solution of rotational diffusers in the absence of an applied electric field (5,6). This paper will demonstrate that direct computation of the expansion coefficients is possible for particle shapes such as the spheroid and that a simple functional form is the result. This form for the coefficients can be used to describe either the alternating electric field light scattering or the intensity autocorrelation function in the absence of an applied field.

## THE SMALL ANGLE APPROXIMATION

For scatterers which can be described as solids of revolution, the scattering amplitude can be expressed as a series in the spherical harmonics $Y_\ell^m(\theta)$:

$$A(\Omega) = \sqrt{4\pi}\, A_o V \sum_{\ell \text{ even}} b_\ell Y_\ell^o(\theta) \tag{1}$$

The particle volume is V and the angle between the symmetry axis of the particle and the scattering vector $\overline{q}$ is $\theta$. The scattering vector is the difference between the incident and scattered wave vectors and is of magnitude $|\overline{q}| = 4\pi/\lambda \sin\Psi/2$ where $\lambda$ is the *in vacuo* wavelength of incident light and $\psi$ is the scattering angle. The scattering intensity is computed by using:

$$I = \int A^*(\Omega)\, A(\Omega)\, \rho(\Omega)\, d\Omega \tag{2}$$

where the integration is over all angles of particle orientation $\theta, \phi$ and $\rho(\Omega)$ is the probability density function for orientation of the particle symmetry axis into the solid angle $\Omega$. The probability density function can be calculated by using the perturbation theory of Kirkwood (7) to obtain a solution of the rotational diffusion equation in the presence of an applied electric field $\bar{E} = \bar{E}_o \sin\omega t$. As a series in the spherical harmonics:

$$\rho(\Omega) = \frac{1}{4\pi}\left\{1 + a_1 \sin(\omega_o t + \varepsilon_1) \cos\gamma \sqrt{\frac{4\pi}{5}}\, Y_1^o(\theta) + \left[a_{20} + a_{22}\sin(2\omega_o t + \varepsilon_2)\right](3\cos^2\gamma - 1)\sqrt{\frac{4\pi}{5}}\, Y_2^o(\theta) + \ldots\right\} \quad (3)$$

where the angle $\gamma$ is between $\bar{E}$ and $\bar{q}$ and the $a_i$, $a_{ij}$, and $\varepsilon_i$ are frequency dependent amplitude and phase factors described by Jennings (2-4).

The electric field strength appears in eq. 3 as $E^n$ in the n-order term of the expansion. Consequently, in the absence of an applied field only the zero order term remains and integration gives:

$$I_o = (A_o V)^2 \sum_{\ell \text{ even}} b_\ell^2 \quad (4)$$

All terms in the probability density function of odd power in field strength vanish upon integration. Therefore, for small applied fields, the additional scattering intensity will be given by inserting the second order term of the density function into eq. 2 and integrating:

$$I_2 = (A_o V)^2 \left\{a_{20} + a_{22}\sin(2\omega t + \varepsilon_2)\right\}(3\cos^2\gamma - 1) \times \sum\sum_{\ell,\ell' \text{ even}} \int b_\ell b_{\ell'} Y_\ell^o(\theta)\, Y_{\ell'}^o(\theta)\, Y_2^o(\theta)\sqrt{\frac{4\pi}{5}}\, d\Omega \quad (5)$$

The integral of the product of three spherical harmonics can be calculated by methods described by Messiah (8) to yield the following result for the additional scattering in the presence of a small alternating field:

$$I_2 = (A_o V)^2 \left\{a_{20} + a_{22}\sin(2\omega t + \varepsilon_2)\right\}(3\cos^2\gamma - 1) \times \sum_{\ell \text{ even} > o} \left\{b_\ell^2\, c(\ell,\ell) + 2b_\ell b_{\ell-2}\, c(\ell,\ell-2)\right\} \quad (6)$$

where $c(\ell_1, \ell_2)$ is given by:

$$c(\ell_1, \ell_2) = \sqrt{\frac{4\pi}{5}} \int Y_{\ell}^{0}(\theta)\, Y_{\ell'}^{0}(\theta)\, Y_{2}^{0}(\theta)\, d\Omega$$

$$= \sqrt{(2\ell_1+1)(2\ell_2+1)} \begin{vmatrix} \ell_1 & \ell_2 & 2 \\ 0 & 0 & 0 \end{vmatrix} \tag{7}$$

and $\begin{vmatrix} \ell_1 & \ell_2 & 2 \\ 0 & 0 & 0 \end{vmatrix}$ is the 3j symbol described by Messiah (8). An expansion of the coefficients $b_\ell(q)$ as a power series in q can be shown to contain only terms of order $q^\ell$ and higher. Therefore, for small scattering angles, $b_\ell \ll b_{\ell+2}$ and the intensity of scattering is given by:

$$I_o = (A_o V)^2 \, b_o^{\,2} \tag{8}$$

$$I_2 = \frac{\sqrt{5}}{15} (A_o V)^2 \left\{ a_{20} + a_{22} \sin(2\omega t + \varepsilon_2) \right\} (3\cos^2\gamma - 1)\, b_o b_2 \tag{9}$$

## CALCULATION OF EXPANSION COEFFICIENTS FOR SPHEROID MACROMOLECULES

Calculation of the expansion coefficients in eqs. 8, 9 is accomplished by using the Rayleigh-Gans-Debye approximation (9) and integrating phase differences over the particle volume. The scattering amplitude is given by:

$$A(\Omega) = (A_o V) \left\{ \frac{1}{V} \int \cos(\bar{q} \cdot \bar{r})\, dV \right\} \tag{10}$$

where $\bar{r}$ is the radius vector to each mass element from the center of mass of the particle. By use of eq. 1 and the addition theorem for spherical harmonics, the equation for computation of the expansion coefficients can be obtained:

$$b_\ell = \frac{1}{\sqrt{2\ell+1}\, V} \int \cos(qr \cos\beta)\, Y_\ell^{0}(\beta)\, Y_\ell^{0}(\alpha)\, dV\, d\Omega_\beta \tag{11}$$

The angle $\alpha$ is taken between $\bar{r}$ and $\bar{\mu}$, with $\bar{\mu}$ as the permanent dipole moment fixed to coincide with the symmetry axis, and the angle $\beta$ is taken between $\bar{q}$ and $\bar{r}$. The volume integration is over all angles $\alpha$. Integration over all angles $\beta$ results in a final expression for the expansion coefficients:

$$b_\ell = \frac{\sqrt{4\pi}}{V} \int j_\ell (qr) \, Y_\ell^o (\alpha) \, dV \tag{12}$$

where $j_\ell$ (qr) is the spherical Bessel function of order $\ell$.

For consideration of spheroid macromolecules, eq. 12 is transformed to prolate or oblate spheroidal coordinates (10). In the case of a prolate spheroid, the expansion coefficients become:

$$b_\ell = \frac{\sqrt{4\pi}}{V} \int_{-1}^{+1} \int_{1}^{e^{-1}} j_\ell (qc(\xi_1^2 + \xi_2^2 - 1)^{\frac{1}{2}}) Y_\ell^o \left| \frac{\xi_1 \xi_2}{(\xi_1^2 + \xi_2^2 - 1)^{\frac{1}{2}}} \right|$$

$$\times \; c^3 (\xi_1^2 - \xi_2^2) \, d\xi_1 \, d\xi_2 \tag{13}$$

where e is the eccentricity of the spheroid and c is the half-focal distance. The low-angle approximation is applied through expansion of the spherical Bessel function in powers of (qr) and truncation of the series with terms of order $(qr)^2$. This procedure gives:

$$b_o = \left[ 1 - \frac{q^2 R_g^2}{6} \right] \tag{14}$$

$$b_2 = \frac{1}{\sqrt{5}} \left[ \frac{q^2 R_g^2}{6} \right] \sigma_p \tag{15}$$

with a squared radius of gyration of:

$$R_g^2 = \frac{1}{5} (3 - 2e^2) \, a^2 \tag{16}$$

and the constant a equal to the length of the semi-major axis of the ellipse. The shape factor $\sigma_p$ is given by:

$$\sigma_p = \frac{2e^2}{3 - 2e^2} \tag{17}$$

The use of oblate spheroidal coordinates yields an identical final result with the exception that the radius of gyration is:

$$R_g^2 = \frac{1}{5} (3 - e^2) \, a^2 \tag{18}$$

and the shape factor is given by:

$$\sigma_o = \frac{2e^2}{e^2 - 3} \tag{19}$$

In either case, by use of the appropriate shape factor and the previous results for light scattering intensities as a function of the expansion coefficients, we obtain:

$$I_o = (A_o V)^2 \left\{ 1 - \frac{1}{3} q^2 R_g^2 \right\} \tag{20}$$

$$I_2 = \frac{1}{15} (A_o V)^2 \left\{ a_{20} + a_{22} \sin(2\omega t + \varepsilon_2) \right\} (3 \cos^2\gamma - 1) \left\{ \frac{q^2 R_g^2}{6} \right\} \sigma \tag{21}$$

DISCUSSION

The description of alternating electric field light scattering has been formulated in terms of an expansion of scattering amplitude in a series of spherical harmonics with coefficients of expansion $b_\ell(q)$. These coefficients are easily described functions of particle shape and size and of scattering angle. In the low angle approximation, only the first two of these coefficients are significant in magnitude and the formulation of the light scattering problem reduces to a simple function of $b_o$ and $b_2$. Since these coefficients are identical to those used in the description of the intensity autocorrelation function for laser light scattering (5,6), useful information might be obtained by experimental determination of $b_o$ and $b_2$ through laser light scattering. In that case, the ratio $b_2/b_o$ could be obtained from the autocorrelation function of the scattered field:

$$g^{(1)}(\tau) \sim 1 + \frac{b_2^2}{b_o^2} e^{-6D_R\tau} + \frac{b_4^2}{b_o^2} e^{-20D_R\tau} + \ldots \tag{22}$$

Additionally, the $b_\ell^2$ have been calculated numerically by Herbert (11) for long thin rods and by Tanaka (12) for prolate spheroids.

Finally, this paper has shown that, in the low angle approximation, an explicit formula can be obtained for computation of the expansion coefficients as functions of radius of gyration and eccentricity of a spheroid. This calculation and future extension to other shapes of macromolecules should allow better comparison of theory and experiment than heretofore possible and accurate determination of the electrical and structural parameters of many macromolecules.

## ACKNOWLEDGEMENTS

Acknowledgement is made to the Donors of the Petroleum Research Fund administered by the American Chemical Society and to the Research Corporation for support of this research. Additionally, the author wishes to thank Professor J. Nearing, Physics Department, University of Miami, for helpful discussions.

## REFERENCES

1 Wippler C, J. Chem. Phys., 53, (1966) 316.
2 Jennings B R and Plummer H, J. Chem. Phys., 50, (1969) 1033.
3 Jennings B R, in "Molecular Electro-Optics, Vol. I", Chapter 8, C T O'Konski, ed., Dekker, New York (1976).
4 Jennings B R and Morris V J, J. Colloid Interface Sci., 49, (1974) 89.
5 Pecora R, J. Chem. Phys. 49, (1968) 1036.
6 Cummins H Z, Carlson R D, Herbert T J and Woods G, Biophys. J. 9, (1969) 518.
7 Kirkwood J G, J. Polymer Sci., 12, (1954) 1.
8 Messiah A, "Quantum Mechanics, Vol. II", John Wiley and Sons, New York (1962).
9 Kerker M, "The Scattering of Light", Academic Press, New York (1969).
10 Arfken G, "Mathematical Methods for Physicists", Academic Press, New York (1966).
11 Herbert T J, Ph.D. Dissertation, John Hopkins University, Baltimore, Maryland (1970).
12 Tanaka T, J. Colloid Interface Sci., 64 (1978) 171.

# A THEORY OF DYNAMIC LIGHT-SCATTERING BY FLEXIBLE MACROMOLECULES IN A FLUCTUATING ELECTRIC FIELD

W. G. Griffin

Cavendish Laboratory, Madingley Road
Cambridge CB3 OHE, U.K.

The effect of a time-dependent electric field on charged macromolecules in solution can be studied using the technique of dynamic light-scattering. In the present theoretical study the effect of intra- and intermolecular hydrodynamic interactions is taken into account for motions of the molecules which satisfy a stochastic equation of motion of specified form. Approximate solutions to the equations of motion are obtained by a regularization procedure. A comparison of these asymptotic results with those obtained by mode-mode coupling analysis is made, from which certain conclusions can be drawn regarding the domain of validity of the so-called 'generalized Brownian equations-of-motion' methods. Finally, an analysis of the properties of other measurables than the scattering cross-section is given, for the particular problem of electrically driven internal motions of flexible macromolecules and the experimental limitations of such many-point correlation measurements are described.

## INTRODUCTION

In the theory here presented we account for some aspects of the behaviour of polyelectrolyte molecules, in solution, in the presence of a fluctuating electric field. Interest centres around a problem of this sort because biopolymers, e.g. DNA, may be subjected to fluctuating external electric fields in vivo and it is therefore important to know how well theory can be set up to account for biologically relevant properties of such molecules. (For example, as in the case of discussions of the kinetics of nucleosome crystallization, DNA replication, etc.).

Even the simplest random walk models of polymer chains present intriguing problems for theorists and problems simpler than that considered in the present work are by no means to be considered as solved. (The problem of calculating the excluded volume of a polymer chain is still under active consideration in the literature (1)). The barest outline of a calculation of the response of a molecule such as DNA to an externally imposed fluctuating electric field is presented here with comments on the special technical difficulties which arise. A detailed treatment of some of the problems mentioned here in passing is complete and will be found in later publications of the author.

## THEORY

Schmitz has suggested (2) that the molecule of DNA having a molecular weight of $\bar{c}$. $10^6$ Daltons can be regarded as being modelled by the simplest form of the theory of stiff chains due to Saito et al. (3) (the so-called STY model). Let us consider briefly what this model involves. The Lagrangian of the model system is taken to be:

$$\mathcal{L} = \frac{1}{2}\int_0^{s'} \rho \left( \frac{\partial \underline{r}}{\partial t} \right)^z ds - \frac{1}{2}\int_0^{s'} ds \left\{ \mathcal{E}\left(\frac{\partial^2 \underline{r}}{\partial s^2}\right)^2 + K\left(\left|\frac{\partial \underline{r}}{\partial s}\right| - 1\right)^2 \right\} \tag{1}$$

in which $\rho$ is the linear density of the stiff chain, $\mathcal{E}$ is the bending modulus, K is the stretching modulus and s is the arc length variable. The deformations accessible to the chain are here considered to be sufficiently small so as to ensure linear response. The STY model is then determined by fluctuation-dissipation considerations, i.e. a Langevin equation of motion is written down. The first variation of the Lagrangian in eq. 1 gives the deterministic dynamical equation of the STY chain:

$$\rho \frac{\partial^2 \underline{r}}{\partial t^2} + \mathcal{E} \frac{\partial^4 \underline{r}}{\partial s^4} - K \frac{\partial^2 \underline{r}}{\partial s^2} = 0 \tag{2}$$

whence we may write down the corresponding Langevin equation as:

$$\rho \frac{\partial^2 \underline{r}}{\partial t^2} + \zeta \frac{\partial \underline{r}}{\partial t} + \mathcal{E} \frac{\partial^4 \underline{r}}{\partial s^4} - K \frac{\partial^2 \underline{r}}{\partial s^2} = F(s, t) \tag{3}$$

in which F(s,t) is a fluctuating force specified by its statistics alone and $\zeta$ is the coefficient of viscosity. Thus, setting

$$\frac{\partial F(s,t)}{\partial s} = \sum_{n=-\infty}^{\infty} G(n, t)\, e^{in\pi s/L} \tag{4}$$

where L is the chain length, we prescribe the coefficients G(n,t) as:

$$\langle G(n,t)\, G(n',t') \rangle = g(n,n')\, \delta\,(t - t') \tag{5}$$

In other words our definition of F(s,t) follows from the $\delta$-correlated property of the harmonic expansion coefficients G. Resolving $\underset{\sim}{u} = \partial \underset{\sim}{r}/\partial s$ into components:

$$\underset{\sim}{u}\,(s, t) = \sum_{n=-\infty}^{\infty} \underset{\sim}{q}\,(n, t)\, e^{in\pi s/L} \tag{6}$$

we obtain the following form of the equation of motion 3:

$$\rho \frac{\partial^2 \underset{\sim}{q}}{\partial t^2} + \zeta \frac{\partial \underset{\sim}{q}}{\partial t} + \lambda_n(\varepsilon, K)\, \underset{\sim}{q} = G(n, t) \tag{7}$$

in which $\lambda_n$ are simply the eigenvalues:

$$\lambda_n \equiv \varepsilon \left(\frac{in\pi}{L}\right)^4 - K \left(\frac{in\pi}{L}\right)^2 \tag{8}$$

The simplest possible form of equation for the above system driven by a random field producing a force F(t,n) on the chain, is obtained by assuming that:

$$\zeta^2 / \lambda_n \rho \gg 1 \tag{9}$$

i.e. that the motion of the chain is highly damped by viscous forces, when the inertia term, $\rho \frac{\partial^2 \underset{\sim}{r}}{\partial t^2}$ in eq. 3 may be neglected to a good approximation.

Then we have:

$$\dot{\phi}_o\,(n, t) = \left\langle \frac{d\underset{\sim}{q}(n,t)}{dt}\, \underset{\sim}{q}\,(n, o) \right\rangle \tag{10a}$$

$$\phi_o\,(n, t) = e^{-(\lambda_n/\zeta)t}\, \phi(n, o) \tag{10b}$$

In the presence of the fluctuating field the approximate equation of motion becomes:

$$\zeta \frac{\partial \underset{\sim}{q}(n,t)}{\partial t} + \lambda_n\, \underset{\sim}{q}\,(n,t) = F(n,t) + G(n,t) \tag{11}$$

in which F(n,t) is specified by its statistical properties alone.

Assume that it is $\delta$-correlated in time:

$$< F(n,t)\, F(n,t') > \;=\; f(n)\, \delta(t - t') \tag{12}$$

Now, if F(n,t) is not correlated with $q(n,t)$:

$$< F(n,t)\, q(m,t') > \;=\; 0 \text{ for all } \{n, m, (t - t')\} \tag{13}$$

Suppose, however, that F(n,t) is not $\delta$-correlated: then a new term appears such that the form of exponentially decaying correlation function given by eq. 10 no longer obtains. If F(n,t) can be considered as having a deterministic part $F_d$:

$$F(n,t) \;=\; F_d(n,t) + F_r(n,t) \tag{14}$$

where now,

$$< F_r(n,t)\, F_r(n,t') > \;=\; f_r(n)\, \delta(t - t') \tag{15}$$

the solution to the equation of motion for the driven chain involves time-integrals over the kernel F (n,t), i.e. a sum-over-histories. Schmitz (2) has calculated the response of the inertia-less STY chain to a sinusoidal field and concludes that if the frequency of the applied field is sufficiently great so as to eliminate centre-of-mass motion from the optical (i.e. scattered-light field) correlation function and at the same time low enough for the driven modes to respond on a spatial scale comparable with the wavelength of the probe radiation, then the driven modes will contribute to the optical spectrum. We wish to comment that a self-consistent calculation for the case of fluctuating external fields is possible even where the field is correlated with the chain motion. In work to appear (4) this calculation is carried out and it emerges that the simple Rayleighean formulation of the STY problem with inertial terms for solvent and polymer is not consistently given in terms of the Langevin (fluctuation-dissipation) approach. More specifically the Oseen tensor is an insufficient approximation for the problem of the STY-backflow model in which hydrodynamic interactions amongst the chain segments are considered in addition to the intrinsic rigidity of the chain. A mode-mode coupling calculation gives results in disagreement with the proper asymptotic analysis of the chain/solvent fluctuations; the proper analysis requires:

(a) that the stiff-chain be regarded as a 'stiff-necklace' with each bead having specified hydrodynamic boundary conditions

(b) that the hydrodynamic equations for the fluctuations be taken in the Boussinesq approximation

(c) that the final equation of motion in self-consistent Fokker-Planck (chain phase-space coordinates) form be solved with the elimination of secularities by averaging (5).

## CONCLUSIONS

It is possible to probe internal relaxations of polymer molecules by intensity-fluctuation spectroscopy, which, in the case where the scattered light field has Gaussian statistics, measures the field-field correlation function:

$$\langle E^{+}(o)\, E(t) \rangle = C(t) \tag{16}$$

provided the spatial scale of the chain fluctuations satisfies the condition that the mean-square amplitude of the scattering mode be comparable with the inverse square scattering vector. Such 'optically-active' modes can be driven, in certain cases, by external fields and it is of interest to consider the case where these fields arise from motions of neighbouring charged molecules. As a result of a detailed analysis to be given elsewhere (4) a revised method of calculating the dynamical properties of intermediate stiffness chains is obtained which illustrates an analytic weakness of the mode-mode coupled Langevin equations-of-motion (6) method characteristic of this latter approach to an arbitrary many-body problem. Finally, we observe that with small scattering volumes, non-Gaussian field statistics may be obtained experimentally and the intensity correlations contain additional information about polymer conformation and dynamics.

## REFERENCES

1 see Edwards S F, Proc. Phys. Soc., 85 (1965) 613.
Daoud M, Cotton J P, Farnoux B, Jannink G, Sarma G, Benoit H, Duplessix C, Picot C and de Gennes P G, Macromolecules, to be published.

2 Schmitz K S, Chem. Phys. Letts., 42, (1976) 137.

3 Saito N, Takahashi K and Yunoki Y, J. Phys. Soc. Japan, 22, (1967) 219.

4 Griffin W G, to be submitted.

5 Bogoliubov N N, Mitroposkii Ju A and Samilenko A M, "Methods of Accelerated Convergence in Non-linear Mechanics", (1976), Springer-Verlag.

6 Kapral R, Ng D and Whittington S G, J. Chem. Phys., 64 (1976) 539.

# ELECTRO-OPTICAL RESPONSES OF CHIRAL SUBSTANCES

Hiroshi Watanabe

Department of Chemistry, College of General Education
University of Tokyo, Komaba, Meguro, Tokyo, Japan

The Jones matrix for chiro-birefringent systems has been obtained. The matrix appears as a function of $\gamma = (\alpha^2 + \beta^2)^{\frac{1}{2}}$, where $\alpha$ is the rotation angle, $\beta$ is a half of the optical retardation due to birefringence. The matrix is quite general and can be used without any limitation, provided that the system is homogeneous and the axis of birefringence can be well determined. By use of the matrix, a general equation for light intensity in an optical system composed of a chiro-birefringent medium sandwiched in between two polarizers is derived. A convenient optical system to measure an anisotropy of optical rotation by use of conventional electro-optical devices is proposed. Effects of slight deviation of the polarizer from a proper orientation of the axis and of stress birefringence in the cell windows on the electro-optical responses of chiro-birefringent substances are discussed.

## INTRODUCTION

Almost all substances become birefringent when external fields (such as electric, magnetic, shear, etc.) which force the constituent molecules to align in some preferential direction, are applied. If the material is chiral, it should therefore become chiro-birefringent when subjected to any of these fields. In order to investigate the electro-optical responses of chiral substances, it is essential to derive a general equation for the effect of chiro-birefringent medium on incident light.

So far, in electric birefringence measurement, the effect of optical rotation has been neglected, simply because we have not had the theory to enable us to account for the effect. The effect of optical rotation must, however, be appreciable particularly when we measure the birefringence of chiral materials which have very small values of the optical anisotropy factor. Besides birefringence measurements, the chirality and its anisotropy is a subject of great interest, because it could provide us with information not given by the other electro-optical experiments. Conformational changes in biopolymers induced by a strong electric field, for instance, can be more directly observed by the change of optical rotation than by the change of birefringence or of dichroism (1,2).

Go developed a general theory of optical activity of anisotropic solutions (3). The theory is described in terms of the Stokes parameters and the Muller matrix. Although the theory is comprehensive, it is not convenient for practical use, as the result is expressed in an exponential function of the matrix. On application, it is necessary to expand the function into an infinite power series of the matrix. The cross terms arising from higher terms are complicated and not easy to analyse, even in the quadratic term. This limits the application for extremely small values of rotation, ellipticity, etc. (4).

In the present work the Jones matrix is derived for chiro-birefringent systems. It requires no restriction on application. The Jones matrix is most conveneint because it is composed of only four elements, each of which includes information of amplitude and phase. Using the Jones matrix, a general equation is derived for the light intensity which passes through a chiro-birefringent substance placed between two polarizers. The equation suggests the best method of measuring the anisotropy of optical rotation from electro-optical responses. Practically, all of the optical elements cannot be considered to be perfect. We therefore discuss the effect of slight imperfection in the optical elements on the electro-optical responses.

## THE JONES MATRIX

Suppose an homogeneous chiro-birefringent material, the rotation angle and the optical retardation of which are $\alpha$ and $\delta$, respectively, is composed of infinite numbers of sets of thin lamellae, one of which is birefringent with optical retardation $\delta/n$, and the other is optically active with rotation angle $\alpha/n$. Here n is the number of a set and therefore should be increased to an infinite value.

Taking the origin of the angular coordinate on the optical axis of the birefringent lamella, the Jones matrix for the birefringent lamella is (5),

$$B(\beta/n) = \begin{bmatrix} \exp(i\beta/n) & 0 \\ 0 & \exp(-i\beta/n) \end{bmatrix}. \tag{1}$$

where $\beta = \delta/2$, and the Jones matrix for the rotative lamella may be

$$A(\alpha/n) = \begin{pmatrix} \cos(\alpha/n) & -\sin(\alpha/n) \\ \sin(\alpha/n) & \cos(\alpha/n) \end{pmatrix}. \tag{2}$$

Accordingly, the Jones matrix of the collection of sets can be expressed as

$$D(\alpha,\beta) = \{A(\alpha/n)\,B(\beta/n)\}^n = \{D(\alpha/n,\beta/n)\}^n \tag{3}$$

Let the matrix be diagonalized by the similarity transformation

$$D(\alpha/n,\beta/n) = T^{-1}\Lambda T. \tag{4}$$

Here $\Lambda$ is the diagonal matrix composed of the eigenvalues $\lambda_1$ and $\lambda_2$ of $D(\alpha/n, \beta/n)$. In an explicit form, the eigenvalues are

$$\lambda_{1,2} = \cos(\alpha/n)\cos(\beta/n) \pm \sqrt{\cos^2(\alpha/n)\cos^2(\beta/n) - 1} \tag{5}$$

The transformation matrices are

$$T = \begin{pmatrix} 1 & 1 \\ (a-\lambda_1)/b^* & (a-\lambda_2)/b^* \end{pmatrix} \tag{6}$$

$$T^{-1} = \frac{1}{\lambda_2 - \lambda_1} \begin{pmatrix} a^* - \lambda_1 & b^* \\ \lambda_2 - a^* & -b^* \end{pmatrix} \tag{7}$$

with abbreviations $a = \cos(\alpha/n)\exp(i\beta/n)$, and $b = \sin(\alpha/n)\exp(i\beta/n)$. The asterisk denotes the complex conjugate. Taking an infinite value of n, we finally have

$$D = \begin{pmatrix} \frac{\beta/\gamma+1}{2} & \frac{i\alpha}{\gamma} \\ \frac{\beta/\gamma-1}{2} & -\frac{i\alpha}{\gamma} \end{pmatrix} \begin{pmatrix} \exp(i\gamma) & 0 \\ 0 & \exp(-i\gamma) \end{pmatrix} \begin{pmatrix} 1 & 1 \\ \frac{i(\beta-\gamma)}{\alpha} & \frac{i(\beta+\gamma)}{\alpha} \end{pmatrix}$$

$$= \begin{pmatrix} \cos\gamma + (i\beta/\gamma)\sin\gamma & -(\alpha/\gamma)\sin\gamma \\ (\alpha/\gamma)\sin\gamma & \cos\gamma - (i\beta/\gamma)\sin\gamma \end{pmatrix} \tag{8}$$

This is the desired Jones matrix. Full details of the derivation will be published elsewhere.

## THE INTENSITY OF LIGHT

By the aid of this Jones matrix, we are able now to elucidate the optical effect of chiro-birefringent substances in full. Considering the commonly used electro-optical procedures, we confine the interest of the present paper to the case where the substance is placed between a polarizer and an analyzer. We take the z axis of the Cartesian coordinates along the light beam. Let the angular coordinate of the polarizer be $\theta_o$ and of the analyzer be $\theta_2$, respectively, as measured from the x axis (fig. 1). Two components of the electric vector X and Y of the light can be expressed by the following equation,

$$\begin{pmatrix} X \\ Y \end{pmatrix} = \begin{pmatrix} C_2^2 & C_2S_2 \\ C_2S_2 & S_2^2 \end{pmatrix} D \begin{pmatrix} C_o \\ S_o \end{pmatrix} \tag{9}$$

where $C_i = \cos\theta_i$ and $S_i = \sin\theta_i$. In the equation, elements of the column vectors $C_o$ and $S_o$ correspond to the components of the output of polarizer in the x and y directions, respectively. The matrix whose elements are composed of $C_2$ and $S_2$ is the Jones matrix for the analyzer.

In practice, the intensity of light which comes out from the analyzer is most important. The intensity can be reduced to a compact form as follows:

$$\begin{aligned} I &= XX^* + YY^* \\ &= \{\cos(\theta_2 - \theta_o)\cos\gamma + \sin(\theta_2 - \theta_o)(\alpha/\gamma)\sin\gamma\}^2 \\ &\quad + \{\cos(\theta_2 + \theta_o)(\beta/\gamma)\sin\gamma\}^2 \end{aligned} \tag{10}$$

It may be instructive to see the equation for several special cases.

Case I. $\alpha = 0$.

$$I = \cos^2(\theta_2 - \theta_o)\cos^2\beta + \cos^2(\theta_2 + \theta_o)\sin^2\beta . \tag{11}$$

This is a general equation of light intensity for merely birefringent substances and reduces to the familiar equation $I = \sin^2\beta$ if we let $\theta_o = 45^o$, $\theta_2 = 135^o$.

Case II. $\beta = 0$.

$$I = \cos^2(\theta_2 - \theta_o - \alpha). \tag{12}$$

This simply indicates that the polarized light at an angle $\theta_o$ is rotated by an angle $\alpha$ by a passage through the chiral medium.

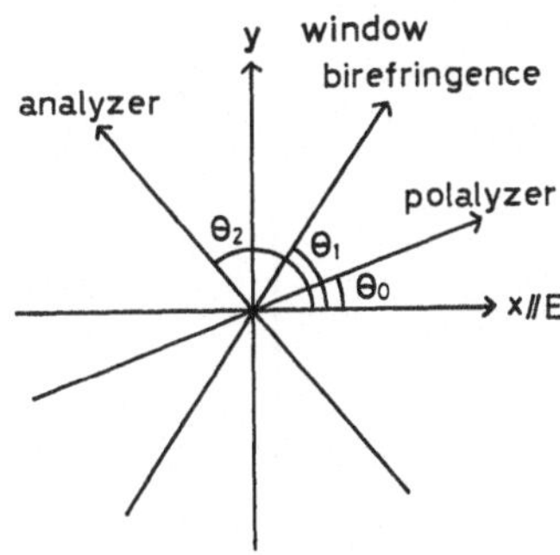


Fig. 1. Angular coordinates of the optical axis of polarizer window birefringence, and analyzer viewed from the photo detector.

Case III. $\theta_o = 0$.

$$I = \cos^2\theta_2 \cos^2\gamma + \frac{\alpha}{2\gamma}\sin 2\theta_2 \sin 2\gamma + \left(\frac{\alpha^2}{\gamma^2}\sin^2\theta_2 + \frac{\beta^2}{\gamma^2}\cos^2\theta_2\right)\sin^2\gamma. \tag{13}$$

This can be further specified

$$\text{for } \theta_2 = 90^\circ,\ I = \frac{\alpha^2}{\gamma^2}\sin^2\gamma, \text{ and} \tag{14}$$

$$\text{for } \theta_2 \pm 45^\circ,\ I = \frac{1}{2} \pm \frac{\alpha}{2\gamma}\sin 2\gamma \tag{15}$$

Special attention should be paid to this case because there has been a misunderstanding that the arrangement of the polarizer at zero angle eliminates the linear birefringence effect leaving only the effect of optical rotation (1). It is worth notice that in the case of eq. 15 the light intensity is a function of $\sin 2\gamma$, suggesting the most sensitive optical system for measuring $\gamma$.

Case IV. $\theta_o = 45^\circ$, $\theta_2 = 135^\circ$.

$$I = \sin^2\gamma \tag{16}$$

This is the commonly used optical arrangement for birefringence measurements. $\beta$, which appears in the case of linear birefringence, is replaced by $\gamma$ in this case.

## EFFECTS OF SMALL DEFECTS ON THE ELECTRO-OPTICAL RESPONSES

The optical elements cannot generally be perfect. In this section we examine the effect of setting the polarizer at a slightly inclined position from the proper orientation of the axis and the effect of stress birefringence in the cell windows on the electro-optical responses of chiro-birefringent substances for the case of eq. 15.

The intensities of light calculated by eq.15 for several small values of $\theta_o$ are plotted against $\gamma$ in fig. 2 and fig. 3, in which

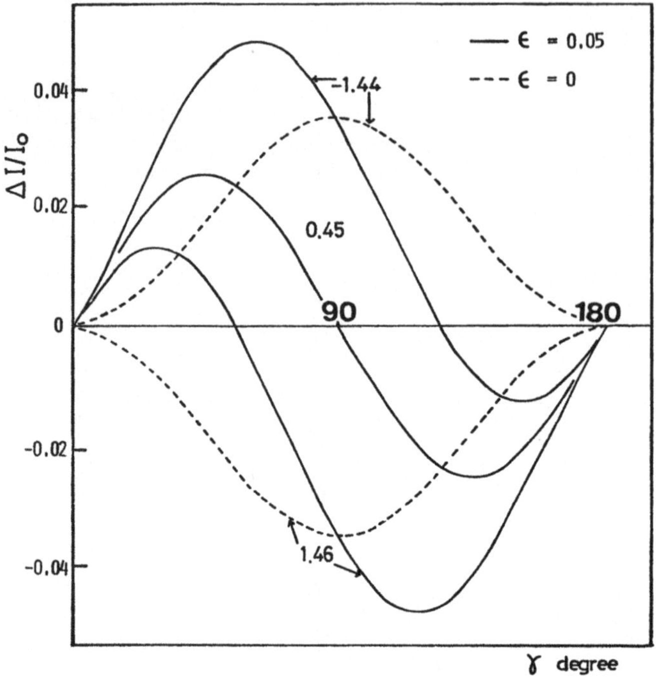


Fig. 2. Dependencies of $\Delta I$ on $\gamma$ by eq. 16 for three different sets of $\theta_o$, $\theta_2$.

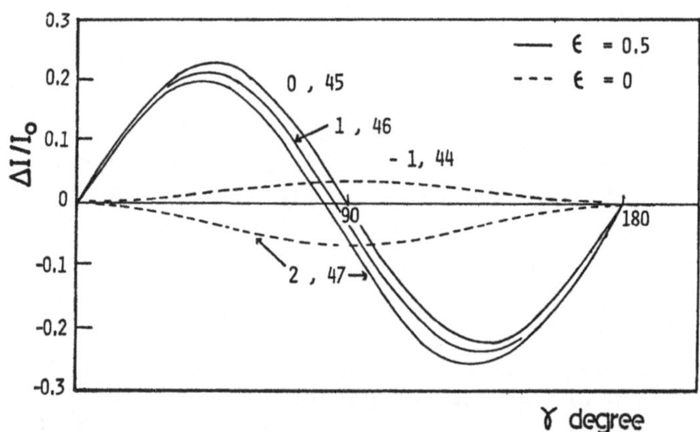


Fig. 3. Dependencies of $\Delta I$ on $\gamma$ by eq. 16 for three different sets of $\theta_o$, $\theta_2$.

$\epsilon = \alpha/\beta$. $\theta_2 - \theta_o = 45°$ for any case. Broken lines in the figures correspond to the case of $\varepsilon = 0$, i.e., of merely birefringent substances. As can be seen, dependence of $\Delta I = I(\gamma) - I(0)$ on $\gamma$ is remarkably deformed by an inclination of the polarizer of 1° when $\varepsilon = 0.05$, whereas the effect of inclination is much smaller when $\varepsilon = 0.5$. The broken lines indicate that we cannot take the negative electro-optical signals in the relevant optical system as an evidence of a change of optical rotation as in ref. 1, unless the exact setting of the polarizer is checked by the electro-optical responses of an achiral substance such as nitrobenzene.

Another possible cause of the negative electro-optical signals in an achiral substance is the window birefringence. The components of light expressed by a product of Jones matrices are

$$\begin{bmatrix} X \\ Y \end{bmatrix} = \begin{bmatrix} C_2^2 & C_2S_2 \\ C_2S_2 & S_2^2 \end{bmatrix} \begin{bmatrix} \exp(i\beta) & 0 \\ 0 & \exp(-i\beta) \end{bmatrix}$$

$$\begin{bmatrix} C_1^2\exp(i\eta) + S_1^2\exp(-i\eta) & 2iC_1S_1\sin(\eta) \\ 2iC_1S_1\sin(\eta) & C_1^2\exp(-i\eta) + S_1^2\exp(i\eta) \end{bmatrix} \begin{bmatrix} C_o \\ S_o \end{bmatrix} \tag{17}$$

where $\eta$ is half of the optical retardation due to the window birefringence. Letting $\theta_o = 0$, we have

$$I = \cos^2\theta_2 - \sin^2 2\theta_1 \cos 2\theta_2 \sin^2\eta + \sin 2\theta_1 \sin 2\theta_2 \{\cos\eta \sin 2\beta + \cos 2\theta_1 \sin\eta \cos 2\beta\} \sin\eta \tag{18}$$

$$\Delta I = \sin 2\theta_1 \sin 2\theta_2 \cos\eta \sin 2\beta + \cos 2\theta_1 \sin\eta\{\cos 2\beta - 1\} \tag{19}$$

Several examples of $\Delta I$ versus $\delta (= 2\beta)$ are shown in fig. 4. In the figure, values of $\theta_1$ and $\eta$ are indicated on each line. As is shown in the figure, we may have negative $\Delta I$ without any change of optical rotation.

## ELECTRO-OPTICAL RESPONSES OF CHIRAL SUBSTANCES

Almost all the naturally occurring biopolymers and their synthetic analogues are chiral. Owing to ordered conformations, they may have an anisotropy of rotatory power as well as anisotropies of electrical and optical properties. In this case an alignment of molecules in solution by an external field such as an electric field causes an anisotropy of optical rotation as well as an anisotropy of refractive index. Assuming that a solute macromolecule has an axis of symmetry

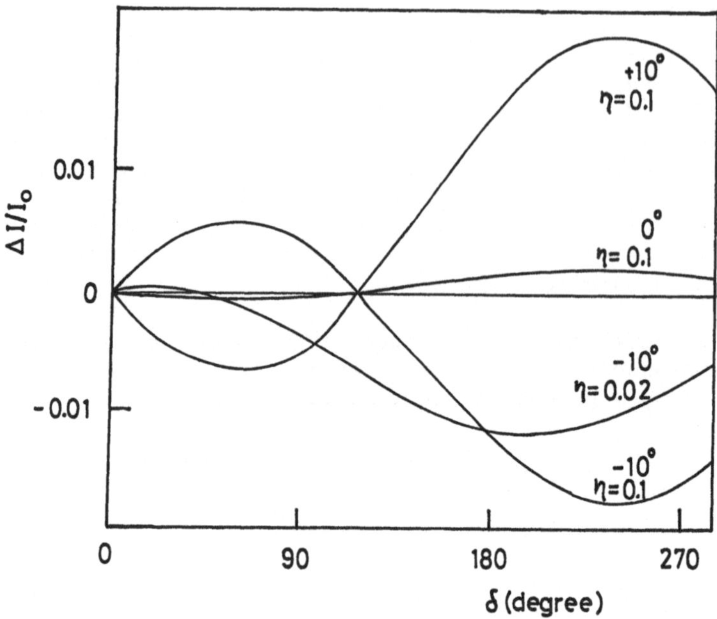


Fig. 4. Dependencies of $\Delta I$ on $\delta$ by eq. 20 for $\eta$ = 0.02 and 0.1. $\theta_0 = 1^\circ$ and $\theta_1 = 20^\circ$. The number of degree on each curve indicates the angle of rotation of the analyzer from the crossed position.

that is the preferential axis of orientation under an external field, and denoting the specific rotation and the optical polarizability along the symmetry axis by $[\alpha_{33}]$ and $g_{33}$, and perpendicular to the axis by $[\alpha_{11}]$ and $g_{11}$, respectively, the values of $\alpha$ and $\beta$ of the solution under an orienting field may be written as follows (6,7);

$$\alpha = A([\alpha_o] + \Delta[\alpha]\Phi), \quad \beta = B\Delta g\Phi, \tag{20}$$

where A and B are constants. In the above equations $[\alpha_o] = \frac{1}{3}([\alpha_{33}] + 2[\alpha_{11}])$, $\Delta[\alpha] = ([\alpha_{11}] - [\alpha_{33}])/3$, $\Delta g = g_{33} - g_{11}$, and $\Phi$ is the orientation factor which is defined as $\Phi = (3\cos^2\theta - 1)/2$ where $\theta$ is the angle between the symmetry axis and the direction of orientating field. The change of the light intensity due to an alignment of solute molecules in solution is therefore

$$\Delta I = \frac{\alpha}{2\sqrt{\alpha^2 + \beta^2}} \sin 2\sqrt{\alpha^2 + \beta^2} - \frac{1}{2}\sin 2\alpha \tag{21}$$

Since the constants A and B are known, and $[\alpha_o]$ and $\Delta g$ are measurable independently, and also $\Phi$ can be determined as a function of the field strength for a given sample, we are now able to determine $\Delta[\alpha]$, at least in principle. Particularly if a strong enough field to attain

$\Phi = 1$ is available, or if an extrapolation to $\Phi \rightarrow 1$ is possible, we have

$$\alpha = A\,[\alpha_{11}], \qquad \beta = B\,\Delta g. \tag{22}$$

If the molecule is oriented perpendicular to the direction of the field, $\Phi = -\frac{1}{2}$ for the complete orientation. Accordingly, we have

$$\alpha = \frac{A}{2}\left([\alpha_{11}] + [\alpha_{33}]\right), \qquad \beta = -\frac{1}{2} B\,\Delta g. \tag{23}$$

NOMENCLATURE

| | |
|---|---|
| A | Jones matrix for chiral lamella |
| a | $= \cos(\alpha/n)\exp(i\beta/n)$ |
| B | Jones matrix for birefringent lamella |
| b | $= \sin(\alpha/n)\exp(i\beta/n)$ |
| $C_i$ | $= \cos\theta_i$ |
| D | Jones matrix for chiro-birefringent system |
| E | external electric field |
| $g_{ii}$ | ii component of optical polarizability |
| I | intensity of light; $\Delta I = I(\gamma) - I(0)$ |
| n | number of set of lamella |
| $S_i$ | $= \sin\theta_i$ |
| T | transformation matrix |
| X | x component of light |
| Y | y component of light |
| $\alpha$ | rotation angle |
| $[\alpha_{ii}]$ | ii component of specific rotation |
| $[\alpha_o]$ | $= ([\alpha_{11}] + [\alpha_{22}] + [\alpha_{33}])/3$ |
| $\gamma$ | $= (\alpha^2 + \beta^2)^{\frac{1}{2}}$ |
| $\varepsilon$ | $= \alpha/\beta$ |
| $\delta$ | optical retardation |
| $\eta$ | optical retardation due to cell windows |
| $\Lambda$ | diagonal matrix |
| $\lambda_i$ | ith element of $\Lambda_i$ |
| $\vartheta$ | angle between external field and molecular axis |
| $\theta_i$ | angular coordinate of ith optical element |
| $\Phi$ | orientation factor |

REFERENCES


1 Cummings A and Eyring E M, Biopolymers, 14, (1975) 2107.
2 Lumry R, Leagare R and Miller W G, Biopolymers, 2, (1945) 489.
3 Go N, J. Phys. Soc. Japan, 23 (1967), 88.
4 Troxell T C and Scheraga H A, Macromolecule, 4,(1971) 519.
5 Schurcliff W A, Polarized Light, Harverd University Press, (1962) pp. 165-170.
6 Tinoco I, Jr. and Hammerle W G, J. Phys. Chem., 60, (1965) 1969.
7 Tinoco I, Jr., J. Am. Chem. Soc., 81, (1959) 1540.

'I pass with relief from the tossing sea of Cause and Theory to the firm ground of Result and Fact.'

SIR WINSTON CHURCHILL (1874-1965)

The Malakand Field Force

# Absorption Phenomena

'The purest and most thoughtful minds
are those which love color the best.'

JOHN RUSKIN (1819-1900)

The Stones of Venice

# AN INSTRUMENT FOR THE MEASUREMENT OF ELECTRIC DICHROISM

William H. Rahe, Robert J. Fraatz, Light K. Sun, Douglas R. C. Priore and Fritz S. Allen

Chemistry Department, University of New Mexico
Albuquerque, NM 87131, U.S.A.

We have constructed an instrument for the measurement of electric dichroism. Through the use of photoelastic modulators we are able to shift the plane of polarization from parallel to the applied electric field to perpendicular, at a cyclic rate of 100 KHz-1 MHz. When an electric pulse is applied to a sample and dichroism arises, an A.C. component appears in the output of the photomultiplier tube. We have devised a digital procedure which is analogous to a lock-in amplifier for processing this A.C. signal.
A waveform recorder captures the dichroism signal and transfers it to a PDP 11/04 minicomputer. The 11/04 can store and/or send the data to the campus IBM machine for processing.

## INTRODUCTION

For the past several years our group has been involved in the construction of an instrument for the more accurate measurement of electric dichroism (ED). We felt it would be best to build a device which would monitor both the parallel and perpendicular components of the dichroism and which would incorporate phase sensitive detection (PSD) techniques to give an improvement in the signal to noise ratio. There have been many problems to overcome but we have nearly completed this task.

The simultaneous measurement of both dichroic components is not a new concept. Allen and Van Holde (1) designed and built an instrument for this purpose. A schematic diagram is shown in fig. 1. This device is essentially two of the more ordinary single beam ED instruments which have the lamp, monochromator, polarizer and electrooptic cell in common. The polarizer produces two beams of opposite polari-

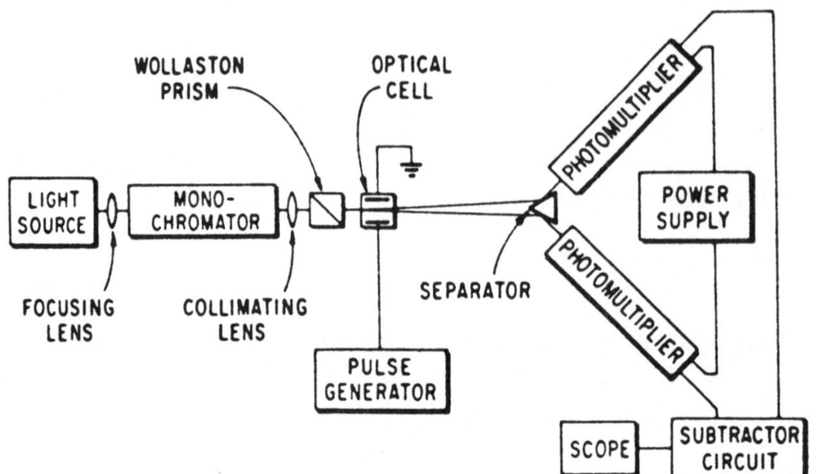


Fig. 1. The electric dichroism instrument of Allen and Van Holde (1).

zation which are passed into separate photo-tubes. The subtraction circuit produces the parallel signal minus the perpendicular signal which can be related to the reduced dichroism parameter, $\Delta\varepsilon/\varepsilon$ by the methods of Allen and Van Holde (1). This instrument is convenient but does not result in a significant improvement in signal to noise ratio (S/N).

The application of PSD to dichroism instrumentation is also not a new idea. Several authors have used this technique in various ways (2,3,4). All previous applications, however, have been for low frequency cases where dichroism is observed in oriented films or static electric field and the sample material is a gas, or an organic solvent solution which will sustain the applied high voltage. Clearly the methods are not applicable to aqueous biopolymer solutions where pulsed electric field techniques must be employed.

The instrument we describe here samples both parallel and perpendicular dichroism by means of a high speed modulating device which alternates the plane of polarization between parallel and perpendicular to the applied field. Thus dichroism appears as an A.C. component in the detector signal. The modulation frequency must be high enough so that many samplings of each polarization form are made during the time of application of the electric field pulse. Since pulse lengths range from $\sim 10\,\mu$sec to several msec, we have chosen a modulation frequency of 1 MHz. This then defines the carrier frequency for the PSD. Since this is a very high frequency for PSD software techniques were used to mimic the behaviour of the PSD instrumentation.

## THE INSTRUMENT

Phase sensitive detection is a means to extract information from a signal where this information is at the same frequency and has same phase relationship to a reference signal. Many companies produce phase sensitive detection equipment but none of these devices perform well at high frequency. We have a Keithley # 822 Phase Sensitive Detector, a # 821 Phase Shifter, and a # 825 Low Noise Amplifier. This system (at 1 MHz) offers one of the highest operating frequencies of any system currently available and yet it produces no noticeable improvement in S/N in our instrumentation even when we operate at 100 KHz. The reason for this is the electronic limitations on the manner in which phase sensitive detection is achieved. It is obvious that if the carrier signal is not modulated at a frequency significantly greater than the maximum frequency at which the information (amplitude) can vary that information loss will occur. Similarly at every point in the processing of the signal, the system must be capable of dealing with a frequency band centered at the carrier frequency and of band-width the order of twice that of the information band. Consequently, when both the carrier and information frequencies are high, a large portion of noise is admitted to the demodulated output signal by the constraints upon the electronic amplification and processing. In addition, several commercial devices have lock-in times which are greater than our events and hence they would miss them entirely.

We have modelled the phase sensitive detection process as shown by the diagram in fig. 2. In order to apply this model to a transient event it is necessary to have a high frequency carrier wave. In general spectroscopic terms, this means a high chopping frequency for the lamp. In our specific dichroism application this means alternation between parallel and perpendicular polarization modes at a high frequency. This frequency must be such that at least ten cycles are contained in the event time interval otherwise there will be excessive loss of amplitude information. In our experiments a short event is the order of 10 $\mu$sec. Hence a carrier frequency of 1 MHz is required. At this frequency, mechanical chopping or exchanging of the beams is not feasible and rapid stress-refraction or electro-refraction methods must be employed. We have used the stress-refraction or photoelastic technique of Kemp (5). In this method the natural piezoelectric properties of crystalline quartz are utilized to drive resonant vibrations in a specially cut fused quartz block which is cemented to the crystal. By varying the dimensions of the crystal and fused silica block the resonant frequency can be varied. The crystal quartz transducer element can be driven with a low voltage signal but it must be within a few Hz of the resonant frequency. For this purpose we have obtained a Harris-PRD # 7828 frequency synthesizer which produces a signal stable in frequency to a few cycles/month. This signal is subsequently amplified to drive the photoelastic modulation system. At this writing we have been able to achieve alternation of the plane of polarization at 1 MHz with ease and have observed the effect as high

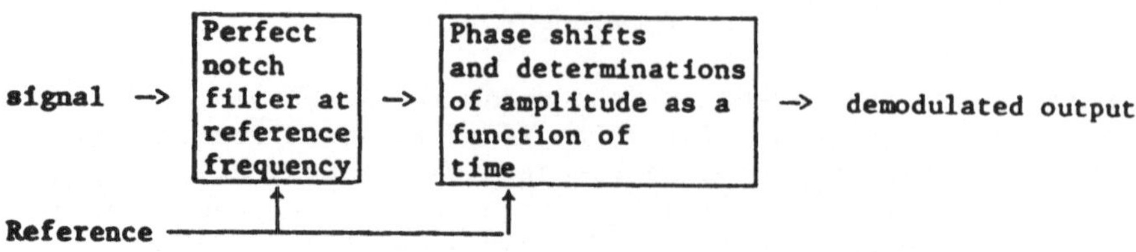


Fig. 2. The phase sensitive detection model.

as 8 MHz. If an additional polarizer is placed after the modulator then the lamp intensity can be modulated at this rate. Thus our modulation system can alternate the plane of polarization or chop the lamp at a high frequency.

We have developed a digital procedure to perform the functions summarized in fig. 2. We begin by postulating that the time dependent signal, which comes from two sources, S(t) is given by S(t) = R(t) + N(t) or S(t) = I(t) + N(t), where R(t) is the reference signal, I(t) is the information signal and N(t) is the noise. We are using a Biomation Model 805 transient recorder. This device samples a single channel at a rate of 5 MHz with 8 bit (0.4%) resolution. We first acquire a reference signal on the 805. We postulate R(t) is A sin $(\omega t + \delta_1)$, where the amplitude A, the frequency $\omega$ and the phase angle $\delta_1$ are all variables. These data are fit in a least square process to determine A, $\omega$ and $\delta_1$. With a large number of data points describing the order of 20-50 periods, noninteractive variables such as amplitude, frequency, and phase angle are very well determined. This process then gives an accurate measure of R(t).

Next we digitize an ED transient. The pulse information portion of the signal is modelled at I(t) = A(t) $\sin(\omega t + \delta_2)$ where A(t) is a time dependent amplitude function and $\delta_2$ is a new phase angle. The value of $\omega$ has been fixed by the reference signal. We allow $\delta_2$ to vary since phase shifts can creep in between the signal and reference. A(t) is a time dependent amplitude function which contains the information of interest. A perfectly general A(t) would be a high order polynomial such as

$$\sum_{i=0}^{10} a_i t^i .$$

However, with a little knowledge about the particular experiment in question, the number of variables in the fitting function (eleven in the above example) can be reduced significantly. In this particular case we are trying only to accurately reproduce the amplitude information, A(t). There may be many sets of values for the fitting variables $\{a_i\}$ which will all give essentially the same information

Table 1·1. Results from computer fitting of ED parameters - <u>SEARCH</u>

| Parameter | Desired Result | ± 0% noise | ± 1% noise | ± 2% noise | ± 5% noise | ± 10% noise | ± 15%* noise | ± 20%* noise |
|---|---|---|---|---|---|---|---|---|
| 1 A | $9.0 \times 10^{-1}$ | $8.9969 \times 10^{-1}$ | $8.9969 \times 10^{-1}$ | $8.9969 \times 10^{-1}$ | $8.9969 \times 10^{-1}$ | $8.9969 \times 10^{-1}$ | $8.9969 \times 10^{-1}$ | $8.9969 \times 10^{-1}$ |
| 2 $\omega$ | $1.0 \times 10^{5}$ | $1.0000 \times 10^{5}$ | $1.0000 \times 10^{5}$ | $1.0000 \times 10^{5}$ | $1.0000 \times 10^{5}$ | $1.0000 \times 10^{5}$ | $1.0000 \times 10^{5}$ | $1.0000 \times 10^{5}$ |
| 3 $\delta_1$ | 0.0 | $1.0304 \times 10^{-5}$ | $1.0304 \times 10^{-5}$ | $1.0304 \times 10^{-5}$ | $1.0304 \times 10^{-5}$ | $1.0304 \times 10^{-5}$ | $1.0304 \times 10^{-5}$ | $1.0304 \times 10^{-5}$ |
| 4 A' | 1.0 | $9.9957 \times 10^{-1}$ | $9.9719 \times 10^{-1}$ | $9.9561 \times 10^{-1}$ | $9.9325 \times 10^{-1}$ | $9.8477 \times 10^{-1}$ | $9.7699 \times 10^{-1}$ | $9.7025 \times 10^{-1}$ |
| 5 $\tau$ | $1.0 \times 10^{-5}$ | $9.9800 \times 10^{-6}$ | $1.0008 \times 10^{-5}$ | $9.9839 \times 10^{-6}$ | $9.9605 \times 10^{-6}$ | $9.9195 \times 10^{-6}$ | $9.9001 \times 10^{-6}$ | $9.8469 \times 10^{-6}$ |
| 6 $\delta_2$ | 1.5707 | 1.5707 | 1.5706 | 1.5707 | 1.5707 | 1.5709 | 1.5709 | 1.5704 |
| 7 $\alpha$ | $7.0 \times 10^{-1}$ | $6.9989 \times 10^{-1}$ | $7.0169 \times 10^{-1}$ | $7.0322 \times 10^{-1}$ | $7.1038 \times 10^{-1}$ | $7.2181 \times 10^{-1}$ | $7.2935 \times 10^{-1}$ | --- |
| 8 $\tau_1$ | $1.0 \times 10^{-6}$ | $9.5588 \times 10^{-7}$ | $1.0796 \times 10^{-6}$ | $1.1473 \times 10^{-6}$ | $1.3900 \times 10^{-6}$ | $1.8220 \times 10^{-6}$ | $2.1234 \times 10^{-6}$ | --- |
| 9 $\tau_2$ | $1.0 \times 10^{-3}$ | $1.0009 \times 10^{-3}$ | $9.8439 \times 10^{-4}$ | $9.8275 \times 10^{-4}$ | $1.0199 \times 10^{-3}$ | $1.0813 \times 10^{-3}$ | $1.1455 \times 10^{-3}$ | --- |
| time (sec) | | 3,619 | 3,404 | 3,465 | 3,476 | 2,973 | 4,649 | 7,126 |
| error 1 | | .003 | .003 | .003 | .003 | .003 | 0.003 | 0.003 |
| 2 | | .007 | .099 | .302 | 1.463 | 5.896 | 13.2 | 22.9 |
| 3 | | .008 | .108 | .300 | 1.253 | 4.453 | 9.478 | |

FITTING FUNCTIONS

Reference $A \sin(2\pi\omega t + \delta_1)$

Pulse $A'(1 - e^{t/\tau})\sin(2\pi\omega t + \delta_2)$

Decay $(\alpha e^{-t/\tau_1} + (1-\alpha)e^{-t/\tau_2})\sin(2\omega t + \delta_2)$

* less stringent fitting criteria

Table 1·2. Results from computer fitting of ED parameters - COMBINED PROGRAM

| Parameter | Desired Result | ± 0% noise | ± 1% noise | ± 2% noise | ± 5% noise | ± 10% noise |
|---|---|---|---|---|---|---|
| 1 A | $9.0 \times 10^{-1}$ | $8.9969 \times 10^{-1}$ | $8.9969 \times 10^{-1}$ | $8.9969 \times 10^{-1}$ | $8.9969 \times 10^{-1}$ | $8.9969 \times 10^{-1}$ |
| 2 $\omega$ | $1.0 \times 10^{5}$ | $1.0000 \times 10^{5}$ | $1.0000 \times 10^{5}$ | $1.0000 \times 10^{5}$ | $1.0000 \times 10^{5}$ | $1.0000 \times 10^{5}$ |
| 3 $\delta_1$ | 0.0 | $-8.4630 \times 10^{-6}$ | $-8.4630 \times 10^{-6}$ | $-8.4630 \times 10^{-6}$ | $-8.4630 \times 10^{-6}$ | $-8.4630 \times 10^{-6}$ |
| 4 A' | 1.0 | $9.9301 \times 10^{-1}$ | $9.9207 \times 10^{-1}$ | $9.9111 \times 10^{-1}$ | $9.8901 \times 10^{-1}$ | $9.8365 \times 10^{-1}$ |
| 5 $\tau$ | $1.0 \times 10^{-5}$ | $1.0015 \times 10^{-5}$ | $1.0035 \times 10^{-5}$ | $1.0008 \times 10^{-5}$ | $9.9824 \times 10^{-6}$ | $9.9238 \times 10^{-6}$ |
| 6 $\delta_2$ | 1.5707 | 1.5704 | 1.5703 | 1.5705 | 1.5705 | 1.5709 |
| 7 $\alpha$ | $7.0 \times 10^{-1}$ | $6.9998 \times 10^{-1}$ | $7.0241 \times 10^{-1}$ | $7.0425 \times 10^{-1}$ | $7.1165 \times 10^{-1}$ | $7.2224 \times 10^{-1}$ |
| 8 $\tau_1$ | $1.0 \times 10^{-6}$ | $7.2991 \times 10^{-7}$ | $1.0208 \times 10^{-6}$ | $1.1251 \times 10^{-6}$ | $1.4244 \times 10^{-6}$ | $1.8371 \times 10^{-6}$ |
| 9 $\tau_2$ | $1.0 \times 10^{-3}$ | $9.4772 \times 10^{-4}$ | $9.4772 \times 10^{-4}$ | $9.5502 \times 10^{-4}$ | $9.9905 \times 10^{-4}$ | $1.0777 \times 10^{-3}$ |
| time | | 3,072 | 1,983 | 2,683 | 3,666 | 2,571 |
| error 1 | | 0.003 | 0.003 | 0.003 | 0.003 | 0.003 |
| 2 | | .262 | 0.413 | 0.646 | 1.892 | 6.084 |
| 3 | | .257 | 0.413 | 0.593 | 1.524 | 4.544 |

FITTING FUNCTIONS

Reference $A \sin(2\pi\omega t + \delta_1)$

Pulse $A'(1 - e^{t/\tau})\sin(2\pi\omega t + \delta_2)$

Decay $(\alpha e^{-t/\tau_1} + (1 - \alpha)e^{-t/\tau_2})\sin(2\omega t + \delta_2)$

function, A(t), if the fitting function is excessively parameterized. Since we are concerned only with reproducing A(t) (the demodulated output), we are not interested in the specific values of the variables in the fitting function. For once we are on the other side of the fence and care nothing about the uniqueness of the fitting function solution; we care only that it models A(t) accurately. The same least squares routine which determined R(t) can now be utilized to determine A(t). Thus we have the envelope of the carrier wave which is the dichroic information of interest.

When the pulse is terminated the oriented system relaxes back to random orientations. This relaxation process is often a multiple exponential. This decay can be modelled in a manner similar to the rise of the steady state dichroism and in this case A(t) (where $A(t) = \alpha e^{-t/\tau_1} + (1-\alpha)e^{-t/\tau_2}$) will contain the relaxation information. Experience has shown, however, that fitting typical relaxation data with more than two components is not easy under any circumstances (6-8). It is at this point that our increase in S/N may be of critical importance.

We have tested our digital technique on synthetic data with varying amounts of random noise added to it. In table 1 we have summarized the results of the fitting from two computational techniques. Search is a sophisticated gradient search method for finding the minimum in the least squares, or $\chi^2$ surface. The combined program involves gradient and simplex methods and Monte Carlo techniques.

The instrument we wish to build is described in a schematic way in fig. 3. We use a 1000 W Hg lamp as a source and a high through-put monochromator to get as high an intensity as possible. The spectral band width of instruments of this type while not considered good by high resolution spectroscopists is excellent for electric dichroism experiments where 1-2 nm is all that is required. The emergent beam for the monochromator will be polarized and then pass through our modulator. The light which is now alternating in the plane of polarization between parallel and perpendicular to the applied field next passes through the electric dichroism cell and then into the detector. The photomultiplier tube is powered by a high voltage supply. To keep the time response of the phototube rapid, a small anode resistor is used and hence, a preamplifier right at the base of the phototube is necessary. The amplified photomultiplier anode signal or reference signal from the modulator driving system can be stored in the Biomation 805.

The Biomation 805 recorder can present both analog and digital output displays. The analog output can be directed to an oscilloscope or graphical recorder for visual inspection of the signal and the digital output can be stored on disk through a digital interface to the PDP 11/04 computer. This machine is not large enough to accommodate the least squares program but it can easily acquire the necessary data arrays. Addition of a serial line interface allows the 11/04 to

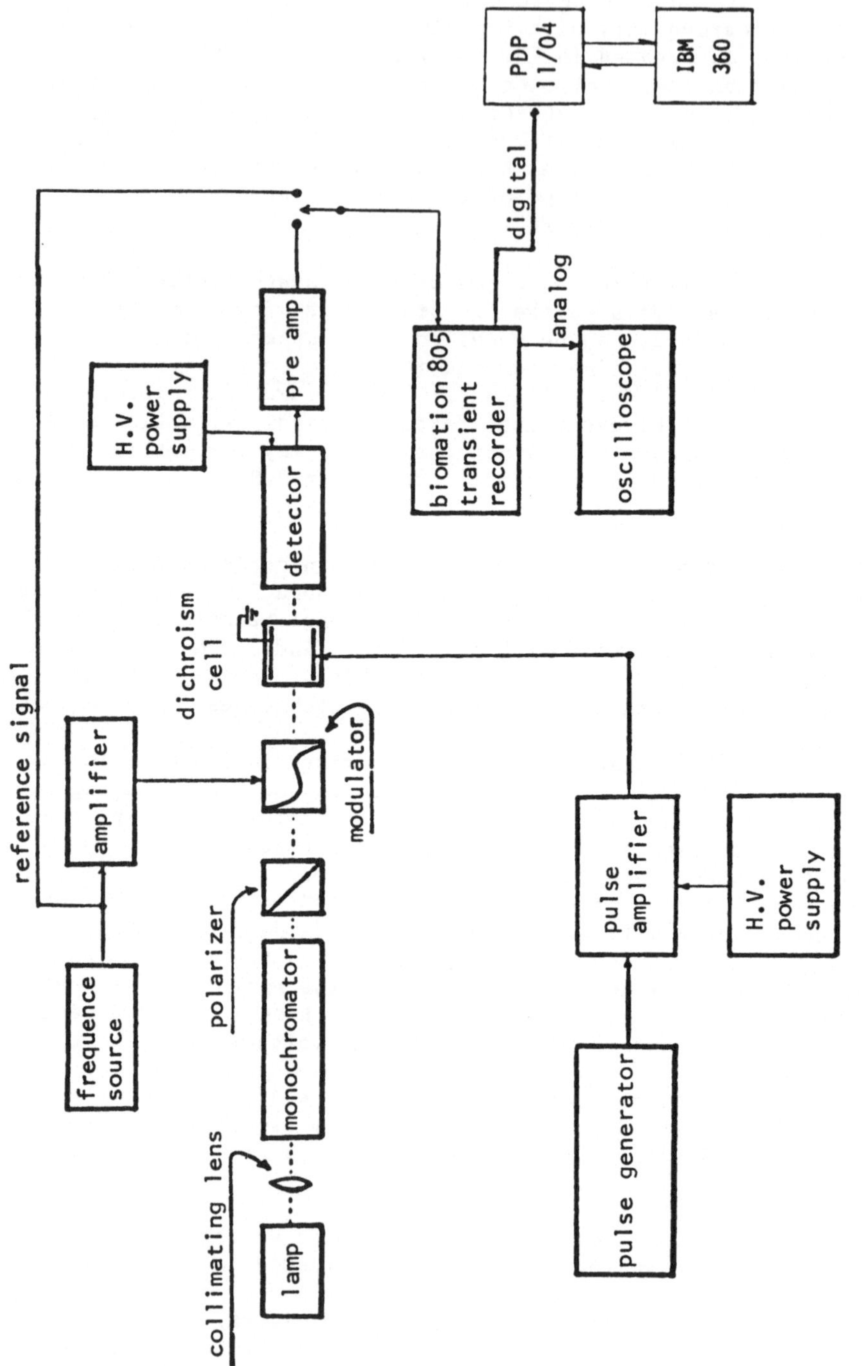


Fig. 3. The electric dichroism instrument block diagram.

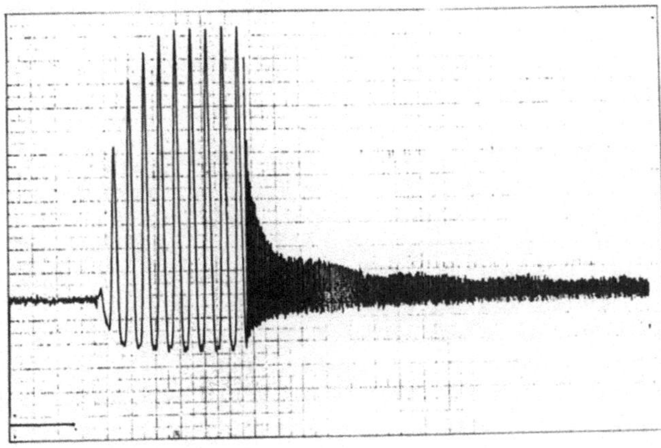

Fig. 4. The electric dichroism of calf thymus DNA in 0.1 mM $MgCl_2$ and 0.1 mM tris-phosphate buffer, pH 7.6 at 267 nm. A field strength of 12 kV/cm was applied for 100 μsec.

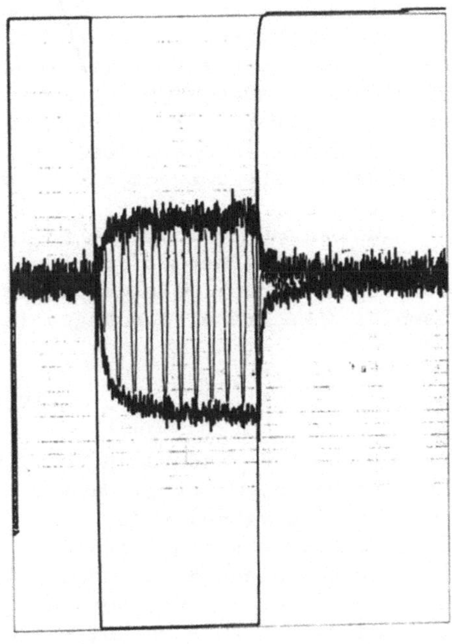

Fig. 5. The electric dichroism of the amylose iodide complex in distilled water. The pulse was 8 kV/cm and lasted 100 μsec.

be interactive with the IBM-360 through a modulator-demodulator (modem) and a phone line. Hence, the data required can be sent to the 360 where all of the necessary programs for analysis are already implemented. The results can be returned to the 11/04 and printed on the teletype. Graphical displays can be presented on oscilloscope or recorder through the digital converter.

Data which we have obtained at a 100 KHz modulation frequency are shown in fig. 4. These data were recorded for calf thymus DNA in 0.1 mM $MgCl_2$ and 0.1 mM tris-phosphate buffer, pH 7.6, at 267 nm. There are three phases to this figure. The first is the base line section which precedes the high voltage pulse which is of the order of 12 kV/cm. The second phase occurs during this pulse and the onset of dichroism can be observed. As the plane of polarization is rotated between parallel and perpendicular to the applied field, the optical signal (due to the electric dichroism) oscillates dramatically. The amplitude of this oscillation is $\Delta A$. The beginning of the third phase is indicated by the termination of the pulse. At this point the display time base changes so as to capture more of the relaxation process. A multi-component exponential decay is very clearly indicated. The cusping which can be seen in the envelope of the relaxation is a beat frequency between the Biomation sampling and the modulation frequencies. It is well known that the magnitude of the parallel component of ED is twice that of the perpendicular component. That is, the signal should go up twice as far as it goes down. This is not seen in this figure because the DNA is melting and there is a strong chemical component in the ED signal. More proper behaviour can be seen in the data for amylose iodide which is shown in fig. 5. Here also we have added the envelope of the signal by measuring and recording the parallel and perpendicular components of the signal without the modulator in operation. This verifies that the modulated signal does define the correct envelope.

We intend to continue to improve our system. The optics, in particular, are in need of replacement and in addition we would like to redesign the high voltage pulser so we may use higher ionic strength samples. We are also engaged in quantifying the signal to noise ratio.

## REFERENCES

1 Allen F S and Van Holde K E, Rev. Sci. Instr., 41 (1970) 211.
2 Brahms J, Pilet J, Damany H and Chandrasekharan V, P.N.A.S., 60 (1968) 1130.
3 Charney E and Halford R S, J. Chem. Phys., 29 (1958) 221.
4 Dows D A and Buckingham A D, J. Mol. Spect., 12 (1964) 189.
5 Kemp J C, J. Opt. Soc. Am., 59 (1969) 950.
6 Allen F S and Van Holde K E, Biopolymers, 10 (1971) 865.
7 Hildebrand F B, "Introduction to Numerical Analysis", McGraw-Hill, New York (1956).
8 Isenberg I and Dyson R D, Biophysical J., 9 (1969) 1337.

# SIMULTANEOUS MEASUREMENT OF DICHROISM AND BIREFRINGENCE ON SUSPENSIONS OF ABSORBING PARTICLES

J. C. Ravey* and C. Houssier**

*Laboratoire de Biophysique, Centre 1er Cycle, Université de Nancy I, C.O. No. 140, 54037 Nancy Cedex, France

**Laboratoire de Chimie Physique, Université de Liège Sart-Tilman, B-4000 Liège, Belgium

The principle of a new method for the simultaneous measurement of the dichroism and birefringence (including the determination of their signs) of absorbing solutions subjected to electric pulses, is described. It involves the measurement of the photocurrent signals as a function of the angle of the polarization directions with respect to the electric field.
Observations on carbon-black suspensions using this procedure are reported.

## INTRODUCTION

In previous studies implying the measurement of the birefringence in absorption regions showing dichroism effects, a preliminary determination of the dichroism was made using an optical arrangement with one polarizer only (1,2). The photocurrent signal detected in the presence of polarizer and analyser was then corrected for the dichroism contribution in order to determine the birefringence or birefringence dispersion curve. In the present study, a procedure is described which does not require any modification of the optical setting whilst allowing the birefringence and dichroism amplitudes and signs to be evaluated. It is applied to the study of carbon-black suspensions and its usefulness is compared to that of other methods.

Carbon-black consists of rigid aggregates made of small isotropic spheres (0.01 $\mu$ about), having a complex refractive index $\hat{n}$. Its value is not perfectly known, but $\hat{n}$ = 1.96 - i 0.66 is the value generally accepted (9). These aggregates are not spherical, and from a statistical point of view, they may be likened to spheroids of ellipticity about 2, and of volume compactness between 0.03 and 0.25 (7). Of course, such anisometric particles may be readily oriented in an electric or hydrodynamic field. The suspensions become at the same time dichroic and birefringent, due to the anisometric form of the particles. Their study necessitates the special experimental set up which will be described in the present paper.

The apparatus uses pulsed electric-fields contrary to the one recently described by Champion et al. (10).

## OPTICAL RELATIONSHIPS

If $\hat{n}_\perp$ and $\hat{n}_\parallel$ are the two complex principal indices of the medium birefringent and dichroic, we define $\delta$ and $\delta'$ as follows:

$$\delta - i\,\delta' = \frac{2\pi \ell}{\lambda_0}(\hat{n}_\parallel - \hat{n}_\perp)$$

where $\ell$ is the pathlength and $\lambda_0$ the wavelength. $\delta$ is then the optical retardation and $\delta'$ the differential absorption related to the more conventionally used dichroism.

$$\Delta A = A_\parallel - A_\perp \quad \text{with} \quad \delta' = 2.3\,\Delta A/2.$$

The expression for the calculation of the birefringence in regions of absorption where a dichroism contribution may be present, has been previously derived for optical arrangements not including (1,2) or including (3-5) a quarter-wave plate between the sample and the analyser. For the latter arrangement, which offers the highest sensitivity, the expression reads (when the slow axis of the quarter-wave plate is at 135° with respect to the field direction):

$$\frac{\Delta I}{I_\alpha} = \frac{I_A - I_\alpha}{I_\alpha} = \frac{\exp(-\delta'/3)\,\{\cosh\,\delta' - \cos(\delta + 2\alpha)\}}{(1 - \cos 2\alpha)} - 1 \tag{1}$$

where $\Delta I$ is the change of light intensity on the photomultiplier with respect to the light intensity "at rest" $I_\alpha$, for a rotation of the analyser through an angle $\alpha$ from the crossed position towards the polarizer.

In the absence of dichroism, eq. 1 becomes:

$$\frac{\Delta I}{I_\alpha} = \frac{\cos 2\alpha - \cos(\delta + 2\alpha)}{(1 - \cos 2\alpha)} \tag{1b}$$

the well-known relationship used for birefringence determinations (2).

For small $\delta'$ values, eq. 1 may be approximated by:

$$(\Delta I/I_{\alpha})\sin^2\alpha \triangleq \sinh^2(\delta'/2) + \sin(\delta/2).\sin(2\alpha+\delta/2) \quad (2a)$$

and for small $\delta$ and $\alpha$ values:

$$(\Delta I/I_{\alpha})\sin^2\alpha \triangleq (\delta'/2)^2 + (\delta/2)^2 + (\delta/2)\sin 2\alpha \quad (2b)$$

The slope of the $(\Delta I/I_{\alpha})\sin^2\alpha$ versus $\sin(2\alpha)$ plot thus gives the birefringence (with its sign) while the intercept at $\alpha = 0^o$ allows one to obtain the square of the dichroism term and thus does not provide the dichroism sign.

If the polarizer is turned through an angle $\alpha'$ towards the electric field, eq. 1 is then replaced by:

$$\frac{\Delta I}{I_{\alpha\alpha'}} = \frac{\exp(-\delta'/3)\{\cosh\delta' - \sin 2\alpha'.\sinh\delta' - \cos 2\alpha'.\cos(2\alpha+\delta)\}}{(1-\cos 2\alpha.\cos 2\alpha')} - 1 \quad (3a)$$

and for $\alpha = 0$

$$\frac{\Delta I}{I_{\alpha'}} = \frac{\exp(-\delta'/3)\{\cosh\delta' - \sin 2\alpha'\sinh\delta' - \cos 2\alpha'\cos\delta\}}{(1-\cos 2\alpha')} - 1 \quad (3b)$$

For low values of $\delta'$ eq. 3b yields the approximate relationships:

$$(\Delta I/I_{\alpha'})\sin^2\alpha' \triangleq \sinh^2(\delta'/2) + \sin^2(\delta/2)\cos 2\alpha' - (1/2)\sinh\delta'\sin 2\alpha' \quad (4a)$$

which further reduce to:

$$(\Delta I/I_{\alpha'})\sin^2\alpha' \triangleq (\delta'/2)^2 + (\delta/2)^2 - (\delta'/2)\sin 2\alpha' \quad (4b)$$

for small values of $\alpha'$ and $\delta$.

The reversed situation is reached as compared to eq. 2b, i.e. the sign of the dichroism is given by the sign of the slope of the $(\Delta I/I_{\alpha'})\sin^2\alpha'$ versus $\sin(2\alpha')$ plot, while the ordinate is the same as that of eq. 2b and does not yield the birefringence sign.

A combination of observations at respective rotations $\alpha'$ and $\alpha$ of the polarizer and analyser and at $\alpha = \alpha' = 0^o$ will thus allow us to solve the problem completely. In the case of negative birefringence measurements, care must be taken to choose values of $\alpha$ sufficiently large so that the requirement $\alpha > -\delta/2$ will always be maintained (2). Alternatively, the experimenter can choose the sign of the uncrossing angle so that the $\Delta I/I_{\alpha}$ will be positive.

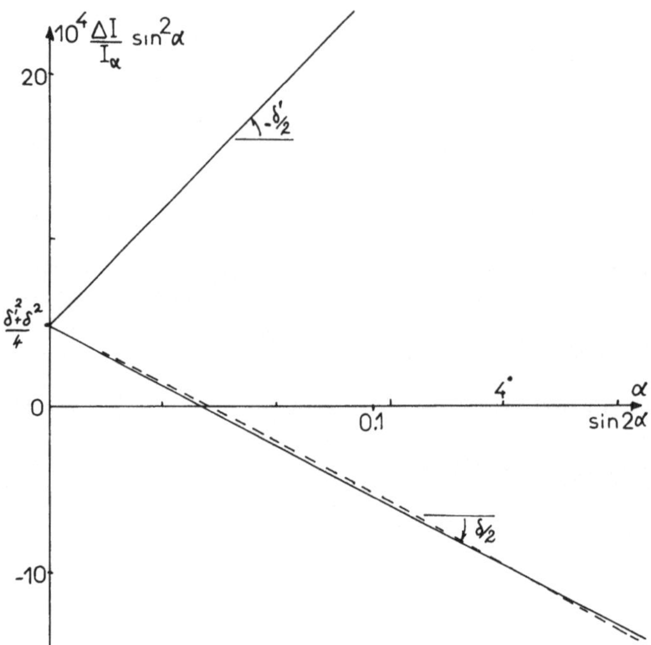


Fig. 1. Graphical representation of eqs. 1 and 3b using artificial data. $\delta$ = -0.02 rad ($\Delta n_{550nm}$ = $-1.75 \times 10^{-7}$); $\delta'$ = 0.04 ($\Delta A$ = -0.035).

In fig. 1 are shown two curves calculated on the basis of the complete eqs. 1 and 3b, which indicate that the approximation of linear relationships (eq. 2b and 4b) is readily applicable in our experimental conditions. Only a small curvature appears when the measurements are extended over the range of 0 to 10$^o$.

Taking into account the stray-light correction and a possible contribution of solvent birefringence ($\delta_s$) and a residual birefringence of the cell ($\delta_o$), eqs. 1, 2b and 4b must be written as:

$$\left(\frac{\Delta I}{I_\alpha}\right)_{corr} = \frac{\exp(-\delta'/3)\{\cosh\delta' - \cos(\delta_t + 2\alpha)\}}{1 - \cos(2\alpha + \delta_o)} - 1 \quad (5a)$$

or, for small $\alpha$, $\delta$ and $\delta'$:

$$\left(\frac{\Delta I}{I_\alpha}\right)_{corr} \cdot \sin^2(\alpha + \delta_o/2) \simeq (\delta'/2)^2 + \left(\frac{\delta_t^2 - \delta_o^2}{4}\right) + \left(\frac{\delta_t - \delta_o}{2}\right) \sin 2\alpha \quad (5b)$$

and

$$\left(\frac{\Delta I}{I}\right)_{\alpha' \mathrm{corr}} = \frac{\exp(-\delta'/3)\left\{\cosh \delta' - \sin 2\alpha' \sinh \delta' - \cos 2\alpha' \cos \delta_t\right\}}{1 - \cos 2\alpha' \cdot \cos \delta_o} - 1 \tag{6a}$$

or

$$\left(\frac{\Delta I}{I}\right)_{\alpha' \mathrm{corr}} \cdot \left(\frac{1 - \cos 2\alpha' \cdot \cos \delta_o}{2}\right) \simeq (\delta'/2)^2 + \left(\frac{\delta_t^2 - \delta_o^2}{4}\right) - (\delta'/2) \cdot \sin 2\alpha' \tag{6b}$$

with $\delta_t = \delta + \delta_o + \delta_s$.

In these relationships, $(\Delta I/I_\alpha)_{corr}$ is the relative change of light intensity corrected for stray-light by the relation (2,4):

$$(\Delta I/I_\alpha)_{corr} = (\Delta I/I_\alpha)_{meas} \left(1 + \frac{K_{SL}}{\sin^2 \alpha}\right) \tag{7a}$$

with $K_{SL} = I_{SL} \sin^2\alpha \,/\, (I_{t,\alpha} - I_{SL})$

where the stray-light constant $K_{SL}$ is determined in the absence of a cell, from the residual light intensity $I_{SL}$ emerging from the analyser in the crossed position, and the total light intensity $I_{t}$, $\alpha$ transmitted for a rotation $\alpha$ of the analyser (or polarizer).

## EXPERIMENTAL PROCEDURE

The electric dichroism and birefringence measurements were performed with the previously described instrumentation (2,5,6) which includes a fast transient recorder (Biomation Model 8100) interfaced to a minicomputer (Modular Computer Systems, II/10) with a $10^6$ words disc unit.

The quarter-wave plate/analyser accessory gives an accuracy of about 0.01$^\circ$ on the $\alpha$ angle setting and allows adjustments at positive or negative $\alpha$ values. The polarizer holder provides $\alpha'$ readings with an accuracy of ca.0.1$^\circ$. The present stray-light constant of the carefully adjusted optical setting amounted to ca. $2.5 \times 10^{-4}$. For $\alpha = 2^\circ$ and 5$^\circ$, the correction amounts respectively to about 20%, and 3.3%.

A quartz-iodine projector lamp was used as light source and the birefringence was measured at 550 nm. The quarter-wave plate is made of mica and has the proper $\pi/2$ retardation in the 500-600 nm region (2). Quartz rectangular cuvettes with a 10 mm pathlength and an electrode spacing of 1.5 mm were used (2). For the present measurements, a cuvette with a particularly low residual birefringence ($\delta_o < 0.2^\circ$) was selected.

The electric pulses (single shot) were delivered by a pulse generator similar to that previously described (2) but using EL 509 tubes, giving a maximum field strength of 13-14 kV/cm.

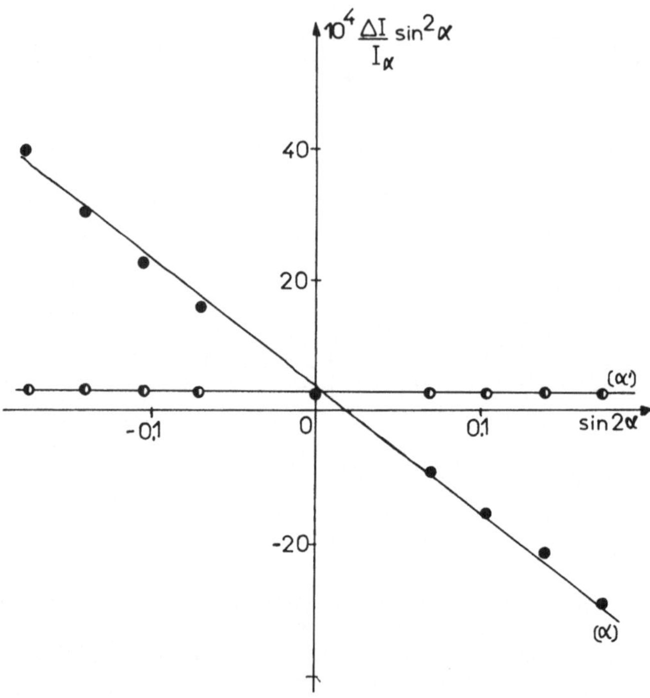


Fig. 2. Check of the polarization directions with respect to the electric field using an aqueous solution of quaternized polyvinylpyridine (0.15 g/l). E = 3.55 kV/cm, $\delta = -38.5 \times 10^{-3}$.

The detailed procedure for the signals recording and processing is described elsewhere (5).

Eqs. 2b and 4b offer an easy means of checking the optical adjustment of the system with respect to the electric field direction. The measurements of $(\Delta I/I_\alpha)\sin^2\alpha$ at positive and negative $\alpha$ and $\alpha'$ values should give two straight-lines which intercept at zero abscissa ($\alpha' = \alpha = 0^o$). The application of this control to our instrumentation using a non-dichroic very stable polymer sample gave the results shown in fig. 2: the slope of $(\Delta I/I_{\alpha'})\sin^2\alpha'$ is effectively zero, within a precision of 0.001. It is also seen that a deviation of less than 20' from the expected orientation could be present. This is an instrument geometrical parameter which has not frequently been controlled; the present procedure appears most adequate for this control and is much more sensitive than other previously described methods.

## RESULTS AND DISCUSSION

The results of the measurements performed on H.A.F. carbon-black suspensions are shown in fig. 3 and fig. 4. The particles studied here are rigid aggregates of spheres (250 Å diameter) with a molecular weight of $2.10^9$, having an equivalent spheroid of ellipticity 1.8 and compactness 0.25. An independent measurement of the dichroism without quarter-wave plate and analyser yielded a reduced dichroism value of 0.054 at 3.5 kV/cm in the cyclohexanone.

The slopes obtained from the straight-lines of fig. 3 and 4 by linear regression give the following values for the birefringence and dichroism (at 3.55 kV/cm):

in cyclohexanone: $\delta = -(42 \pm 1) \times 10^{-3}$ rad ($\Delta n = 3.7 \times 10^{-7}$)

($A_{550}^{10} = 1.22$) $\delta' = (81.2 \pm 0.6) \times 10^{-3}$ ($\Delta A = 0.07$; $\Delta A/A = 0.0575$)

$\delta'/\delta = -1.95$

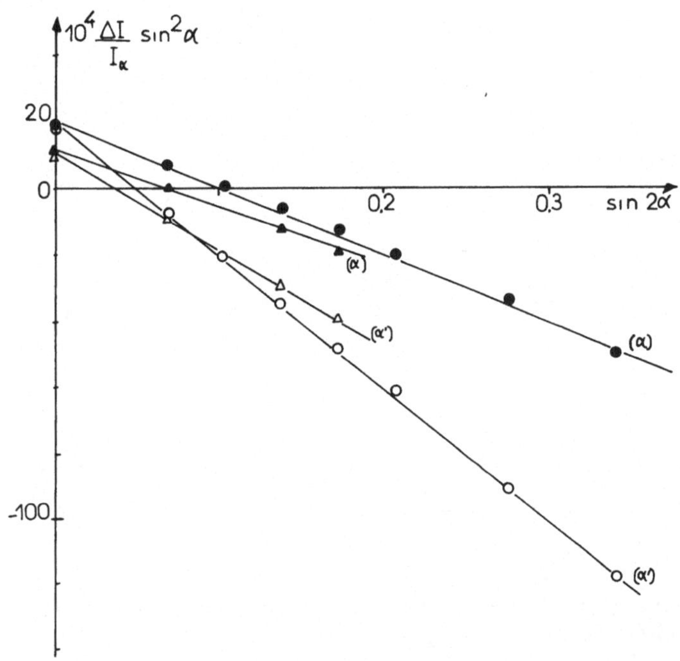


Fig. 3. Electro-optical behaviour of H.A.F. carbon-black suspensions. E = 3.55 kV/cm; pulse duration 5 ms. Circles: suspension in cyclohexanone ($A_{550nm}^{10mm} = 1.22$); triangles: suspension in acetophenone ($A_{550nm}^{10mm} = 0.88$). Measurements as a function of the uncrossing angles $\alpha$ and $\alpha'$ of the analyser and polariser, respectively.

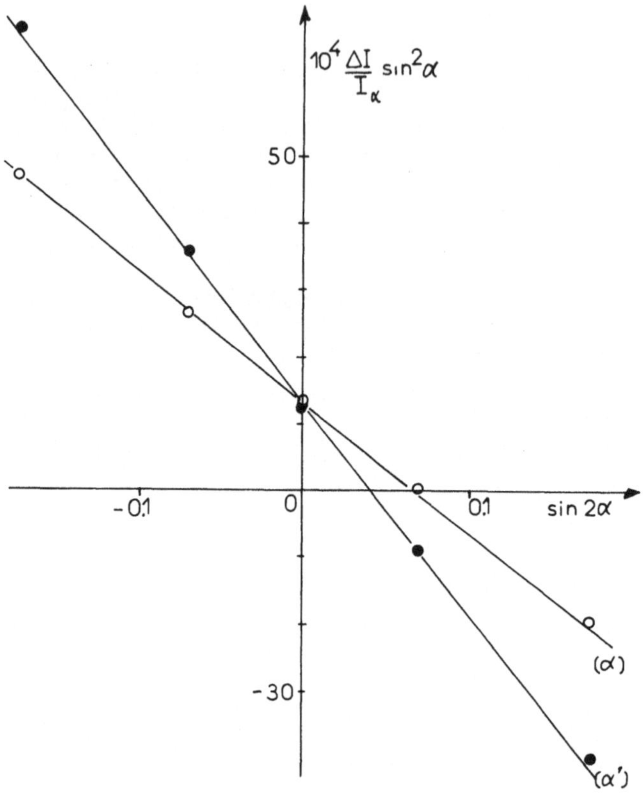


Fig. 4. Symbols as in fig. 3, for carbon-black suspension in acetophenone. Measurements covering positive and negative $\alpha$ and $\alpha'$.

in acetophenone: $\delta = -(43.6 \pm 0.6) \times 10^{-3}$ rad ( $\Delta n = 3.82 \times 10^{-7}$ )

( $A_{550}^{10} = 0.88$ ) $\delta' = (63.4 \pm 1.4) \times 10^{-3}$ ( $\Delta A = 0.055$; $\Delta A/A = 0.0625$ )

$\delta'/\delta = -1.45$

The intersection of the two lines at zero abscissa confirms the good setting of the optical arrangement.

Of course, full curves are not necessary, since in principle, $\delta$ and $\delta'$ may be obtained with only two measurements, i.e. for two pairs of $\alpha$, $\alpha'$ values, by inversion of the formulae 3a. In fact, it is much simpler to add a third measurement with $\alpha$ or $\alpha' = 0$ and use the conditions ( $\alpha$, o), (o, o), (o, $\alpha'$) which also allow a check on the validity of the results, for which the accuracy cannot be so good.

In that case, we have:

$$\frac{\delta'}{\delta} = -\frac{\sin 2\alpha}{\sin 2\alpha'} \cdot \frac{y_{\alpha'} - y_o}{y_\alpha - y_o}$$

where $y_\alpha = (\Delta I/I_\alpha)\sin^2\alpha$.
The value obtained by this way with $\alpha = \alpha' = 5^o$ is $\delta'/\delta = -2.04$ in cyclohexanone. If $\hat{\gamma}_1$ and $\hat{\gamma}_2$ are the complex principal optical polarizabilities of the particles, then

$$\frac{\delta'}{\delta} = \frac{\mathrm{Im}\,(\hat{\gamma}_1 - \hat{\gamma}_2)}{\mathrm{Re}\,(\hat{\gamma}_1 - \hat{\gamma}_2)},$$

independent of the degree of their orientation. As found previously (3), $\delta$ is a negative form birefringence, due to the complex value of the carbon-black and the actual value of the compactness. Extensive study of this unusual effect will be reported elsewhere.

The most important advantage of this method is the fact that all the parameters can be obtained without any requirement to alter the optical arrangement.

The sensitivity is not better than that reached by the usual procedure for dichroism measurements. For the results presented in this paper, the relative change of light intensity at 3.55 kV/cm observed in the dichroism measurement with polarizer only was ca. 10% for the parallel polarization direction With the arrangement including the quarter-wave plate and analyser, the corresponding relative change of light intensity in the condition of maximum $\Delta I/I$ was of the order of (0.2 - 0.3)%. However, this new method is nearly a null method which allows much larger amplifications for low absorbing and weakly dichroic samples: values of $\delta'$ less than 0.001 should be still measurable.

The most drastic limitations of this method are the following: (i) the system must be equipped with graduated holders for the polarizer and analyser, allowing $\alpha$ and $\alpha'$ to be set with an accuracy of at least 0.1$^o$; (ii) the quarter-wave plate must be either adjustable (Babinet-Soleil compensator) or achromatic (8) in order to allow measurements over an extended wavelength region; (iii) for measurements in the ultraviolet region, the polarizer, analyser and quarter-wave plate must have the highest possible transmittance (with a subsequent large increase of the cost of the instrumentation); (iv) corrections must always be applied for stray-light, Kerr cell and solvent birefringence contributions, all of which are absent when the polarizer is used alone.

## REFERENCES

1 Houssier C and Kuball H G, Biopolymers 10 (1971) 2421-2433.
2 Fredericq E and Houssier C, Electric dichroism and electric birefringence, Clarendon Press, Oxford (1973).

3 Ravey J C, Colloid and Polymer Sci., 253, (1975) 292-305.
4 Tricot M and Houssier C, in Polyelectrolytes (Frisch K C, Klempner D and Patsis A V, Eds.) Technomic Publishing Co., Westport (1976) p. 43-90.
5 Houssier C, Thèse d'Agrégation de l'Enseignement Supérieur, Université de Liège (1977).
6 Houssier C, Laboratory Practice (1974) 562-563.
7 Ravey J C, Premilat S and Horn P, Eur. Polymer J. 6 (1970) 1527-1537.
8 O'Konski C T in Encyclopedia of Polymer Science and Technology (Bikales, N Ed.) Vol. 9, p.555, Wiley N.Y.
9 Senftleben H and Benedict E, Ann. Phys. 60, (1919) 297.
10 Champion J V, Downer D, Meeten G H and Gate L F, J. Phys. E., Scientific Instruments, 10, (1977) 1137.

# ELECTRIC DICHROISM OF PURPLE MEMBRANE

K. Tsuji and K. Rosenheck

Department of Membrane Research, The Weizmann Institute of Science, Rehovot, Israel

The electric dichroism of bacteriorhodopsin has been studied by subjecting suspensions of purple membrane fragments (mean diameter $\sim 0.5\,\mu m$) to pulsed electric fields. The reduced dichroism as a function of varying orientation of the purple membrane sheets was calculated, taking the $P_3$ symmetry of the arrangement of bacteriorhodopsin molecules into account. The results for this symmetrical trimer model are different from those derived from the monomeric model used by others, showing that the latter underestimates the angle between the normal to the membrane plane and chromophore transition moment by several degrees. Analysis of the experiments in terms of the trimer model restricts the allowed transition moment orientations to either one or the other of two ranges of angles, according to whether the planes of the membrane fragments orient parallel or perpendicular to the imposed electric field.

## INTRODUCTION

The purple membrane is one of the cell membrane fragments of the halophilic bacterium Halobacterium halobium (1). Its main, and perhaps only, function is that of a light-driven proton pump (2,3). The light-absorbing pigment, bacteriorhodopsin, has a broad absorption band with a maximum at 570 nm, which is attributed to a retinal molecule bound by a protonated Schiff base linkage to one of the lysine residues in the protein chain (1,4). Bacteriorhodopsin constitutes about 75 weight percent of the membrane and is arranged in a crystalline lattice of space group $P_3$ with three protein molecules per unit cell (5-7). Each protein molecule consists of seven $\alpha$-helical segments oriented nearly perpendicular to the plane of the membrane, and almost totally immersed within the lipid layer (5-7).

Recently experiments of transient dichroism and linear dichroism have been carried out to determine the direction of the chromophore transition moment in the purple membrane (8-10). The single chromophore model (11) has been applied for the calculation of the angle between the direction of the chromophore transition moment and the normal to the membrane plane. However, the purple membrane does not fit this model because of the $P_3$ symmetry of the bacteriorhodopsin molecules. In this article the dichroism for the more realistic trimer model, with the chromophores in $P_3$ symmetry will be derived theoretically and applied to the electric dichroism of the purple membrane.

It has been reported that purple membrane fragments are oriented in the presence of magnetic or electric fields (12,13). Neugebauer et al. (12) suggested that the membrane plane orients perpendicularly to the magnetic field and that the degree of the orientation is strongly dependent on their state of aggregation. Shinar et al. (13) reported two different kinds of relaxation in the decay process of the electric dichroism of purple membrane suspensions, one being due to the rotational motion of the membrane fragments, the other to that of bacteriorhodopsin molecules inside the membrane. Rotational motion of bacteriorhodopsin has been observed when the rigid lattice of the purple membrane is disrupted by organic solvents (14) or when bacteriorhodopsin is incorporated into lipid vesicles (15). On the other hand, according to Razi Naqvi et al. (16), bacteriorhodopsin is immobilized in the native purple membrane. The above work notwithstanding, the orientation direction of purple membrane fragments, as well as the direction of the chromophore transition moment, in the presence of an electric field, are as yet open questions. Electric dichroism measurements, giving information with respect to the possible ranges of the fragment orientation and the direction of the chromophore transition moment, will be discussed in the present work.

## THEORY

Let us consider the Cartesian coordinate system $0\text{-}\xi\eta\zeta$ which is fixed on the disk as shown in fig. 1; $0\zeta$ is the direction of the normal with respect to the plane of disk, 0 is the point of intersection of the $P_3$ symmetry axis with the plane of the disk, the chromophore transition moment $\mu$ makes an angle $\alpha$ with $0\zeta$, and the direction of one of the projections of the transition moments on the plane of the disk $(\xi 0 \eta)$ is $0\xi$.

The components of the transition moments $\mu$ along $\xi$, $\eta$ and $\zeta$ are described as following:

$$\vec{\mu}_1 \; (\mu_{\xi 1} = \mu\sin\alpha; \quad \mu_{\eta 1} = 0; \quad \mu_{\zeta 1} = \mu\cos\alpha) \tag{1a}$$

$$\vec{\mu}_2 \; (\mu_{\xi 2} = -\tfrac{1}{2}\mu\sin\alpha; \quad \mu_{\eta 2} = \frac{\sqrt{3}}{2}\mu\sin\alpha; \quad \mu_{\zeta 2} = \mu\cos\alpha) \tag{1b}$$

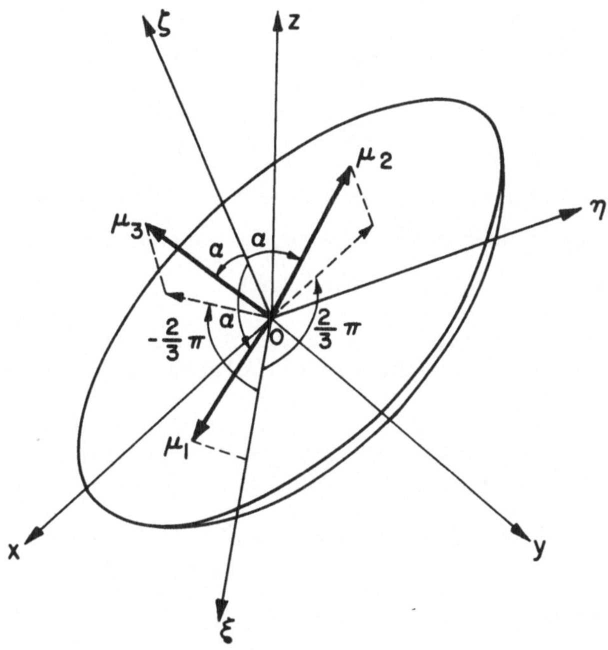


Fig. 1. Geometrical representation of a disk-like particle containing three chromophores in $P_3$ symmetry.

$$\vec{\mu}_3 \,(\, \mu_{\xi 3} = -\tfrac{1}{2}\mu\sin\alpha\,; \quad \mu_{\eta 3} = -\frac{\sqrt{3}}{2}\mu\sin\alpha\,; \quad \mu_{\zeta 3} = \mu\cos\alpha \,) \tag{1c}$$

where $0 \leq \alpha \leq \frac{\pi}{2}$. Since the contribution of a vector $\mu$ to the absorbance is equivalent to that of a vector $-\mu$, the components of the total transition moment of the trimer which contribute to the absorbance are:

$$\mu_\xi = \sum_{i=1}^{3} |\mu_{\xi i}| = 2\mu\sin\alpha \tag{2a}$$

$$\mu_\eta = \sum_{i=1}^{3} |\mu_{\eta i}| = \sqrt{3}\mu\sin\alpha \tag{2b}$$

$$\mu_\zeta = \sum_{i=1}^{3} |\mu_{\zeta i}| = 3\mu\cos\alpha \tag{2c}$$

In order to calculate the absorbance, the other Cartesian coordinate O-xyz which is fixed in the space is introduced. The direction Oz is parallel to the electric field, Ox is the direction of the incident light beam, and Oy is perpendicular to both Ox and Oz. The relation between the coordinate O-xyz and O-$\xi\eta\zeta$ is expressed by the Euler's angles $\theta$, $\phi$ and $\psi$. $\theta$ is the angle between the z and $\zeta$ axes, $\phi$ is the angle of rotation around the z axis and $\psi$ defines the rotation in the plane of the disk, i.e. around $\zeta$ axis. The transformation matrix S from the O-$\xi\eta\zeta$ to the O-xyz coordinate system is given by (17):

$$S = \begin{Bmatrix} \cos\phi\cos\psi-\cos\theta\sin\phi\sin\psi & -\cos\phi\sin\psi-\cos\theta\sin\phi\cos\psi & \sin\theta\sin\phi \\ \sin\phi\cos\psi+\cos\theta\cos\phi\sin\psi & -\sin\phi\sin\psi+\cos\theta\cos\phi\cos\psi & -\sin\theta\cos\phi \\ \sin\theta\sin\psi & \sin\theta\cos\psi & \cos\theta \end{Bmatrix} \quad (3)$$

By using the transformation matrix S, the x, y and z components of the total transition moment are calculated from eqs. 2a, 2b and 2c as follows:

$$\mu_x = \mu\Big[\big\{2(\cos\phi\cos\psi-\cos\theta\sin\phi\sin\psi) - \sqrt{3}(\cos\phi\sin\psi +\cos\theta\sin\phi\cos\psi)\big\}\sin\alpha + 3\sin\theta\sin\phi\cos\alpha\Big] \quad (4a)$$

$$\mu_y = \mu\Big[\big\{2(\sin\phi\cos\psi+\cos\theta\cos\phi\sin\psi) - \sqrt{3}(\sin\phi\sin\psi -\cos\theta\cos\phi\cos\psi)\big\}\sin\alpha - 3\sin\theta\cos\phi\cos\alpha\Big] \quad (4b)$$

$$\mu_z = \mu\big\{(2\sin\theta\sin\psi + \sqrt{3}\sin\theta\cos\psi)\sin\alpha + 3\cos\theta\cos\alpha\big\} \quad (4c)$$

The absorbances along Ox, Oy and Oz are given by:

$$A_x = k\,\mu_x^2 \quad (5a)$$

$$A_y = k\,\mu_y^2 \quad (5b)$$

$$A_z = k\,\mu_z^2 \quad (5c)$$

k is a proportionality constant.

Let us suppose now that the disk remains at angle $\theta$ with respect to the electric field but that rotation about the field direction and about the O$\zeta$ axis is free. Averages over all possible orientations of $\phi$ and $\psi$ yield:

$$\bar{A}_x = \bar{A}_y = \frac{1}{(2\pi)^2}\int_0^{2\pi}\int_0^{2\pi} A x\, d\phi\, d\psi$$

$$= A\mu\Big\{\frac{7}{4}(1+\cos^2\theta)\sin^2\alpha + \frac{9}{2}\sin^2\theta\cos^2\alpha\Big\} \quad (6a,b)$$

$$\bar{A}_z = \frac{1}{2\pi}\int_0^{2\pi} A_z d\psi = A\mu(\frac{7}{2}\sin^2\theta\sin^2\alpha + 9\cos^2\theta\cos^2\alpha) \quad (6c)$$

with $A\mu = k\mu^2$, where $A\mu$ is the absorbance along the transition moment direction. Here, the parallel component of the absorbance with respect to the electric field $A_{\parallel}$, corresponds to $\bar{A}_z$, the perpendicular components, $A_{\perp}$, to $\bar{A}_x$ $(=\bar{A}_y)$.

For the random orientation, the average of the direction of $\theta$ is required. Since the probability of finding $\mu$ between $\theta$ and $\theta + d\theta$ is equal to $\sin\theta \, d\theta/2$ the mean value is given by:

$$A = \frac{1}{2}\int_0^{\pi} \bar{A}_z \sin\theta \, d\theta = \frac{1}{2}\int_0^{\pi} \bar{A}_x \sin\theta \, d\theta = \frac{1}{2}\int_0^{\pi} \bar{A}_y \sin\theta \, d\theta$$

$$= \frac{A\mu}{3}(7 + 2\cos^2\alpha) \qquad (7)$$

Consequently, the parallel and perpendicular dichroism $\Delta A_{\parallel}$ and $\Delta A_{\perp}$ are:

$$\Delta A_{\parallel} = A_{\parallel} - A = \frac{A\mu}{6}(3\cos^2\theta - 1)(25\cos^2\alpha - 7) \qquad (8a)$$

$$\Delta A_{\perp} = A_{\perp} - A = -\frac{A\mu}{12}(3\cos^2\theta - 1)(25\cos^2\alpha - 7) \qquad (8b)$$

and the reduced dichroism is:

$$\frac{\Delta A}{A} = \frac{A_{\parallel} - A_{\perp}}{A} = \frac{3}{4} \cdot \frac{(3\cos^2\theta - 1)(25\cos^2\alpha - 7)}{7 + 2\cos^2\alpha} \qquad (9)$$

Note that the dichroism is zero when $\cos^2\alpha = \frac{7}{25}$, i.e. $\alpha = 58°$, or $\cos^2\theta = \frac{1}{3}$, i.e. $\theta = 54.7°$.

In the case of the single transition moment, the same calculation is done by using the equation (1a) for the transition moment instead of equation (2), resulting in:

$$A_{\parallel} = A\mu\left(\tfrac{1}{2}\sin^2\theta\sin^2\alpha + \cos^2\theta\cos^2\alpha\right) \qquad (10a)$$

$$A_{\perp} = A\mu\left\{\tfrac{1}{4}(1 + \cos^2\theta)\sin^2\alpha + \tfrac{1}{2}(\sin^2\theta\cos^2\alpha)\right\} \qquad (10b)$$

$$A = \frac{1}{3}A\mu \qquad (11)$$

Therefore,

$$\Delta A_{\parallel} = \frac{A\mu}{6}(3\cos^2\theta - 1)(3\cos^2\alpha - 1) \qquad (12a)$$

$$\Delta A_{\perp} = -\frac{A\mu}{12}(3\cos^2\theta - 1)(3\cos^2\alpha - 1) \qquad (12b)$$

and:

$$\frac{\Delta A}{A} = \tfrac{3}{4}(3\cos^2\theta - 1)(3\cos^2\alpha - 1) \qquad (13)$$

Here, the dichroism is zero when $\cos^2\alpha = \frac{1}{3}$, i.e. $\alpha = 54.7°$ or $\cos^2\theta = \frac{1}{3}$, i.e. $\theta = 54.7°$.

Let us consider two extreme cases; when $\theta = 0°$, the disk orients with its normal parallel to the field, and when $\theta = 90°$, the disk orients with the normal perpendicular to the field. For $\theta = 0^{\circ}$, the reduced dichroism is:

$$\frac{\Delta A}{A} = \frac{3}{2} \cdot \frac{25\cos^2\alpha - 7}{7 + 2\cos^2\alpha} \qquad \text{for } P_3 \text{ trimer} \tag{14a}$$

$$\frac{\Delta A}{A} = \frac{3}{2}(3\cos^2\alpha - 1) \qquad \text{for single transition moment} \tag{14b}$$

For $\theta = 90^{\circ}$:

$$\frac{\Delta A}{A} = -\frac{3}{4} \cdot \frac{25\cos^2\alpha - 7}{7 + 2\cos^2\alpha} \qquad \text{for } P_3 \text{ trimer} \tag{15a}$$

$$\frac{\Delta A}{A} = -\frac{3}{4} \cdot (3\cos^2\alpha - 1) \qquad \text{for single transition moment} \tag{15b}$$

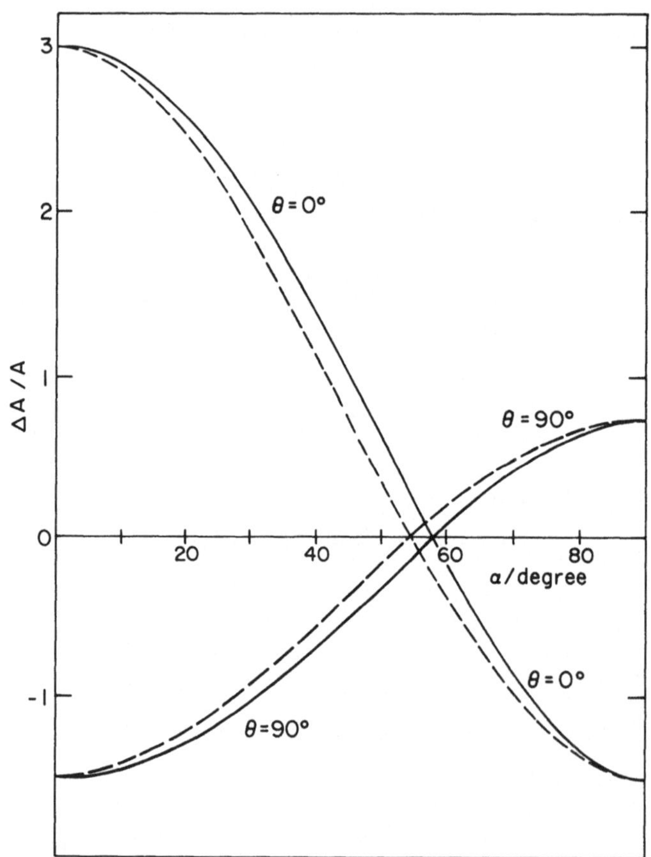


Fig. 2. Comparison of the reduced dichroism as function of $\alpha$ for trimer model (____) with that for single chromophore model (----) $\theta = 0^{\circ}$ and $90^{\circ}$

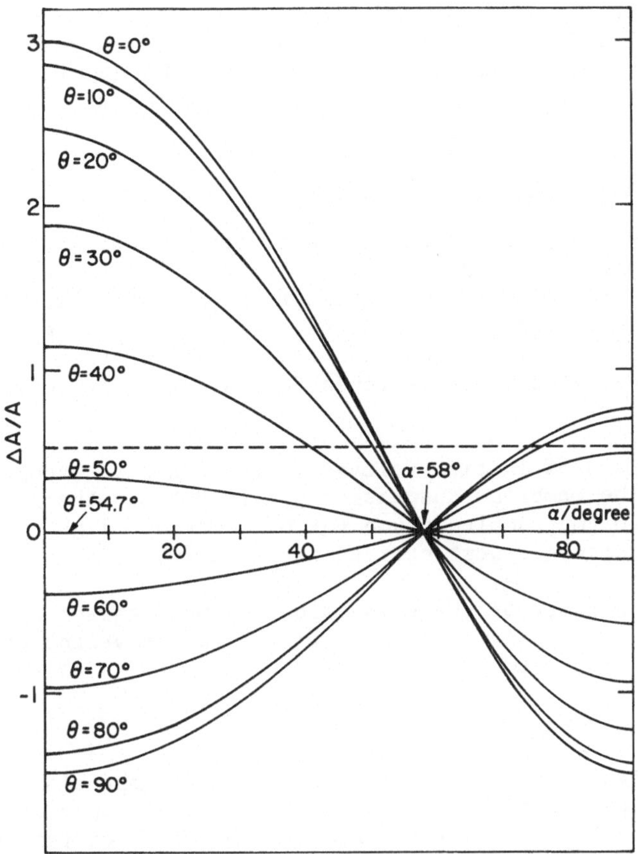


Fig. 3. Variation of the reduced dichroism with $\alpha$ for trimer at varying values of $\theta$.

Fig. 2 shows the variation of the reduced dichroism with $\alpha$ in these two cases both for the trimer and for the single transition moment. Thus, the relation between the reduced dichroism and the angle $\alpha$ is slightly different for $P_3$ trimer and the single transition moment.

The plots of the reduced dichroism versus $\alpha$ at various $\theta$'s for the $P_3$ trimer are shown in fig. 3. When $\cos^2\theta = 1/3$ the reduced dichroism is zero at any angle of $\alpha$, and when $\cos^2\alpha = 7/25$ the reduced dichroism is zero at any orientation.

## EXPERIMENTAL

### Materials

Purple membranes were isolated from Halobacterium halobium as described by Oesterhelt and Stoeckenius (18). The molecular weight of the bacteriorhodopsin was estimated to be 24,000 by sodium dodecyl sulfate-polyacrylamide gel electrophoresis. Bacteriorhodopsin concentration in the purple membrane suspension was determined on the basis of a molar extinction coefficient of 63,000 $M^{-1}$ $cm^{-1}$ at 570 nm (19).

## Measurements of the Electric Dichroism

A temperature jump apparatus (20) was used for the measurements of the electric dichroism. A polacoat polarizing filter was inserted between the exit slit of the high-intensity Bausch and Lomb monochromator and the sample cell. The transient signal from the photomultiplier was stored in the fast transient recorder (Biomation Model 8700), displayed on an oscilloscope (Tektronix 546) or recorded on an XY recorder.

Electric pulses of initial voltage Vo which decays exponentially were applied to the membrane suspension placed between the electrodes of a cell, d = 1.0 cm, sample volume v = 0.9 $cm^3$, and light path l = 0.7 cm. All measurements were carried out at 20°C.

The optical density of the suspension in the presence of the electric field and for polarized light with electric vector parallel to the field, $A_{\parallel}$ is given by (11):

$$A_{\parallel} = A - \log (I_{\parallel} / I) \tag{16}$$

where I is the light intensity transmitted by the suspension in the absence of the electric field, $I_{\parallel}$ is that in the presence of the electric field for light polarized parallel to the electric field, and A is the optical density of the suspension at rest. A was measured with a Cary 15 spectrophotometer.

The reduced dichroism is defined as (11):

$$\Delta A/A = (A_{\parallel} - A_{\perp}) / A \tag{17}$$

where $A_{\perp}$ is the optical density of the suspension in the presence of the electric field with light polarized perpendicular to the electric field. The concentration of the purple membrane was adjusted such that the optical density at rest was between 0.28 and 0.30 ($\sim$4.7 nanomole bacteriorhodopsin in 1 $cm^3$).

Scattering effects were corrected by recording the absorption spectrum of the purple membrane suspension in 61% Urografin (a mixture of the sodium and methylglucamine salts of 3,5-bis-acetamido-2,4,6-triiodobenzoic acid, Schering AG, Berlin), the refractive index of which is 1.42.

<u>Measurements of the Flow Dichroism</u>

A Cary 15 spectrophotometer was used for the measurements of the flow dichroism. The sample cell consisted of two concentric cylinders. The sample was placed in the annular space between cylinders. When the outer cylinder was set in motion by an air jet a laminar flow was set up across the gap. The resulting shearing forces produce an orientation of the suspended particles. A polacoat polarizer was inserted between the exit slit of the monochromator and the cell.

## RESULTS

<u>Electric Dichroism</u>

Fig. 4 shows a typical signal of the change in transmission with monitoring light polarized parallel and perpendicular to the electric field, when the electric pulse was applied to the purple membrane suspension. The maximum value of the reduced dichroism at 560 nm calculated from these data was 0.53. A very fast decrease of the transmission ($\sim$100 $\mu$s), i.e. increase of the optical density, followed by a slow decay consisting of several relaxation times was observed, when the light was polarized parallel to the electric field. The sign of the change in transmission was reversed for perpendicularly polarized light.

The average relaxation time in the decay process, $\bar{\tau}$, calculated by the area method (21) is 100 ms in 0.05M NaCl aqueous solution, and 150 ms and 310 ms in 0.05M NaCl - 13.5% glycerol and 0.05M NaCl - 27% glycerol, respectively. The sensitivity of the average relaxation time to medium viscosity suggests that the membrane fragments themselves are oriented by the electric field. According to Perrin's equation for oblate ellipsoids (22):

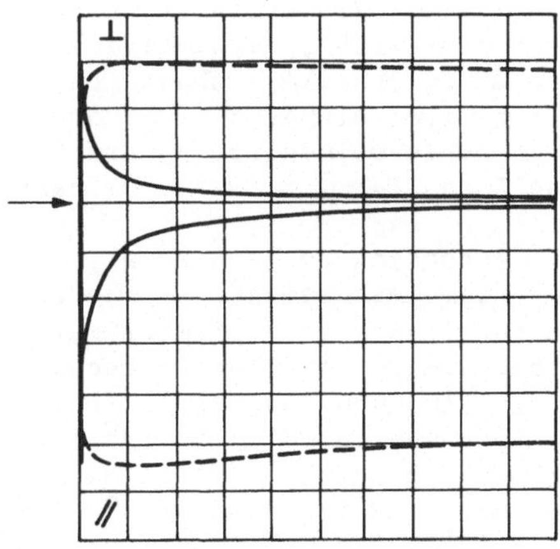


Fig. 4. Typical transient signals of transmission change with polarized light parallel and perpendicular to the electric field; ordinate, 200 mV/div; abscissa, —— 200 m sec/div, ---- 50 $\mu$sec/div, initial output voltage of photomultiplier, 8.0 V. The base line is indicated by an arrow.

$$\tau = \frac{\eta . 2\pi r^3}{3kT} \tag{18}$$

a diameter of the orienting unit r can be estimated. Here $\eta$ is the viscosity of the medium, k is the Boltzmann constant and T is the absolute temperature. The average diameters thus calculated are 0.58 $\mu$m in aqueous, 0.61 $\mu$m in 13.5% glycerol and 0.67 $\mu$m in 27% glycerol. These values are in good agreement with the observed diameter of purple membrane fragments (0.5 - 1.0 $\mu$m (4,6)).

Flow Dichroism

Laminar flow in the Couette type flow device described in the experimental section can be reasonably assumed to orient the membrane fragments with their planes aligned parallel to the stream lines and perpendicular to the direction of the incident light beam. It, therefore, provides a simple means to ascertain qualitatively the chromophore orientation (in plane versus out of plane). The optical density recorded in membrane suspensions during flow was always larger ($\sim$ 15%) than that at rest, irrespective of the state of polarization of the incident light.

## DISCUSSION

From the value of the reduced dichroism, $\Delta A/A = 0.53$, we can estimate the possible range of orientation of the fragments and the angle between the chromophore transition moment and the normal to the plane of the membrane by using fig. 2. From the points at which the line $\Delta A/A = 0.53$ crosses the calculated curves corresponding to $\Delta A/A$ at varying angles of orientation $\theta$ of the fragments with respect to the electric field, the dependence of $\alpha$ on $\theta$ can be derived. It is found that the possible values of $\alpha$ are $0^o < \alpha < 51.4^o$ (case I) and $73.8^o < \alpha < 90^o$ (case II), corresponding to $47.8^o > \theta > 0^o$ and $90^o > \theta > 71.8^o$, respectively. If we define the degree of orientation $\Phi$ by $\Delta A/\Delta A_s$, where $\Delta A_s$ is the dichroism for complete orientation, i.e. either $\theta = 0^o$ or $\theta = 90^o$, the above mentioned two regions of possible values for $\alpha$ are seen to correspond to $0.18 < \Phi < 1$ for case I and $1 > \Phi > 0.71$ for case II.

According to Heyn et al. (8), the angle $\alpha$ is $78^o \pm 3^o$ and $71^o \pm 4^o$ from measurements of the transient dichroism of large liposomes and the linear dichroism of multilayers of oriented purple membranes, respectively. The latter technique has also been used by Bogomolni et al. (10) who reported $\alpha$ to be $67.5^o$, as well as by Korenstein and Hess (9) who found $\alpha \geqslant 63^o$. The angles were derived from the single chromophore model. Using the more realistic trimer model, these values are corrected to be $80^o \pm 3^o$, $74^o \pm 4^o$ (Heyn et al.'s), $70.5^o$ (Bogomolni et al.'s) and $\geqslant 66^o$ (Korenstein and Hess), by referring to the data in fig. 2.

The positive values of the flow dichroism in our experiments indicate that the angle $\alpha$ is larger than $58^{\circ}$, according to eq. 9. This is consistent with the results just mentioned above. All these suggest that the chromophore transition moment is more in-plane. However, case I cannot be excluded, if it is assumed that the orientation of either the entire bacteriorhodopsin molecule, or parts of it within the membrane is changing under the influence of the electric field. From the narrow range of possible values of $\Phi$ for case II, the purple membrane fragments seem to be oriented almost completely with the membrane plane parallel to the electric field. This orientation would require an electric dipole moment lying in the plane of the membrane. In view of the hexagonal structure it seems improbable that this could be a permanent dipole since it is very difficult to visualize a permanent charge distribution giving rise to such a dipole. The direction of the permanent dipole moments associated with the helical residues is along the helical axis (23) and therefore any net permanent dipole of this type would be perpendicular to the plane of the membrane. It must hence be concluded that for case II the major contribution to the membrane orientation comes from an induced type dipole moment. The latter could arise from the polarization of the counterions within the double layer surrounding the membrane, giving a net resultant moment along the long axis of the particles (24). If, on the other hand, case I is assumed to represent the real situation, this could be attributed to the permanent dipole moments of the helical regions in bacteriorhodopsin.

Let us consider these two models in terms of the kinetic behaviour shown in fig. 1. A prominent feature of the transient changes in transmission is the very fast rise time, of the order of $\sim 100$ $\mu$s. Such a fast rise time was observed under conditions somewhat different from ours, also by Shinar et al. (13). In order to achieve the high degree of particle orientation, nearing saturation, that are required by case II, an extremely large dipole moment would be required (25). If, on the other hand, case I applies, the degree of orientation can vary within a much larger range. The very fast rise time, could then be explained by assuming that the observed maximum in the transients does not correspond to saturation of the electric field effect, simply because the pulse is too short to achieve a true stationary state.

So far, none of the two cases described above can be rigorously excluded. Further investigation, using rectangular electric field pulses with longer duration times, is necessary.

ACKNOWLEDGEMENT

We thank Prof. I. R. Miller and Dr. R. Korenstein for many useful discussions. This work was supported by a grant from the Stiftung Volkswagenwerk.

## REFERENCES


1 Oesterhelt D and Stoeckenius W, Nature New Biol., 233, (1971) 149.
2 Oesterhelt D and Stoeckenius W, Proc. Natl. Acad. Sci. U.S., 70, (1973) 2853.
3 Danon A and Stoeckenius W, Proc. Natl. Acad. Sci. U.S., 71, (1974) 1234.
4 Blaurock A E and Stoeckenius W, Nature New Biol., 233, (1971) 152.
5 Henderson R and Unwin P N T, Nature, 257, (1975) 28.
6 Henderson R, J. Mol. Biol., 93, (1975) 123.
7 Blaurock A E, J. Mol. Biol., 93, (1975) 139.
8 Heyn M P, Cherry R J and Muller U, J. Molec. Biol., 117, (1977) 607.
9 Korenstein R and Hess B, FEBS Lett., 89, (1978) 15.
10 Bogomolni R A, Hwang S B, Tseng Y W, King G I and Stoeckenius W, Biophys. J., 17, (1977) 98a.
11 Fredericq E and Houssier C, "Electric Dichroism and Electric Birefringence", Oxford University Press, London (1973).
12 Neugebauer D-C, Blaurock A E and Worcester D L, FEBS Lett., 78, (1977) 31.
13 Shinar R, Druckmann S, Ottolenghi M and Korenstein R, Biophys. J., 19, (1977) 1.
14 Heyn M P, Bauer P J and Dencher N A, "Biochemistry of Membrane Transport", Semenza G and Carafoli E, eds., Springer-Verlag, Heidelberg (1977) p.96.
15 Cherry R J, Muller U and Schneider G, FEBS Lett., 80, (1977) 465.
16 Razi Naqvi K, Gonzalez-Rodriguez J, Cherry R J and Chapman D, Nature New Biol., 245, (1973) 249.
17 Corben H C and Stehle P, "Classical Mechanics", John Wiley & Sons, Inc., New York (1950) p.174.
18 Oesterhelt D and Stoeckenius W, Meth. Enzymol., 31, (1974) 667.
19 Oesterhelt D and Hess B, Eur. J. Biochem., 37, (1973) 316.
20 Rigler R, Rabl C R and Jovin T M, Rev. Sci. Instrum., 45, (1974) 580.
21 Yoshioka K and Watanabe H, "Physical Principles and Techniques of Protein Chemistry, Part A", Leach S J, ed., Academic Press, New York (1969) p.344.
22 Perrin F J, Phys. Radium., 7, (1926) 390.
23 Wada A, "Poly-$\alpha$-Amino Acids", Fasman G, ed., Marcel Dekker, New York (1967) p.369.
24 Neumann E and Katchalsky A, Proc. Natl. Acad. Sci. U.S., 69, (1972) 993.
25 O'Konski C T, Yoshioka K and Orttung W H, J. Phys. Chem., 63, (1959) 1558.

# ELECTROCHROMISM OF ORGANIC DYES IN POLYMER MATRICES

E. E. Havinga and P. Van Pelt

Philips Research Laboratories, Eindhoven,
The Netherlands

Results of electrochromic measurements on several organic dyes dissolved in polymer matrices are presented. The advantages of polymers over common liquids (higher breakdown field strengths and the existence of a glass point) are discussed. Moreover, the possibility of studying the softening of polymers around their glass transitions with an electrochromic dye as a probe is demonstrated.

## INTRODUCTION

Electrochromism comprises all reversible variations in the absorption (or emission) spectra that are induced by an applied electric field. For solutions of organic molecules, two different physical processes lead to field-induced variations of their spectra, viz. a reorientation of the molecules and a change in their energy levels.

For the purpose of electrochromic measurements polymer matrices are in two ways superior to common liquid solvents. First, the electric breakdown fields in polymers ($\geqslant 10^6$ V/cm) are much higher than those typically met in liquid dielectrics ($\approx 10^5$ V/cm). The high fields attainable in polymers give rise to large electrochromic effects, and moreover, in favourable cases, the saturation of these effects can be studied. Secondly, amorphous polymers possess a glass transition temperature, separating a temperature region in which solute molecules are practically fixed in a rigid matrix from a higher temperature region in which the matrix behaves more or less liquid-like such that reorientation of the solute molecules can occur if electric fields are applied.

In the present paper we will demonstrate how to take advantage of polymer matrices to measure electrochromic properties of dissolved organic molecules, and reversely how such molecules can be used as a probe to study the softening of the polymer matrix around its glass transition.

## THEORY

Liptay (1) has worked out a theory for the electrochromic effects occurring in dilute liquid solutions of absorbing molecules subject to low applied fields. An extension of this theory to high-field effects has been given by Yamaoka and Charney (2). The equations of these theories can readily be adapted to cover our experiments on polymer matrix samples. As our measurements were carried out on samples in which the direction of the applied field is parallel or anti-parallel to that of the Pointing vector of the unpolarized light, we will give below only equations pertinent to this case.

The simplifications usually made to reduce the number of parameters in the interpretation of broad-band spectra amount to the assumption that both the ground state $|0\rangle$ and the excited state $|1\rangle$ of an absorbing molecule are independent of its specific surroundings in the solution and its rotational and vibronic quantum numbers. The latter influences are taken into account by blurring the sharp transition into a band, whose shape remains invariant during the experiment. For such an "averaged" molecule, one can define the dipole moment in the ground state, $\mu_0 = \langle 0 | e\bar{r} | 0 \rangle$, that in the excited state, $\bar{\mu}_1 = \langle 1 | e\bar{r} | 1 \rangle$, and the transition matrix element $\bar{m} = \langle 1 | e\bar{r} | 0 \rangle$ (e is the electronic charge). The influence of applied fields on these moments may be taken into account by introducing polarizabilities, hyperpolarizabilities, etc. The molecule absorbs light of frequency $\nu (h\nu = E_1 - E_0)$ with a probability proportional to $(\bar{m}.\bar{E})^2$ ($\bar{E}$ is the vector of the electric field of the light beam).

Reorientation of molecules, e.g. caused by an electric field, will change the average value of $(\bar{m}.\bar{E})^2$ for an ensemble of molecules and hence change the intensity of the absorption band without affecting its shape or position. In liquid solutions field-induced reorientations of molecules can only be sustained by holding the orienting field (short relaxation times). Consequently on top of the influence of the reorientation of the molecules one always measures a contribution of the field-dependence of the transition matrix element. In polymer matrices, on the other hand, the field-induced reorientation effect can be measured separately from other electrochromic effects by comparing the absorbance of poled and virgin samples at temperatures well below the glass transition of the polymer and measured without applying an electric field. In a poled sample a field-induced reorientation has been accomplished by applying an electric field at a temperature $T_1$ (about equal to the glass transition temperature of the polymer), followed by cooling down rapidly to a much lower measuring temperature while holding the field. The rigidity of the polymer matrix at this temperature prevents any appreciable relaxation to random orientations of the solute molecules when the field is released.

The reorientation effect, measured by the ratio of the absorbances (A) of a poled sample (p) and a virgin sample (v) is given by:

$$A(p)/A(v) = 1 - \tfrac{1}{2}(3\cos^2\psi - 1)(1 - 3\coth(x)/x + 3/x^2) \qquad (1)$$

where

$$x = |\bar{\mu}_0| \cdot |\bar{F}_i| / kT_1. \qquad (1a)$$

In deriving this equation we retained the dipolar terms only, having verified that contributions to the reorientation of the molecules in the field due to anisotropic ground state polarizabilities have, for all molecules in this investigation, a negligible influence on the measured effect. In eq. 1 $\psi$ denotes the angle between $\bar{\mu}_0$ and $\bar{m}$ and $\bar{F}_i$ is the internal field acting on the solute molecules.

We approximate $\bar{F}_i$ by the Onsager field:

$$\bar{F}_i = \bar{F}\,\varepsilon\,(D\,\varepsilon_\infty + 1 - D)/(D\,\varepsilon_\infty + \varepsilon.(1 - D)) \qquad (2)$$

Here $\bar{F}$ is the applied field, $\varepsilon$ is the static dielectric constant of the matrix, whilst the solute molecule is represented by an ellipsoid of molecular dimensions (obtainable from standard bond lengths and angles) with optical dielectric constant $\varepsilon_\infty$ and depolarization factor D.

The second effect of an applied electric field is to change the position of the energy levels of the molecules, shifting the transition frequency of the individual molecules. For molecules having large ground state dipole moments the electrochromic effect due to reorientation of the molecules in the field is much larger than this effect. Therefore, the rigidity of the polymer matrix at low temperatures is also particularly useful for measuring the latter effects as orientation is strongly suppressed.

The field dependence of the absorbance of a sample at low temperature may be analyzed with the help of the function

$$A(\nu,F)/\nu = (1+\delta)\{A(\nu,o)/\nu\} + \langle\Delta\nu\rangle \cdot \frac{\delta[A(\nu,o)/\nu]}{\delta\nu} + \tfrac{1}{2}\langle(\Delta\nu)^2\rangle \frac{\delta^2[A(\nu,o)/\nu]}{\delta\nu^2} \qquad (3)$$

Here $\nu$ is the frequency of the incident light beam, and $\delta$ is a parameter, much smaller than 1, that provides for the field dependence of the transition moment of the absorption band as well as for small reorientational effects, which will still occur in the glass phase of the polymer matrix. Our measurements did not allow a quantitative separation of both effects. The parameter $\langle\Delta\nu\rangle$ describes a shift of the absorption band as a whole; in samples with randomly distributed orientations of dye molecules it is in general very small (only polarizability terms), but in poled samples its magnitude is governed by

$$h\langle\Delta\nu\rangle = (\cos\psi \; \cos\varphi - 2\cos\eta) \, . \, xy/5 + \ldots . \qquad (4)$$

where

$y = |\bar{\mu}_1 - \bar{\mu}_0| \, . \, |\bar{F}_i| = |\Delta\bar{\mu}| \, . \, |\bar{F}_i|$, the maximum value of the difference in the field induced shifts of the ground state energy level and the excited state energy level, $\varphi$ is the angle between $\bar{\mu}_0$ and $\Delta\bar{\mu}$ and $\eta$ the angle between $\bar{m}$ and $\Delta\bar{\mu}$. The parameter $\langle(\Delta\nu)^2\rangle$ describes a field-induced broadening of the absorption band; both in virgin and poled samples the dipolar contribution amounts to

$$h^2 \langle(\Delta\nu)^2\rangle = (2 - \cos^2\eta) \, . \, y^2/5 + \ldots . \qquad (5)$$

and polarizability terms contribute only fourth order terms in $\bar{F}$.

## EXPERIMENTAL

The samples used in our present investigation of electrochromism were prepared as follows: A solution of both dye and polymer (polymethyl methacrylate, Elvacite* 2009) in acetone or chloroform was poured on to a teflon* plate and the solvent was allowed to evaporate slowly. The resulting polymeric film containing the dye molecules in solution was heated to about 120°C in vacuum to remove the solvent more completely and then separated from the teflon* plate. Two or three layers of such a film are hot-pressed between glass or quartz plates provided with transparent electrodes (indium oxide or evaporated gold). A different preparation method was used to obtain samples with an epoxy resin (Epikote+ 1001 or 1004) as polymer matrix: the resin was rubbed in a mortar into a fine powder, and ground together with the dye. The powder was melted under vacuum on to a glass plate with electrode, and a similar glass plate placed on top of it. Then the sample was allowed to cool down slowly. The concentration of dye molecules in the matrix was always chosen to be around $3.10^{-4}$ mol/kg, the layer thickness about 100 $\mu$m and the electrode surface about $2 \times 1$ cm$^2$.

Absorbance curves were measured in a Cary 17 spectrometer with the light beam perpendicular to the electrode surfaces. The tempera-

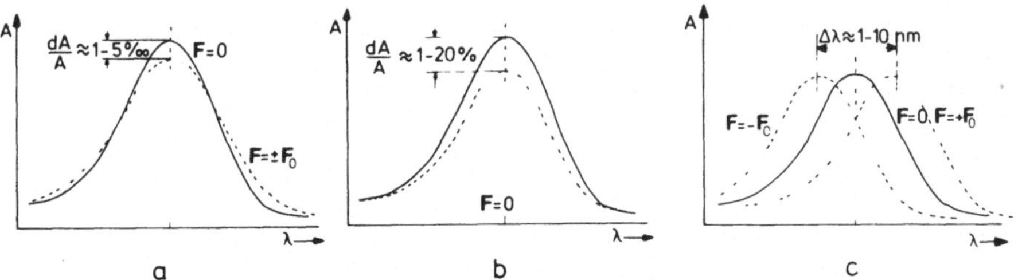


Fig. 1. Measuring cycle (Absorption, A, versus wavelength, $\lambda$):
a. Virgin sample, field-induced broadening.
b. Poled vs virgin sample, no applied field.
c. Poled sample, field-induced broadening and shift.

ture of the sample was controlled between -100°C and +150°C by way of a steady flow of cooled or heated nitrogen. The sample holder, moreover, allowed the application of high field strengths over the sample. A full measuring cycle is sketched in fig. 1. First the broadening effect (eqs. 3 and 5) was measured on a virgin sample well below the glass transition temperature (in most cases room temperature) applying high DC voltages up to 10 kV. Then the sample was poled and its absorption remeasured, giving the reorientation effect (eq. 1). Finally the field dependence of the absorbance of the poled sample was determined and analyzed with the help of eqs. 3, 4 and 5. If the angles $\psi$, $\delta$ and $\eta$ are fixed by the symmetry of the dye molecule (two orthogonal symmetry planes) the latter measurements merely give a check on the values obtained for $\bar{\mu}_0$ and $\Delta\bar{\mu}$; if the molecule has a lower symmetry (only a single symmetry plane) the three measurements may be used to calculate the angles and $\Delta\bar{\mu}$, once the ground state dipole moment is known from a dielectric measurement.

For the electrochromic measurement of the softening of the polymer matrices, using the dyes as a probe, the sample was placed in the same sample holder, as used in the Cary 17, but now situated between a monochromator and a PM tube, a lock-amplifier serving as a phase-selective detector of the changes in the transmitted light intensity when the sample was subjected to an AC square wave voltage (up to 1 kV with waveform ⊓⊔ ).

## RESULTS

The structure formulas of the dyes selected in the present investigation are given in fig. 2. The results of the evaluation of the electrochromic measurements expressed as values of $\bar{\mu}_0$, $\bar{\mu}_1$ and the direction of $\bar{m}$ are indicated on the left of each molecule. In the first instance we have used a spherical molecular model to evaluate the internal fields (eq. 2 with D=1/3). For compound I such a model is in fair agreement with the molecular dimensions. For

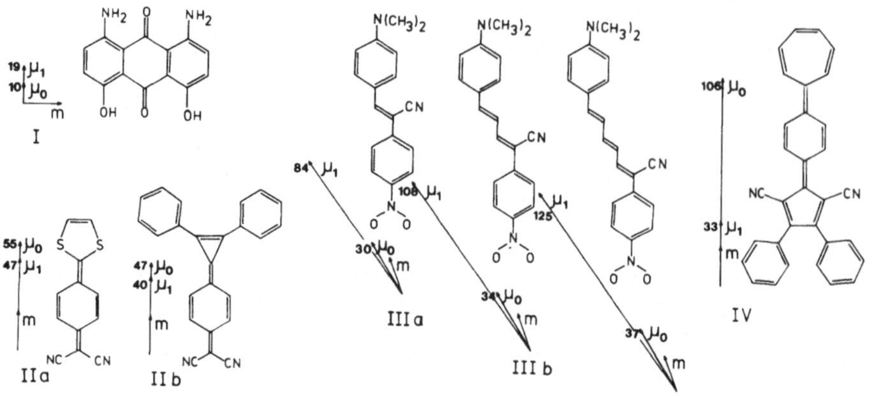

Fig. 2. Structure of the measured dyes, dipole moments of ground state, $\bar{\mu}_0$, and excited state, $\bar{\mu}_1$, (in $10^{-30}$ C.m) and direction of transition moments. For compound IIa a transition with $\bar{m}$ perpendicular to $\bar{\mu}_0$ is also present (see text).

the compounds IIIa, IIIb and IIIc, where the symmetry does not fix the direction of the dipole moments, the electrochromic measurements were combined with dielectric measurements of the ground state dipole moment. It turned out that the results of both types of measurements were contradictory if spherical molecules were assumed. With ellipsoidal shapes of the molecules (D = 0.165, 0.142 and 0.124, derived from standard molecular dimensions) the contradiction was removed. A detailed discussion of these compounds will be given elsewhere (3). For compound IIb the shape of the molecule is rather different from ellipsoidal so that no realistic improvement can be expected by choosing an ellipsoidal model for the calculation of the internal field. Finally, for the compounds IV and IIa the accuracy of the measurements was so low that any refinement of the internal field calculation seemed to be superfluous. The low accuracy of the measurements for compound IV was caused by difficulties encountered in dissolving of this dye in a polymer matrix (large $\bar{\mu}_0$); in the case of compound IIa the reason for the low accuracy was the complicated spectrum (four bands, each consisting of overlapping bands with mutual orthogonal polarization). For compound IIb, which has a large ground state dipole moment and which could be measured quite accurately, the deviation of the orientational effect from a simple quadratic function in F (eq. 1) could be observed and was found to be in agreement with the theoretical formula.

The results of the measurements shown in fig. 2 may be summarized as follows: Compound I possesses a small moment both in the ground state and in the excited state; its $\mu_0$-value agrees more or less with the sum of local group moments. Compounds IIa and IIb have a rather large dipole moment (much in excess of local moments) both in the ground

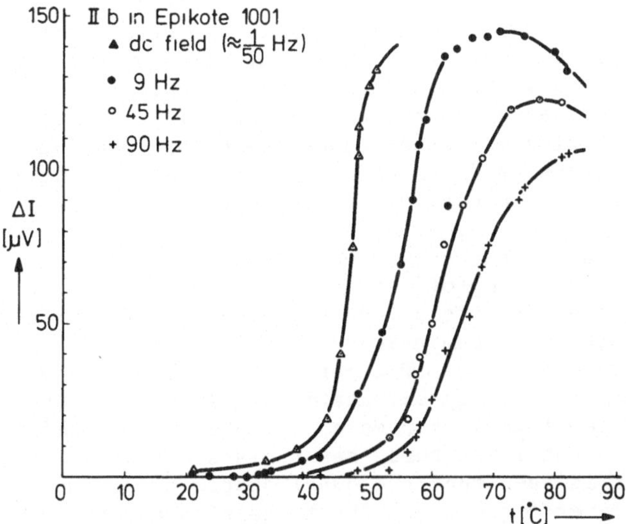


Fig. 3. Field-induced change in transmitted light intensity as a function of temperature for a solution of dye IIb in Epikote+ 1001.

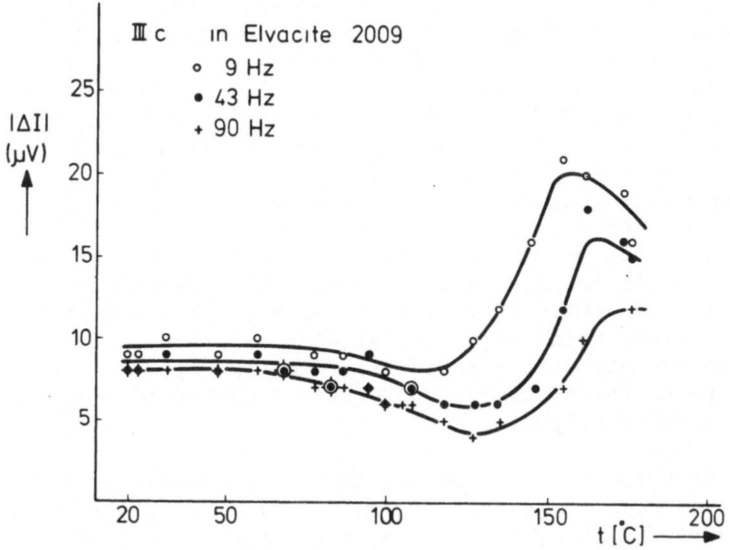


Fig. 4. Field-induced change in transmitted light intensity as a function of temperature for a solution of dye IIIc in Elvacite* 2009.

state and in the excited state. They may be described as "Zwitterions". Compounds III a, b, c have dipole moments in the ground state only slightly larger than expected from the local moments present; upon excitation, however, a large electron transfer takes place, leading to more Zwitterionic character in the excited state. Finally, compound IV is clearly Zwitterionic in the ground state, but upon excitation such a large back-transfer of electrons takes place that in the excited state only a small dipole moment is left.

In fig. 3 and 4 we have shown some results of electrochromic measurement of the softening of the polymer matrix around its glass transition by plotting the field-induced change in transmitted light intensity as a function of temperature. In fig. 3 dye IIa, which is very well suited for this type of experiment (small value of $\Delta\bar{\mu}$ and large value of $\bar{\mu}_0$ i.e. large orientational effects and only small other electrochromic effects) serves as a probe in Epikote+ 1001, which, according to the literature, has a glass transition temperature of about 40°C. It is seen that the quasi-static measurement reproduces this value; at higher frequencies the transition to the liquid-like behaviour occurs at higher temperatures and is also more gradual. The decrease of the measured effect with increasing temperature is partly due to the dependence of the absorbance change on T and partly to the variation of the absorbance with temperature (about 1% decrease for an increase of 10°C). In fig. 4 the softening of the PMMA matrix is followed using the dye IIIc as a probe. This dye has a very large $\Delta\mu$, accounting for a non-zero effect at temperatures well below the glass transition at about 120°C. Apart from the temperature effects, mentioned above, the curves show minima just before the rise due to the orientational effect. This complication arises because the measurements were carried out at a wavelength slightly different from that of maximum absorption. Here the bandshift effect is comparable in magnitude with the orientation effect, but with opposite sign. Although information about the microscopic viscosity of the polymer can still be obtained from this measurement, it is clear that this dye is much less suited than IIa for quantitative study of the softening of the polymer matrix. We remark that with the help of this technique one can supplement dielectric relaxation studies with low and very low frequency measurements and, moreover, by choosing dyes of different molecular shape and/or size, one can obtain detailed information about the softening of the polymer.

We are indebted to Mr. J. Boven and Dr. J. van der Veen for the synthesis of the dyes and to Mr. J. van Hoof and Mr. J. van Vledder for their assistance during part of the measurements.

* Du Pont trade name
\+ Shell trade name

## REFERENCES

1 Liptay W, in "Excited States", Vol. I, Lim E C ed., Academic Press, New York/London (1974) 129-229.
2 Yamaoka K and Charney E, J. Amer. Chem. Soc. 94 (1972) 8963.
3 Havinga E E and Van Pelt P, to be published.

# ELECTRICALLY INDUCED FLUORESCENCE CHANGES FROM SOLUTIONS OF DYE TAGGED POLYRIBONUCLEOTIDES

P. J. Ridler and B. R. Jennings

Electro-Optics Group, Physics Department, Brunel University, Uxbridge, Middlesex, U.K.

Transient changes have been recorded in the polarised components of fluorescence from aqueous solutions of poly r A, Poly r C, poly r G, poly r I and poly r U complexed with the dye acridine orange upon alignment of the polymers with pulsed electric fields. The novel data reveal the nature of the fluorescent dye ordering. For poly r A and poly r C the effects are found to depend strongly upon the polymer concentration.

## INTRODUCTION

The absorption of light and its re-emission as fluorescence at longer wavelengths are processes associated with electronic transition dipole moments that have fixed directions within the framework of a fluorescent molecule. Polarised incident light mostly will be absorbed when the molecule is oriented so that the transition moment of absorption is parallel to the polarisation direction. The emitted light will be predominantly polarised parallel to the transition moment of emission.

If fluorescent polyelectrolytes in solution are aligned by the application of an electric field, then provided the fluorescent groups have ordering with respect to the alignment axes of the polymers and provided the electric field does not shift the absorption spectrum or change the quantum yield and does not affect motions which depolarise the fluorescence within the lifetime of the excited state, then a change may be induced in both the intensity and polarisation of fluorescence due respectively to the ordering of the absorption and emission transition moments.

Such effects have been measured recently on solutions of dye tagged DNA (1-5) and polymers (6) by applying pulsed electric fields. When pulsed fields are applied, changes in the fluorescence components are transient in nature. From the amplitudes of the transient changes estimates of the azimuths of the transitions with respect to the alignment axes have been obtained and hence the geometry of the fluorescent dye attachment. The times of decay of fluorescence changes upon removal of the field are characteristic of the time taken for the polyelectrolyte to revert to random orientations in the solution.

We report herein preliminary experiments using the transient electric fluorescence method for the first time to study aqueous solutions of various dye tagged homopolyribonucleotides. Data are presented as functions of electric field strength and polymer concentration at a fixed monomer-to-dye ratio.

## EXPERIMENTAL

### Materials

The various polyribonucleotides were obtained as salts from Miles Laboratories Ltd. of Stoke Poges, U.K., and stored at minus 20°C until used. The maker's median sedimentation coefficients of the polymers in 0.05 M $PO_4^{3-}$ buffer and 0.1 M NaCl at pH 7.0 are as follows; poly r A, 15.4 S; poly r C, 8.0 S; poly r G, 17.6 S; poly r I, 4.8 S and poly r U, 7.6 S. The dye acridine orange (AO) was a commercially purified sample obtained from G.T. Gurr Ltd., London. Stock solutions were prepared by dissolving the polynucleotides in de-ionised and distilled water which had a specific conductivity of less than 2 M $\Omega\,cm^{-1}$. Solutions of the dye, dissolved in similar water, were then added to give stock solutions with a polynucleotide concentration typically of $9.0 \times 10^{-5}$ gm $cm^{-3}$ and a monomer-to-dye ratio of 250. These stock solutions were diluted as required for the experiments. Final solutions at pH 6.8 ± 0.2 and 23°C were used within two hours of preparation.

### Method

The apparatus used has been previously described (4,5). In principle a beam of 488 nm wavelength light, which is either vertically or horizontally plane polarised, is incident upon the solution which is held in a glass cell fitted with electrodes. Any fluorescent light which is emitted at 90° to the incident beam direction is analysed into its vertically and horizontally polarised components and recorded. Fig. 1 shows the geometrical arrangement when an electric field of amplitude E is applied in the vertical direction. Four polarised components of the fluorescence can thus be measured. These are designated $V_V$, $V_H$, $H_V$ and $H_H$ where the capital and subscripts refer to the polarisation states of the incident and analysed beams respectively. Supplementations of these components in the presence of electric fields are prefixed $\Delta$.

Using rectangular pulsed electric fields the changes appear as transients. They are recorded photographically from an oscilloscope. Fig. 2 provides a typical example.

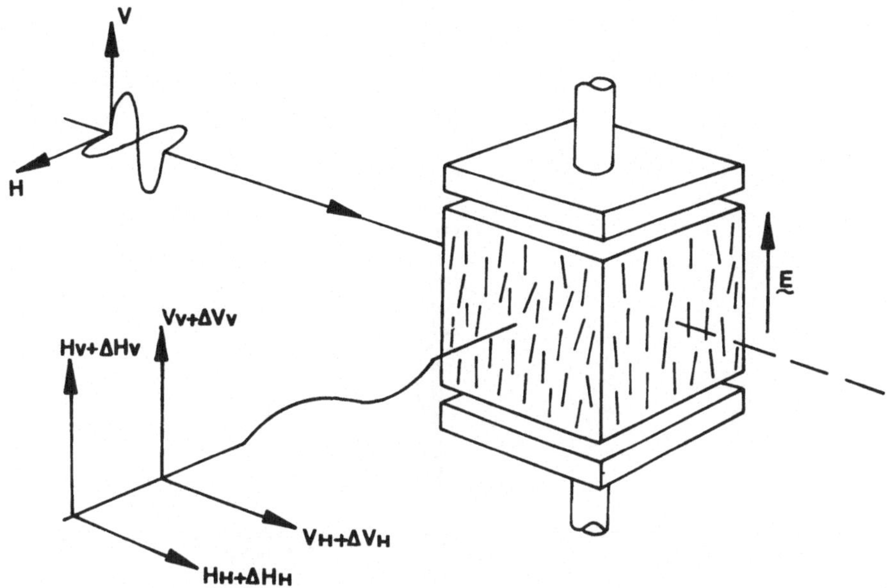


Fig. 1. Schematic diagram of the experimental arrangement.

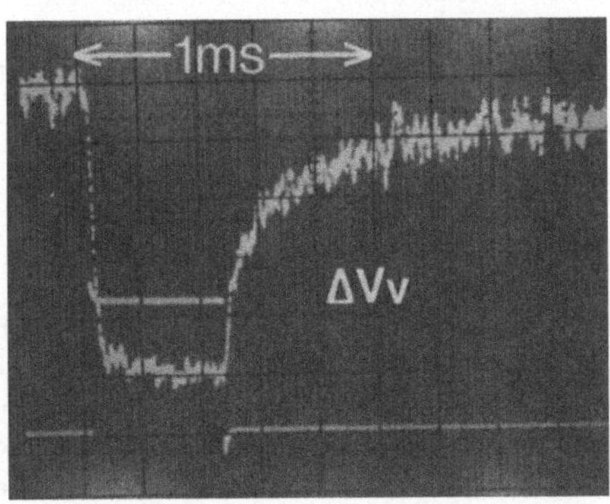


Fig. 2. A typical electrically induced transient change in the $V_V$ component of fluorescence for poly r A + AO. The lower trace is the applied field of 6.5 kV $cm^{-1}$.

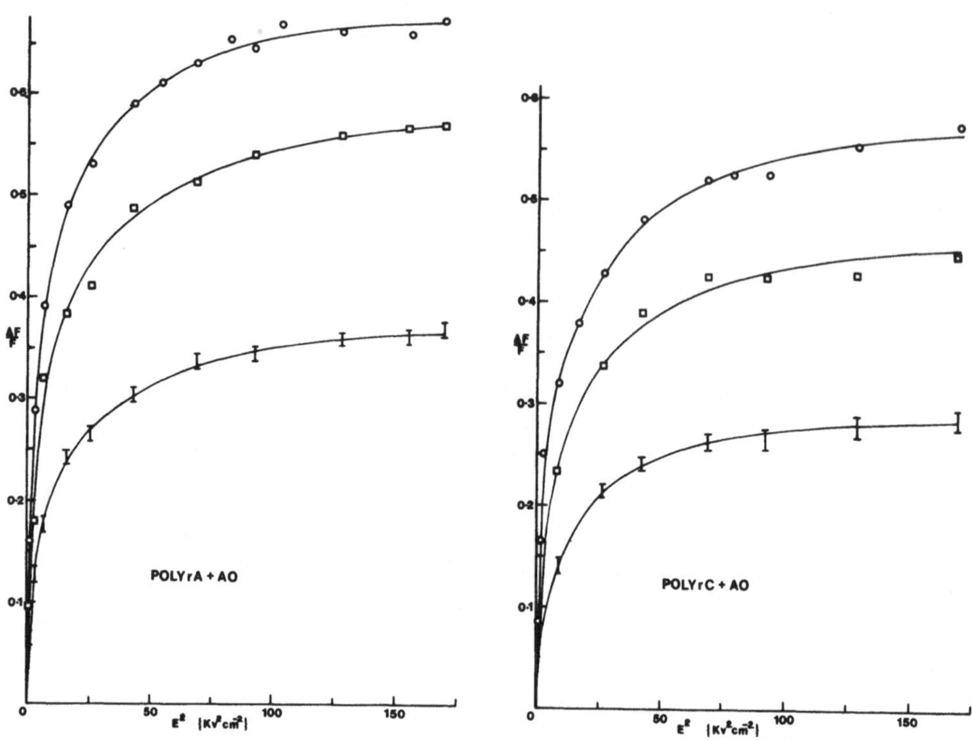


Fig. 3. Variation of the relative amplitudes of the electrically induced fluorescence changes with the square of the applied field for polyribonucleotides complexed with acridine orange. Data are for polymer concentrations of 7.5 x $10^{-6}$ gm $cm^{-3}$ and $^P/_D$ ratios of 250. The fluorescence was excited at 488 nm and detected at wavelengths exceeding 515 nm. Circles represent $-\frac{\Delta V_V}{V_V}$, squares $+\frac{\Delta H_H}{H_H}$, and bars $-\frac{\Delta V_H}{V_V}$ and $-\frac{\Delta H_V}{H_V}$.

## RESULTS

Figs. 3 and 4 show how the relative component changes of fluorescence $\Delta V_V/V_V$, $\Delta V_H/V_H$, $\Delta H_V/H_V$ and $\Delta H_H/H_H$ vary with the square of the applied field. Only poly r U + AO showed no measurable effects under these conditions. The other three systems showed similar, but not identical, behaviour. For these, $\Delta H_H/H_H$ was found always to be positive and $\Delta V_V/V_V$, $\Delta V_H/V_H$ and $\Delta H_V/H_V$ always appeared to be negative. The values of $\Delta V_H/V_H$ and $\Delta H_V/H_V$ are equal within the range shown by the vertical bars. The average time $\tau$ taken for the effects to decay to $^1/_e$ of their pre-field values upon removal of the electric field are shown in table 1. The times are similar to those found

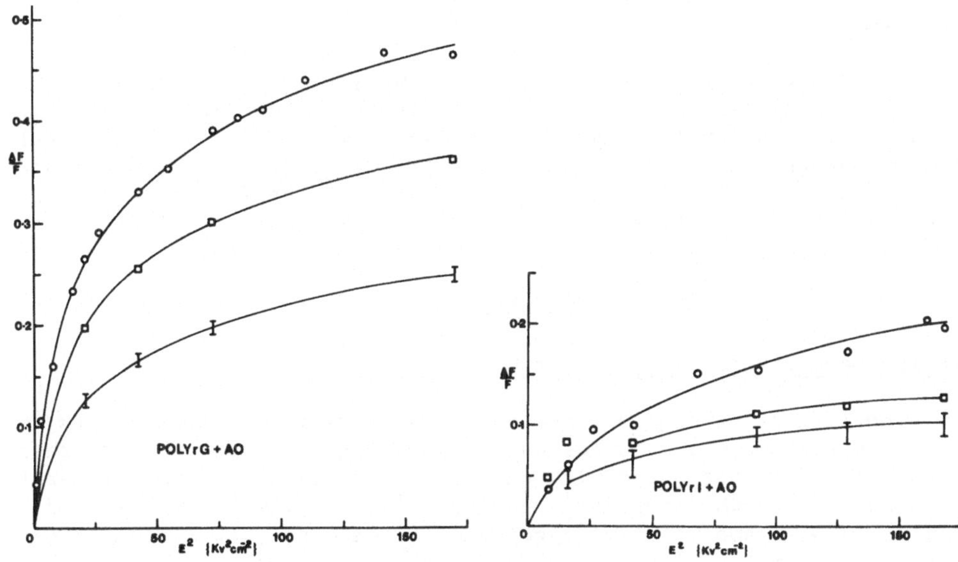


Fig. 4. Caption as for fig. 3.

Table I

| Polynucleotide - Dye Complex | Polymer concentration gm $cm^{-3}$ | Relative Component Changes of Fluorescence | | | | Average Relaxation time $\tau$ $\mu s$ | Polarisation $\frac{V_V - V_H}{V_V + V_H}$ |
|---|---|---|---|---|---|---|---|
| | | $\frac{\Delta V_V}{V_V}$ | $\frac{\Delta V_H}{V_H}$ | $\frac{\Delta H_V}{H_V}$ | $\frac{\Delta H_H}{H_H}$ | | |
| POLY(rA) + A.O. | $4.5 \times 10^{-5}$ | - 0.27 | - 0.11 | - 0.11 | + 0.23 | 50 | 0.24 |
| | $7.5 \times 10^{-6}$ | - 0.58 | - 0.30 | - 0.30 | + 0.45 | 400 | 0.35 |
| POLY(rC) + A.O. | $4.5 \times 10^{-5}$ | - 0.15 | - 0.06 | - 0.06 | + 0.11 | 50 | 0.23 |
| | $7.5 \times 10^{-6}$ | - 0.48 | - 0.24 | - 0.24 | + 0.38 | 350 | 0.33 |
| POLY(rG) + A.O. | $4.5 \times 10^{-5}$ | - 0.39 | - 0.20 | - 0.20 | + 0.29 | 30 | 0.42 |
| | $7.5 \times 10^{-6}$ | - 0.33 | - 0.17 | - 0.17 | + 0.25 | 40 | 0.41 |
| POLY(rI) + A.O. | $4.5 \times 10^{-5}$ | - 0.12 | - 0.06 | - 0.06 | + 0.09 | 20 | 0.22 |
| | $7.5 \times 10^{-6}$ | - 0.12 | - 0.07 | - 0.07 | + 0.11 | 20 | 0.24 |
| POLY(rU) + A.O. | $4.5 \times 10^{-5}$ | + 0.08 | + 0.04 | + 0.03 | + 0.03 | 200 | 0.10 |
| | $7.5 \times 10^{-6}$ | - | - | - | - | - | 0.11 |
| D.N.A. + A.O. | $4.5 \times 10^{-5}$ | - 0.63 | - 0.36 | - 0.36 | + 0.48 | 350 | 0.35 |
| | $7.5 \times 10^{-6}$ | - 0.63 | - 0.34 | - 0.34 | + 0.47 | 350 | 0.35 |

using electric birefringence (7,8) and electric dichroism (8,9) to study polynucleotides in solution. They are much longer than the lifetime of the excited state, which for bound acridines is at most some tens of nano seconds (10).

Table I also shows the influence of polymer concentration on the foregoing parameters. The final column contains the field-free polarisation of fluorescence, defined by the ratio $(V_V - V_H) / (V_V + V_H)$. All data in the table are for a field strength of 6.5 kV $cm^{-1}$ and a monomer to dye ratio of 250. By way of comparison, the foot of the table displays comparative data for a calf thymus DNA ($M_w = 6 \times 10^6$) sample.

## DUSCUSSION

### Binding Geometry for Poly r A, r C, r G and r I

The effects of figs. 3, 4 and table I arise from the electrically induced orientation of the polynucleotides and the dyes bound to them because the dyes are bound with uniform orientations to the polymers' backbone. Since the component changes of figs. 3, 4 and table I are similar in sign and relative magnitude for each polymer system, the binding geometry of fluorescent acridine orange is also likely to be similar.

The orientation of the transitions of absorption and emission, and hence the dyes with respect to the electric dipoles of the polynucleotides, may be estimated as follows. In conducting media such as water, the majority of polyelectrolytes orient with their major dimensions along the applied electric field lines (11). The mechanism for this is the anisotropic polarisability associated with the polarisation of the cloud of counterions surrounding the polyelectrolyte. With increasing field strength, the macromolecules extend and orient with the chain backbone drawing more nearly parallel to the applied field vector. With our experimental array, the polynucleotides tend to orient with their long axes increasingly parallel to the vertical direction. With vertically polarised incident light, the changes $\Delta V_V$ and $\Delta V_H$ are negative. Hence the total intensity of fluorescence, which is proportional to the sum $\{(V_V + \Delta V_V) + 2\ (V_H + \Delta V_H)\}$, decreases upon application of the field. This indicates that the absorption process is less efficient when the electric field is applied and that the absorption transitions of the bound fluorescent dyes must therefore align predominantly away from the direction of the electric vector of the incident light beam. In our experimental arrangement this is the horizontal plane at right angles to the long polyelectrolyte axis.

The predominant emission transition moment may be deduced from the <u>relative</u> amplitudes of the component changes. Whatever the polarisation state of the incident beam it will be noted that the change in the horizontally emitted fluorescence component is greater (i.e. more positive) than its vertically polarised counterpart. In fact $\Delta V_V$ is less than $2\Delta V_H$ when the signs of the changes are considered. The emission is thus more polarised towards the horizontal plane when the electric field is applied. Like the absorption moment, this relates to the polymer backbone. With acridine orange the transitions lie in the plane of the molecule (12,13) containing the three membered rings. Hence the dyes must be predominantly ordered so that their planes are more or less perpendicular to the chain axes of the polynucleotides.

## Calculations for Poly r A

For poly r A + AO it can be seen from fig. 3 that the effects change little with field strength at the highest fields attained. Under such circumstances the orientation of the polyelectrolytes may be considered to be complete and a more exact caclulation of $\chi$ and $\chi'$ (the azimuthal angles of the absorption and emission transitions with respect to the orientation axes) of the polymers may be made. For this preliminary analysis, it is assumed that the extended molecules are approximately rod-like under high fields.

(i) The average absorption transition azimuth $\psi$ may then be calculated from the intensity of fluorescence with and without the field applied as follows:

For a randomly oriented collection of absorption transitions, the number excited at an azimuthal angle $\theta$ to the electric vector of vertically polarised incident light is proportional to (a) the probability of excitation at $\theta$, and (b) the number at $\theta$, i.e. $\cos^2\theta \sin\theta \, d\theta$. Therefore, the total number excited is proportional to

$$\int_{\pi}^{0} \cos^2\theta \sin\theta \, d\theta = \frac{2}{3}$$

With the polyelectrolytes completely aligned in the vertical direction the absorption transitions are concentrated at $\theta = \psi$. Then the total number excited by vertically polarised light is proportional to

$$\cos^2\psi \int_{\pi}^{0} \sin\theta \, d\theta = 2\cos^2\psi$$

The fluorescence intensity is proportional to the number of excited absorption transitions. The proportionality constant will be the same for the random and ordered cases provided the average quantum yield does not vary with the transition azimuth (in the case of more than one type of binding site). This can be assumed at the high monomer-to-dye ratios used in these experiments. Thus,

$$\frac{\text{Fluorescence intensity without field}}{\text{Fluorescence intensity with field}} = \frac{V_V + 2V_H}{V_V + \Delta V_V + 2(V_H + \Delta V_H)}$$

$$= \frac{{}^2/_3}{2\cos^2\psi}$$

substituting the data for poly r A + AO at saturation gives $\psi$ = 66°

This value is remarkably similar to the absorption transition azimuth of the planar adenine bases of poly r A recently determined in an electric dichroism study (9). It is consistent with a face to face or intercalative mode of dye binding.

(ii) The average emission transition azimuth $\psi'$. Here, the calculation is less rewarding. The value of $\psi'$ is simply obtained from the ratio of the components of fluorescence polarised in the vertical direction and in the horizontal plane at full orientation, thus

$$\frac{2(V_H + \Delta V_H)}{V_V + \Delta V_V} = \tan^2\psi'$$

Substituting values at saturation gives $\psi' = 54°$. We would have expected $\psi'$ to equal $\psi$. The origin of this discrepancy is uncertain. Reasons may include inadequacy of the current theory, non-parallel transition moments when the dye is bound, or motion of polymer or dye during the excited state lifetime. Current studies are being directed towards the resolution of this problem.

Poly r U

The absence of effects for the poly r U + AO complex at the lower concentration probably indicates an absence of discrete ordering of the dye molecules on this polynucleotide. The low polarisation of fluorescence may indicate the presence of unbound dye and, or, a large mobility of the bound dye within the lifetime of the excited state. At the higher concentration electrically induced transient fluorescent changes are observed which are all positive. The relatively long relaxation times of these suggests the effects are results of the macromolecular orientation. It is however improbable that an anisotropy of dye ordering on the polynucleotide would lead to four positive fluorescence changes with both vertically and horizontally polarised light. A possible explanation is that the macromolecules are more compact at the higher concentration and the application of the electric field causes the structures to orient and extend. This would cause a change in the environment of the bound dyes, which in turn could induce a change in the extinction coefficient or quantum yield giving rise to an increase in the fluorescence with both polarisation states of the incident light.

## Concentration Dependence

It is interesting to note from the data that the fluorescence parameters for the poly r A, poly r C and poly r U dye complexes are strongly dependent upon the polymer concentration, whereas those for poly r G and poly r I dye complexes are not.

We note a remarkable similarity of the data for poly r A and poly r C at the low concentrations with that for native DNA complexed with AO (table I). This suggests a similarity between these structures.

At the higher concentrations the reduced relaxation times and reduced magnitude of the relative components suggest that at these concentrations the polynucleotides are more compact, flexible, have a less anisotropic ordering of the dyes and have a lower electric polarisability. The lower polarisation of fluorescence indicates a greater mobility of the bound dyes within the time between the obsorption and re-emission processes.

## ACKNOWLEDGEMENTS

The Science Research Council is thanked for a grant to one of us (B.R.J.) which funded a fellowship for P.J.R.

## REFERENCES

1 Weill G and Hornick C, Biopolymers, 10, (1971) 2029.
2 Weill G and Sturm J, Biopolymers, 14, (1975) 2537.
3 Jennings B R and Ridler P J, Chem. Phys. Lett., 45, (1977) 550.
4 Ridler P J, Ph.D. Thesis, Brunel University (1977).
5 Ridler P J and Jennings B R, J. Phys. (E), 10, (1977) 558.
6 Ridler P J and Jennings B R, Polymer, 19, (1978) 627.
7 Jakabhazy S Z and Flemming S W, Biopolymers, 4, (1966) 793.
8 Bradley D F et al., Biopolymers, 11, (1972) 645.
9 Charney E and Milstein J B, Biopolymers, 17, (1978) 1629.
10 Duportail G, Mauss Y and Chambron J, Biopolymers, 16, (1977) 1397.
11 Fredericq E and Houssier C, Electric Dichroism & Electric Birefringence, Oxford University Press, London (1973).
12 Zanker V Z, Z. Phys. Chem., 2, (1954) 52.
13 Ballard R E, McCaffery A J and Mason S F, Biopolymers, 4, (1966) 97.

# POLARIZED FLUORESCENCE IN AN ELECTRIC FIELD : THEORETICAL CALCULATION AT ARBITRARY FIELDS, EXPERIMENTAL COMPARISON WITH OTHER ELECTROOPTICAL EFFECTS, SATURATION OF THE INDUCED DIPOLE MOMENT IN POLYELECTROLYTES

S. Sokerov and G. Weill

Centre de Recherches sur les Macromolécules
CNRS, Strasbourg, France

The molecular properties accessible from electrooptic effects are stressed and polarized fluorescence in an electric field is then introduced in view of its specific properties. The need to calculate the fourth moment of the orientation function at high fields is emphasized for its use in the determination of the mechanism of orientation. The results of such a calculation for permanent and induced dipole orientation are given. They are applied in an experimental comparison with electric birefringence and dichroism on fragments of DNA labelled with intercalated Acridine Orange. The significance of an apparent permanent dipole orientation mechanism is discussed in terms of saturation of the induced dipole moment in polyelectrolytes.

## INTRODUCTION

The development of the measurement of electrooptical effects in solution or suspensions of rigid particles, is directed towards the determination of three types of molecular properties:

(a) the size L and the distribution of sizes;

(b) the mechanism of orientation of the particles and the value of their permanent electric dipole moment $\mu$ and/or of the anisotropy of their polarizability $\Delta\alpha$ ;

(c) some structural information such as the optical anisotropy or the direction of some transition moments with respect to the orientation axis.

A simultaneous determination of these quantities is generally desirable for the unambiguous measurement of any one of them. As an example, the type of average mean size obtained from the field free relaxation of the effect at low degrees of orientation depends upon the mechanism of orientation because of the difference in the L dependence of $\mu$ and $\Delta\alpha$ (1). Similarly, the determination of information of type (c) requires an extrapolation of the effect at full orientation, the form of which can be critically dependent upon properties of type (a) and (b).

Another approach consists of using several different electro-optical effects, each of them having a specific advantage: electric birefringence is very sensitive, but an a priori determination of the anisotropy tensor is very difficult if total orientation is hard to reach. It is generally simpler then to use electric dichroism where the orientation of the absorption transition moment with respect to the orientation axis may be known from other structural determination. In electric light scattering the maximum change of scattered intensity at full orientation can in principle be calculated from the scattering in the absence of an electric field, but the result is very sensitive to the polydispersity.

Our proposal for the use of the changes in the intensity of the polarized components of fluorescence upon orientation of the molecules in an electric field arose from two specific problems connected with the physical chemistry of DNA :

(a) in the study of dye - DNA interactions, it is often found that the fluorescence of a part of the bound molecules is completely quenched while another part still fluoresces. Electric dichroism will give information on the average geometry of the two types of binding sites, while polarized fluorescence in an electric field (PFEF) can give information on the geometry of only the unquenched bound molecules (2).

(b) PFEF is dependent not only on the second but also on the fourth moment of the orientation function. This can make this effect more sensitive to differences with the mechanism of orientation and help to solve this controversial problem in the case of DNA.

PFEF consists in the observation of the change in the four polarized components of fluorescence - $V_v$, $H_v$, $V_h$ and $H_h$ - observed at right angles (fig. 1) upon orientation of the molecules in an electric field. This orientation produces a change in the absorption probability related to the orientation of the absorption transition moment $\mu_a$ with respect to the incident light polarization direction. This effect is observed in electric dichroism. The orientation also produces a redistribution of the emitted light into the different polarized components related to the orientation of the emission transition moment $\mu_e$ with respect to the directions of the emitted light polarization direction.

The calculation of these intensities of the polarized components has been performed for a rodlike particle bearing a fluorescent

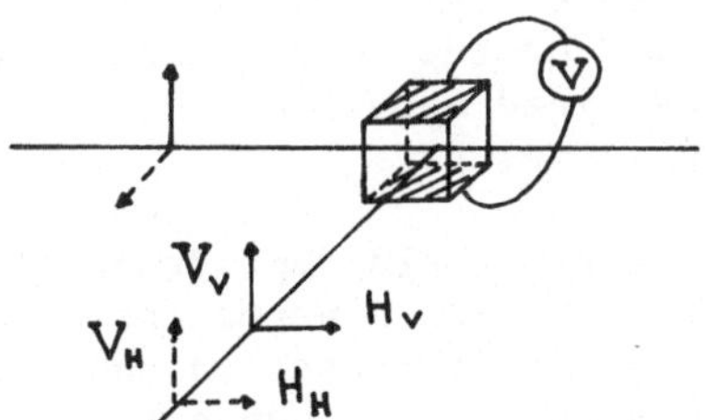


Fig. 1. Definition of the polarized components of fluorescence. The arrows indicate the direction of the polarizer and analyzer. V (vertical); H (horizontal).

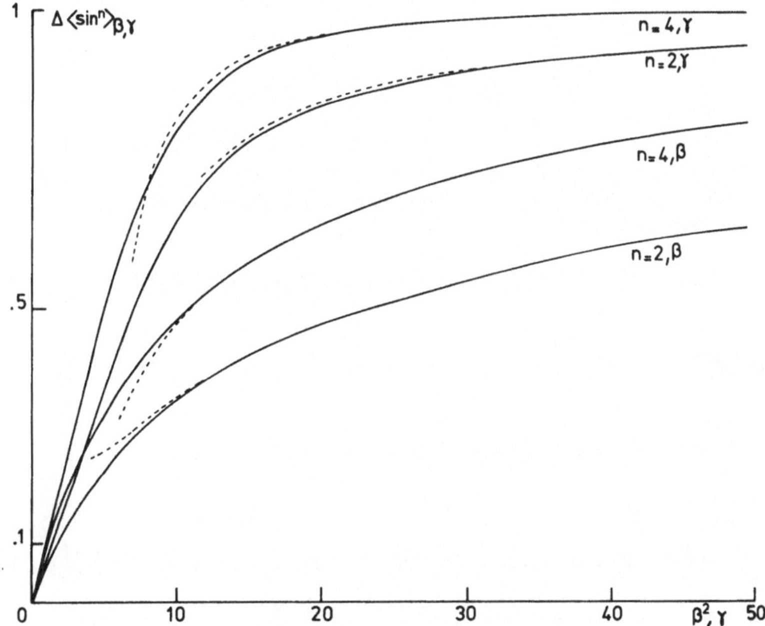


Fig. 2. Relative changes of the second and fourth moments of the orientation function for induced ( $\gamma = \Delta\alpha E^2/kT$) and permanent ( $\beta = \mu E/kT$) dipole moment orientation.

group as a function of the angles $\chi$, $\chi'$ and $\psi$ which fix the position of $\mu_a$ and $\mu_e$ with respect to the rod axis, and of the angle $\theta$ which gives the orientation of this axis with respect to the orientation field. The final results are expressed as a function of the two mean values $\langle \sin^2\theta \rangle$ and $\langle \sin^4\theta \rangle$ the fourth moment appearing because of the double projection described above and the squaring of the amplitudes to obtain the intensities (2).

The interpretation of the field dependence of the effect at large degrees of orientation in terms of a permanent dipole or an induced dipole orientation mechanism of orientation thus requires the calculation of the field dependence of these two averages at arbitrary fields for the two cases. The results are well known for $\langle \sin^2\theta \rangle$ from the classical calculation of O'Konski (3) of the orientation factor in Kerr effect. We had to carry out a similar calculation for $\langle \sin^4\theta \rangle$ before using it in a comparison of the field dependence of electric birefringence, dichroism and PFEF in an effort to characterize the mechanism of orientation of dye labelled fragments of sonicated DNA fragments.

## CALCULATION OF $\langle \mathrm{SIN}^4\theta \rangle$

Both the steady state and the transient values of $\langle \sin^4\theta \rangle$ when the field is turned on and off have been obtained for a permanent dipole and an induced dipole mechanism, the dipole being oriented along the molecular axis.

The steady state calculation starts from the definition

$$\langle \sin^4\theta \rangle = \frac{\int_0^{\pi} \sin^4\theta \ f(\theta) \ \sin\theta \ d\theta}{\int_0^{\pi} f(\theta) \ \sin\theta \ d\theta} \tag{1}$$

where the unnormalized orientation distribution function is

$$f(\theta) = \exp - U(\theta) / kt \tag{2}$$

the energy $U(\theta)$ of the molecule in the electric field being given by

$$\begin{aligned} U(\theta) &= - \mu E \cos\theta - 1/2 \ \Delta\alpha E^2 \cos^2\theta \\ &= - kT \left( \beta \cos\theta + 1/2 \ \gamma \cos^2\theta \right) \end{aligned} \tag{3}$$

All our results will be given as a function of $\beta$ and $\gamma$ (note that our $\gamma$ is twice that chosen by O'Konski).

The most interesting results are for the low field and high field expansions:

$$\langle \sin^4\theta \rangle_{\beta,\gamma\to 0} = \frac{8}{15}\left(1 - \frac{2}{21}(\beta^2 + \gamma) - \frac{1}{315}(\gamma^2 + 2\beta^2\gamma - 3\beta^4)\right) \tag{4}$$

$$\langle \sin^4\theta \rangle_{\beta,\gamma\to\infty} = (\beta^8 + \gamma)^2 = 2\ \langle \sin^2\vartheta \rangle^2_{\beta,\gamma\to\infty} \tag{5}$$

The relative change of $\langle \sin^4\theta \rangle$ with $\beta^2$ and $\gamma$

$$\Delta\langle \sin^4\theta \rangle = \frac{\langle \sin^4\theta \rangle_{E \neq 0} - \langle \sin^4\theta \rangle_{E=0}}{\langle \sin^4\vartheta \rangle_{E=0}} \tag{6}$$

is given in fig. 2 together with the equivalent quantity $\Delta\langle \sin^2\vartheta \rangle$. It is seen that PFEF will reach saturation more rapidly than do birefringence or dichroism which depend only on $\langle \sin^2\theta \rangle$. For an induced dipole moment mechanism one predicts an $E^2$ dependence of the effect up to rather high degrees of orientation while curvature arises at low degrees of orientation for a permanent dipole moment mechanism.

A plot of $\langle \sin^4\vartheta \rangle_{E \neq 0} / \langle \sin^4\theta \rangle_{E=0}$ as a function of $\langle \sin^2\theta \rangle_{E \neq 0} / \langle \sin^2\theta \rangle_{E=0}$ in fig. 3 shows that their relation is not very dependent upon the mechanism of orientation. They remain proportional up to a high degree of orientation where the electrooptical effect itself no longer follows the low field $E^2$ dependence (Kerr region). The relation between the relative changes in the different polarized components of fluorescence established at low fields (2) and based on this proportionality

$$\frac{\Delta H_v - \Delta V_h}{V_v + 2\,H_v} = \frac{\Delta H_h + 2\Delta H_v}{V_v + 2\,H_v} = -\frac{\Delta H_h + 2\Delta V_h}{V_v + 2\,H_v} \tag{7}$$

should therefore remain valid up to high fields. This can help in checking some important experimental corrections (see next paragraph).

## SOME EXPERIMENTAL PROBLEMS

The geometry of an instrument devoted to the observation of the changes in the polarized components of fluorescence is obvious from fig. 1. Since the requirements of stability and sensitivity are exactly similar to those of a Temperature Jump apparatus, it is very easy to modify such an instrument, equipped with fluorescence detection, by interposition of polarizers and analyzers (2) and the replacement of the cell by a square cuvette equipped with electrodes on a special Teflon holder (5). A specially designed instrument, using an Argon ion laser and later on a dye laser as a light source has been described recently by Ridler and Jennings (6). Introducing some experimental data into their theoretical expression of the signal to noise (S/N) ratio, these authors conclude that "the sensitivity of the transient fluorescence method is equal, if not greater than that of electric birefringence".

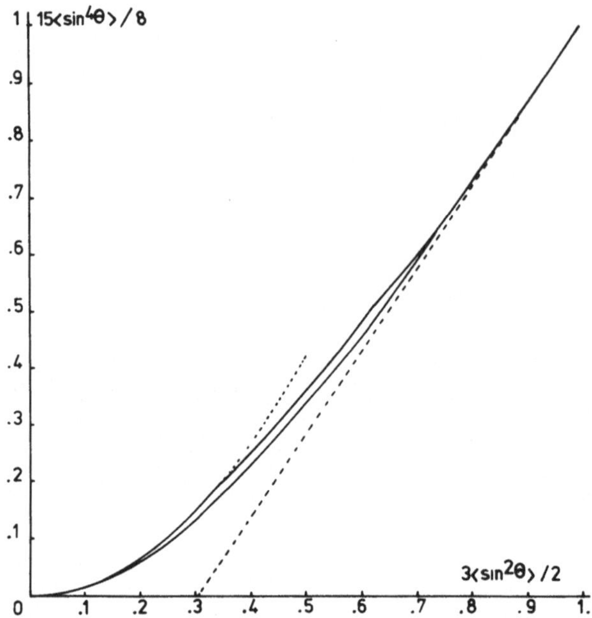


Fig. 3. $<\sin^4\theta>$ as a function of $<\sin^2\theta>$ for $\beta$ and $\gamma$ orientation.

This statement is in contradiction with the fact that a powerful laser beam, rather high electric fields and very large molecular weight DNA (where the rod approximation is in fact no longer valid) had to be used to obtain very nice oscilloscope recordings of the transients, in contrast with the noisy recordings previously recorded by us with a conventional Xenon-Mercury source and short rodlike fragments of sonicated DNA.

We have traced an error, leading to such a false conclusion, in the estimation of the S/N ratio. What has been forgotten in the proportionality factor $\sigma$ between the fluoresced intensity I and the incident intensity $I_0$ is the ratio of the solid angle $\Omega$ of the collecting optic to the total space solid angle $4\pi$. Thus $\sigma$ is the product of the quantum yield Q, the fraction of absorbed photons in the observed volume and the ratio $\Omega/4\pi$. This last factor can be of the order of $10^{-5}$, therefore $\sigma$ should be of the order of $10^{-6}$. Since what is observed is $\Delta I/I = \Delta\sigma/\sigma$ and the theoretical S/N ratio is approximately equal to $\sigma/(\sigma+\Delta\sigma)^{\frac{1}{2}}$ this brings a factor of $10^{-2}$ - $10^{-3}$ into the final result. This does not take into account the increase in dark current when measuring a slight change in intensity from an already high flux of photons as compared to essentially no photons on the photomultiplier for the birefringence case.

It is therefore fair to say that PFEF is $10^{-2}$ - $10^{-3}$ less sensitive than electric birefringence. This implies for relatively small molecules the use of high electric fields and gives another interest to our calculation of $<\sin^4\theta>$ at high fields.

A second important remark concerning the experimental difficulties has to be made. A good definition of the electric field acting on the solution requires the electrodes to be larger than the observed volume. Moreover, a compromise has to be made between the concentration and the length of this volume which govern the fraction of absorbed photons. One must then consider that the calculation of the changes in intensities has been performed at constant $I_o$, i.e. for a thin layer of solution subjected to the electric field. In the real set up two factors are important. Firstly, in order to obtain a uniform electric field the electrodes are larger than the observed volume. Therefore the intensity incident on the volume will vary with the electric field as a result of the dichroism of the external layers submitted to the electric field. Secondly, the observed volume is not a thin layer and the same effect must be taken into account inside the observed volume.

Approximate corrections have been proposed (7). They are most simply evaluated if electric dichroism can be measured on the same cell and instrument. Since these are of different sign for the two polarizations of the incident beam, their consistency can be checked using eq. 7. As a simple rule one should not work with optical densities higher than 0.1, especially at high fields, to limit this correction.

## A COMPARISON OF ELECTROOPTICAL EFFECTS ON ACRIDINE ORANGE LABELLED DNA FRAGMENTS

From a comparison of their rotational relaxation time, as measured by the field free decay of electric birefringence, with the length L of the double helix deduced from the molecular weight, DNA fragments are known to behave as rigid rods at low ionic strength. We have used fragments with $L \sim 1200$ A labelled with Acridine Orange at a phosphate to dye ratio $> 100$. In these conditions the dye is quantitatively intercalated with $\chi = \chi' = \pi/2$ (ref. 8). The quantities directly available from the electrooptical measurements are simple functions of $\Delta <\sin^2\theta> = \phi$ and $\Delta <\sin^4\theta>$ (refs. 9,2).

Electric Birefringence:

$$\Delta n = \frac{2}{n_o} c \bar{v} (g_1 - g_2) \qquad (8)$$

where $n_o$ is the solvent index of refraction, c the concentration, $\bar{v}$ the molar partial volume (0.57 for DNA), $g_1$ and $g_2$ the principal polarizabilities per unit volume of the solute.

Electric Dichroism:

$$\log (1 + \Delta I/I) = (\log I_o/I)\,\phi \qquad (9)$$

$I_o$, I and I + $\Delta$I are respectively the incident intensity and the transmitted intensity in the absence and presence of an electric field parallel to the polarization direction.

PFEF

$$V_v \propto 1/8 \langle \sin^4\theta \rangle \; (1 + 2 \cos^2 \psi) \qquad (10)$$

$$H_v = V_h \propto 1/4 \langle \sin^2\theta \rangle \; - 1/16 \langle \sin^4 \theta \rangle (1 + 2 \cos^2\psi) \qquad (11)$$

$$V_h \propto 24/64 \, (1 - \langle \sin^2\theta \rangle + \langle \sin^4 \theta \rangle) + \cos^2 \psi /64 \, (16 - 16 \langle \sin^2 \theta \rangle - 2 \langle \sin^4\theta \rangle) \qquad (12)$$

$\psi$ can be determined from the anisotropy of fluorescence in the absence of an electric field:

$$(V_v - H_v) / (V_v + 2H_v) = (3 \cos^2 \psi - 1) /5 \qquad (13)$$

It is however not needed to calculate $\Delta\langle \sin^4\theta \rangle$ from the change in $V_v$.

The values of $\phi$ and $\Delta \langle \sin^4 \theta \rangle = \Delta V_v/V_v$ deduced from electric dichroism and PFEF are given in fig. 4 as a function of $E^2$. Electric birefringence data can be matched to the dichroism data using $(g_1 - g_2) = 20\ 10^{-3}$.

Since the field free decay does not present a tail which could be attributed to the existence of remaining large particles, the curvature observed at low degrees of orientation excludes a fitting with an induced dipole orientation mechanism. A fit with a permanent dipole moment $\mu = 4\ 10^3 D$ gives a very satisfactory account of the variation of both $\phi$ and $\Delta V_v/V_v$. Using the same $\mu$ to recalculate the changes of the other components gives a less satisfactory agreement but the experimental variation shows even more curvature. This can be partly explained by the importance of the dichroism correction at high field but also by the fact that the other components depend on the difference between different moments of the orientation function.

## A DISCUSSION OF THE SATURATION OF THE INDUCED DIPOLE MOMENT IN DNA

A field dependence of the orientation factor $\phi$ following closely that predicted for a permanent dipole moment orientation mechanism has

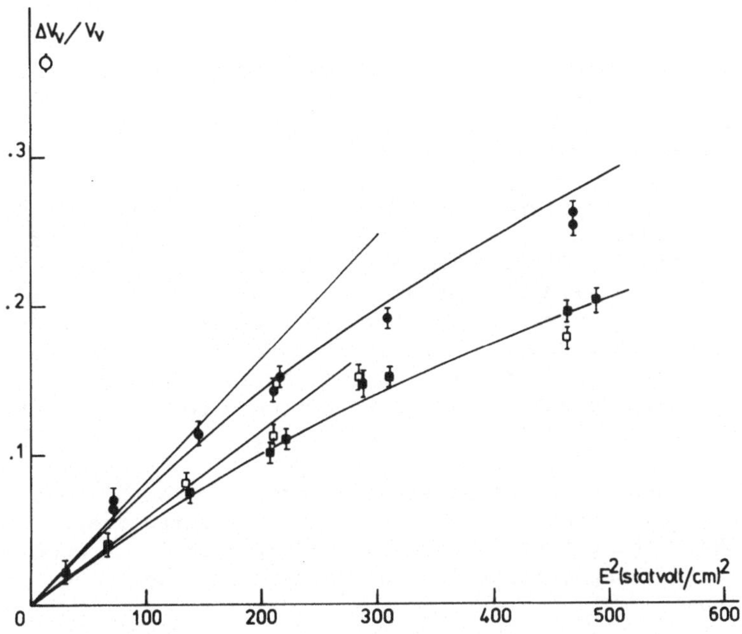


Fig. 4. The orientation factor and the relative change in the fluorescent component $V_v$ as a function of $E^2$: $\phi$ from electric dichroism (■) and from electric birefringence (□) $V_v/V_v$ from PFEF (●). Solid lines are calculated for permanent moment orientation with $\mu = 4\ 10^3$ Debye.

already been observed in electric dichroism measurements on DNA fragments (10) and in electric birefringence of other polyelectrolytes (11). Our comparative measurements of several electrooptical effects, depending opon different moments of the orientation function, bring further evidence for such an orientation mechanism.

DNA however, due to its double helical antiparallel structure, cannot have such a large true permanent dipole moment. The apparent contradiction can be explained by the saturation of the induced dipole moment due to counterion polarization at low enough electric fields for the orientation to be still very small.

Yoshioka (12) has performed a calculation of the orientation factor for a saturating induced moment based on Mandel's model of counterion polarization which completely neglects counterion-counterion repulsion. For a polyelectrolyte of length L with n condensed counterions of valency z, the anisotropy of polarizability is (13)

$$\Delta\alpha = (n\, z^2\, e^2\, L^2) \,/\, 12\, k\, T \qquad (14)$$

and the moment at saturation

$$\mu = n z e L/2 \quad (15)$$

For n large, $\Delta\alpha$ is large and the orientation saturates before the induced dipole. The reverse is true for small n; for very small n the field variation of $\phi$ cannot be distinguished from what is predicted for permanent dipole orientation.

A similar result has been obtained by Shirai (14) starting from a very different model in which the field induced flow of counterions is balanced by diffusion (15). The only difference is in the numerical factor in eq. 14 which slightly depends on the polyelectrolyte charge per unit length. One interesting feature in the calculation is the remark that since we are dealing with a saturated induced dipole which is reversed by reversing the field, the calculation of the averages (eq.1) should be carried out by integration from 0 to $\pi/2$. This brings a linear dependence in E at intermediate fields.

A linear dependence of the orientation factor $\phi$ with E has been observed recently by Crothers (16) on small monodisperse restriction fragments of DNA. The effect is however quadratic in L. A somewhat ad hoc model based on an orientation due to an anisotropic ion flow has been developed.

Going back to eqs. 14 and 15 one can calculate from the initial part of our results in fig. 4 a value of $\Delta\alpha = 3.5\ 10^{-16}\ cm^3$ and compare the corresponding value of n with that calculated from $\mu$ (n = 4) and with the theoretical number of condensed counterions N $\sim$ 700.

From the polarizability one gets n $\sim$ 50, i.e. n/N = 0.07. Such a ratio has already been observed and explained (17) by the influence of counterion-counterion repulsion on the linear response of the counterion atmosphere, following the arguments of Oosawa (18).

There is not at the present time any theory which can predict how counterion-counterion repulsion would affect the value of the saturated dipole moment, outside of some numerical calculations of Gibbs and McTague (19). A further tenfold decrease in the apparent value of n is not unreasonable. From our results, such a saturation would be obtained in electric fields as low as 500 V/cm.

## REFERENCES


1 Coles H and Weill G, Polymer, 18, (1977) 1235.

2 Weill G and Sturm J, Biopolymer, 14, (1975) 2537.

3 O'Konski C T, Yoshioka K and Orttung W H, J. Phys. Chem., 63, (1959) 1558.

4 Sokerov S and Weill G, Biophys. Chem., in press.

5 Sokerov S and Weill G, Biophys. Chem., in press.

6 Ridler P J and Jennings B R, J. Phys. E., Sc. Instr., 10, (1977) 558.

7 Hornick C and Weill G, Biopolymer, 10, (1971) 2029.
8 Hornick C and Weill G, Biopolymer, 10, (1971) 2345.
9 Fredericq E and Houssier C, Electric Dichroism and Electric Birefringence, Clarendon Press, Oxford (1973).
10 Ding D W, Rill R and Van Holde K E, Biopolymer, 11, (1972) 2109.
11 Kikuchi K and Yoshioka K, J. Phys. Chem., 17, (1973) 2101.
12 Kikuchi K and Yoshioka K, Biopolymer, 15, (1976) 583.
13 Mandel M, Mol. Phys., 4, (1961) 489.
14 Shirai M and O'Konski C T, unpublished results.
15 Schwarz G, Z. Physik. Chem., Neue Folge, 19, (1959) 286.
16 Hogan M, Dattagupta N and Crothers D M, P.N.A.S., 75, (1978) 195.
17 Oosawa F, Biopolymer, 9, (1970) 677.
18 McTague J and Gibbs J H, J. Chem. Phys., 44, (1966) 4295.

# *Nucleic Acids and Polynucleotides*

'Twist ye, twine ye! even so
Mingle shades of joy and woe,
Hope and fear, and peace and strife,
In the thread of human life.'

SIR WALTER SCOTT (1771-1832)

Guy Mannering

# DIELECTRIC PROPERTIES OF LOW-MOLECULAR WEIGHT DNA IN AQUEOUS SOLUTIONS AT LOW IONIC STRENGTH

Th. Vreugdenhil, F. van der Touw and M. Mandel

Department of Physical Chemistry, Gorlaeus Laboratories, University of Leiden, Leiden, The Netherlands

The dielectric properties of sonicated calf-thymus Na-DNA ($M_w \simeq 3 \times 10^5$ g mol$^{-1}$) have been investigated in the frequency range 5 kHz - 100 MHz, both in pure water and in $3 \times 10^{-4}$ M NaCl. These properties have been found to be qualitatively analogous to those observed for other, synthetic polyelectrolytes both in their frequency and concentration dependence. No specific effects could be detected. The experimental results, after extrapolation to zero concentration, have been interpreted in terms of the theory proposed previously by Van der Touw and Mandel.

For many years the dielectric properties of charged macromolecules in aqueous solution have been investigated in this laboratory (1-5). We have developed special methods enabling us to measure these properties of conducting solutions in a frequency range of a few kHz to 100 MHz, covering thus nearly five decades (6-8). These methods ensure inter alia that at frequencies below 1 MHz parasitic electrode effects are properly corrected for. In most cases we have been dealing with simple, synthetic polyelectrolytes. However DNA, a typical charged biopolymer, has attracted our attention as conflicting mechanisms have been proposed so far to explain its dielectric behaviour and no systematic study has been performed covering the two dispersion regions that have been found previously, one in the kHz-region or below, one in the MHz-region (9-25). We wish to report here some results of our DNA-investigation. A more detailed paper will be published elsewhere.

Preliminary measurements have shown that the commercially available calf-thymus Na-DNA of molar mass $5 \times 10^6$ g mol$^{-1}$ (provided as samples of protein content smaller than 0.5%) exhibits in aqueous solution the two dispersion regions mentioned but with a lower critical frequency which falls outside the frequency range covered by our equipment (1).

Therefore we have applied the ultrasonic degradation method (sonication) to obtain DNA of lower molar mass ($\sim 3 \times 10^5$ g mol$^{-1}$) as has often been described in the literature. Sonication has been applied to the commercial calf-thymus Na-DNA either in pure water (DNA I) or in the more usual buffered 0.2 M NaCl solution (DNA II). Both samples have been characterised by their weight average molar mass using the light scattering and viscosimetric technique and their average dimensions at high ionic strength (0.2 M NaCl) from the former. With electron microscopy it could be shown that the sonicated samples were heterodisperse with a contour length ($L_c$) distribution in fair agreement with the experimentally found weight-average molar mass, the z-averaged square radius of gyration and intrinsic viscosity (assuming for the two last quantities that DNA may be represented as a wormlike chain with a persistence length of ($55 \pm 10$) nm.

Dielectric measurements were performed with DNA I in pure water and with DNA II in $3 \times 10^{-4}$ NaCl over the whole frequency range available, a lower limit being imposed however by the conductance of the solution. Several concentrations of Na-DNA between $10^{-4}$ and $3 \times 10^{-3}$ monomol $\ell^{-1}$ were investigated. Under all circumstances at the lowest frequency attainable the electric permittivity has not yet reached a constant value but the critical frequency of the lower dispersion region could be clearly observed. Extrapolation to zero-frequency was performed by a least squares fit to a superposition of two Cole-Cole dispersion functions with constant parameter $\beta = 0.75$ (for $\beta = 1$ a Debye-type dispersion function would have been obtained). This yields the static electric permittivity assuming that no third dispersion occurs at still lower frequencies. No evidence for such an additional dispersion has been presented so far (15). At the high frequency end the value of the electric permittivity of the solvent is reached within experimental accuracy. This shows that the dispersions observed should be attributed to the DNA. Some of the experimental results are represented in figs. 1 and 2.

All the results exhibit the same features as also observed with other, synthetic, polyelectrolytes (2,3): a) as already mentioned two dispersion regions are found, the low-frequency one being characterised by a molecular-weight dependent mean relaxation time, in contrast to the high-frequency one; b) the static dielectric increments with respect to the solvent as well as the amplitudes $\Delta\epsilon_s$ of the high-frequency dispersion increase with increasing polyelectrolyte concentration $C_p$ (expressed in monomol $\ell^{-1}$), but the corresponding specific quantities $\Delta\epsilon_s/C_p$ or $\Delta\epsilon_2/C_p$ decrease with increasing concentration. This concentration behaviour can be represented empirically, just as for other polyelectrolytes investigated, by a linear dependence of $C_p/\Delta\epsilon_s$ or $C_p/\Delta\epsilon_2$ on $C_p$, according to

$$\Delta\epsilon_i/C_p = \frac{(\Delta\epsilon_i/C_p)_{C_p=0}}{1 + B_i C_p} \qquad (i = s,2) \qquad (1)$$

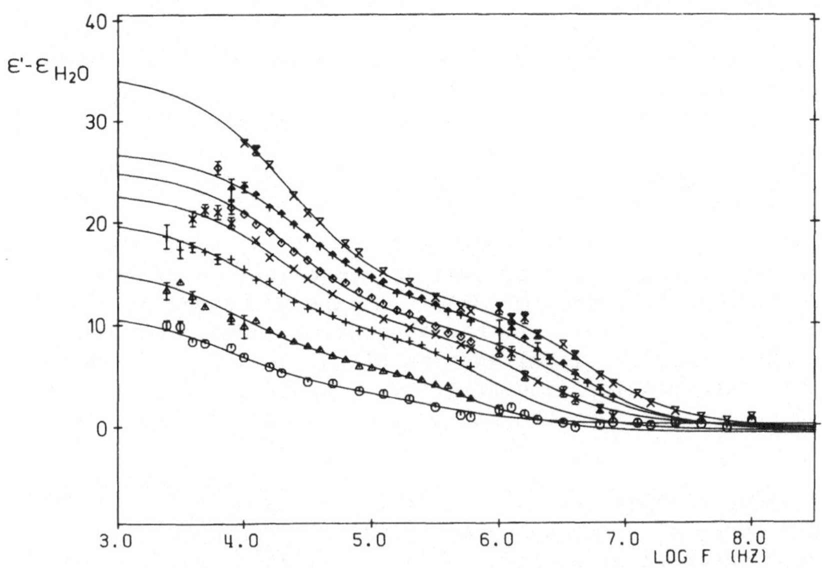


Fig. 1. Frequency dependence of the electric permittivity increment of DNA I in $H_2O$. The concentrations are as follows, for the bottom curve to the top; $1.0 \times 10^{-4}$, $2.5 \times 10^{-4}$, $5.0 \times 10^{-4}$, $7.5 \times 10^{-4}$, $1.50 \times 10^{-3}$, $2.00 \times 10^{-3}$ and $2.88 \times 10^{-3}$ monomol $\ell^{-1}$. For clarity much high-frequency data are omitted, and Duplo's are not shown.

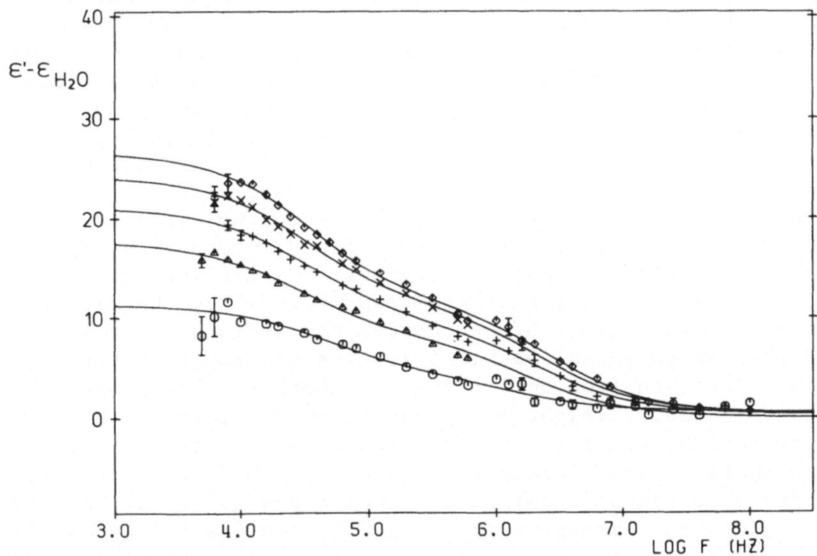


Fig. 2. Frequency dependence of the electric permittivity increment of DNA II in $3 \times 10^{-4}$ M NaCl. The concentrations are as follows, for the bottom curve to the top; $3.0 \times 10^{-4}$, $6.0 \times 10^{-4}$, $9.0 \times 10^{-4}$, $1.20 \times 10^{-3}$ and $1.50 \times 10^{-3}$ monomol $\ell^{-1}$. For clarity much high-frequency data are omitted.

c) the larger mean relaxation time $\tau_1$ shows within experimental accuracy only little concentration dependence but the smaller $\tau_2$ decreases with increasing concentration. This concentration dependence can be represented by a linear increase of $\tau_2^{-1}$ with $C_p$

$$\tau_2 = \frac{(\tau_2)_{C_p=0}}{1 + B_2' C_p} \qquad (2)$$

It should be noted that there was no appreciable difference between the dielectric behaviour of DNA I in pure water and of DNA II in $3 \times 10^{-4}$ M NaCl. According to the absorption coefficient at $\lambda = 260$ nm the former should already be slightly denatured, at least at the lowest concentrations investigated. Apparently this slight denaturation does not affect the dielectric properties in an appreciable way.

The striking analogy between the dielectric properties of DNA solutions and those of other polyelectrolytes, has prompted us to apply the same theoretical analysis (26) as has been proposed previously by Van der Touw and Mandel for charged macromolecules in aqueous solutions. The contribution of the macromolecules to the polarisation of the solution is assumed not to arise from a permanent or conventionally defined induced dipole moment, but to an <u>induced dipole moment arising from a perturbation of the ion atmosphere by the electric field</u>, in particular a perturbation in the distribution along the macromolecular chain of the counterions which are known to be closely associated or "bound" to the charged, elongated polyion. In the absence of the field these counterions are assumed to be distributed uniformly along the chain. For synthetic polyelectrolytes of corresponding degree of polymerisation the polyion was not modelled as a rigid rod but, due to its intrinsic flexibility, as a sequence of rigid rodlike subunits in a time-independent configuration. The perturbed distribution of the associated counterions is assumed to relax on two different time scales after disappearance of the electric field in order to explain the occurrence of the two dispersion regions. On the shortest one the distribution of these counterions becomes uniform along the different subunits only, but the number of associated counterions on each subunit may still differ as potential barriers are assumed to prevent the passage of a counterion from one subunit to another. Therefore the high-frequency molecular weight independent, mean relaxation time $\tau_2$ will be determined by the rate of diffusion of associated counterions along the molecular weight independent subunits. Only on a larger time scale the complete uniformization becomes possible. It has been shown that, within the framework of the model, the low-frequency molecular weight dependent mean relaxation time $\tau_1$ should be determined either by the rate of diffusion of the associated counterions over the complete macromolecular domain or the rate of rotational diffusion of the polyion (or by both mechanisms simultaneously) depending on the values of the corresponding relaxation times which both depend strongly on the macromolecular dimensions.

The theoretical expressions derived for systems without polyion-polyion interactions (in which also counterion-counterion interactions are neglected) show that from $\Delta\epsilon_s/C_p$ the weight-averaged mean square radius of gyration $R_g^2$ of the polyion may be calculated and from $\Delta\epsilon_2/C_p$ the square of the length of a subunit $b^2$, provided the fraction $\bar{f}$ of associated or "bound" counterions is known as well as the ratio $\gamma$ of the effective field acting on these counterions to the Maxwell field. The internal consistency may be checked by comparing the experimental value of $\tau_2$ to the theoretical expression for the diffusional relaxation time for an ion along a subunit of length b according to Wyllie (27)

$$\tau_{2,W} = \frac{b_2}{u\, kT\, \pi^2} \tag{3}$$

Here u represents the mobility of a counterion along a subunit for which the value $u_o$ of the counterion's mobility in an electrolyte solution at infinite dilution may be taken as a first approximation. For $b^2$ the value derived from $\Delta\epsilon_2$ should be used in eq. 3. It has been shown that for several synthetic polyelectrolytes reasonable values for $R_g$ and b are obtained in such a way and that there is reasonable agreement between $\tau_2$ and $\tau_{2,W}$ if the dielectric parameters extrapolated to infinite dilution according to eqs. 1 and 2 and the assumption $\gamma = 1$ are used (2,3,5).

If such an analysis is applied to the results of Na-DNA I and Na-DNA II in water and $3 \times 10^{-4}$ M NaCl, using for $\bar{f}$ values derived from activity measurements with the help of a specific $Na^+$ electrode, again reasonable values for $R_g$ and b are found (Table I). In particular it should be noted that the weight averaged root-mean-square of gyration of DNA I in water is larger than the corresponding quantity of DNA II in the salt solution where the repulsive electrostatic forces along the chain may already be screened off to some extent. Also the value of b, roughly a measurement of the stiffness of the chain, is found to be larger for the former as compared to the latter. Also the values of $(\tau_2)_{C_p=0}$ and $\tau_{2,W}$ agree satisfactorily. However it is clear that the absolute values $R_g$ are probably not representing the true values for that quantity at infinite dilution. This follows from comparison with the values of the weight averaged root-mean-square radius of gyration at infinite polyelectrolyte dilution calculated for DNA I and DNA II in 0.2 M NaCl from light scattering and electron microscopy (contour-length distribution) data. These values are (44 $\pm$ 2) nm and (38 $\pm$ 2) nm respectively and definitely <u>larger</u> than the corresponding values derived from the dielectric measurements at much lower ionic strength. It should be kept in mind however, that the latter values are only rough estimates in view of all the approximations involved in the theoretical Van der Touw - Mandel treatment (such as $\gamma = 1$, neglect of counterion-counterion interactions, etc.). Also the extrapolation procedure to $C_p = 0$ according to eqs. 1 and 2 does not necessarily yield dielectric parameters for a truly infinitely diluted Na-DNA solution. It cannot be

Table I. Dielectric parameters extrapolated to infinite dilutions and average dimensions at infinite dilution from electric permittivity for DNA.

| | DNA I/$H_2O$ | DNA II/$3x10^{-4}$M NaCl |
|---|---|---|
| $(\Delta\varepsilon_s/C_p)_{C_p=0}$ (in $10^3$ l $monomol^{-1}$) | 140 ± 30 | 51 ± 9 |
| $(\Delta\varepsilon_2/C_p)_{C_p=0}$ (in $10^3$ l $monomol^{-1}$) | 60 ± 10 | 25 ± 5 |
| $R_g$ +) (in nm) | 38 ± 4 | 23 ± 2 |
| b +) (in nm) | 86 ± 8 | 55 ± 6 |
| $(\tau_2)_{C_p=0}$ (in $10^{-6}$s) | 1.3 ± 0.9 | 0.15 ± 0.07 |
| $(\tau_{2,W})$ *) (in $10^{-6}$s) | 0.6 ± 0.1 | 0.25 ± 0.06 |
| $\bar{L}_c^w$ †) (nm) | 194 | 167 |

*) calculated according to eq.3 with $u=u_o = 3.1 \times 10^{11}$ m $s^{-2}$ $N^{-1}$ for $Na^+$ and b taken from the table.

+) calculated with $\bar{f}=0.68\pm0.06$ from activity measurements and $\gamma=1$.

†) weight-average contour length from electron microscopy.

excluded that at very low $C_p$, where measurements cannot be performed due to lack of experimental sensitivity, deviations with respect to eqs. 1 and 2 may occur such that the extrapolated dielectric parameters rather refer to a hypothetical state where some polyion-polyion interaction effects have disappeared but the conformational state of the polyions is not that prevailing at infinite dilution. Therefore the average dimensions determined from the dielectric data may be smaller than expected in a truly infinite dilution of vanishingly small ionic strength. In this respect our conclusion, that the dielectric behaviour of sonicated Na-DNA in solution of very low ionic strength is qualitatively the same as other polyelectrolytes and that no specific effects have to be taken into consideration in understanding the frequency or concentration dependence, will not be affected.

Finally it may be interesting to compare the electric polarisability due to the perturbation of the ion atmosphere calculated from the theoretical expressions for the dielectric increments with the experimental values determined from electric birefringence measurements (28,29). From the former, two different polarisabilities may be estimated, $\alpha_s$ referring to the static dielectric increment $\Delta\epsilon_s$ and $\alpha_2$ corresponding to the polarisability in the higher frequency region as determined from $\Delta\epsilon_2$. From the Van der Touw - Mandel theory these polarisabilities are, for a polydisperse macromolecule with monovalent counterions only, given by the following expressions valid only for infinite polyelectrolyte dilution and $\gamma = 1$.

$$\alpha_s = \frac{3\epsilon_o \Delta\epsilon_s}{C_M} = \frac{\bar{f} e^2 \bar{N}^n}{kT} R_g^2 \tag{4}$$

$$\alpha_2 = \frac{3\epsilon_o \Delta\epsilon_2}{\bar{p}^n C_m} = \frac{\bar{f} e^2}{12\ kT\ L_m} b^3 \tag{5}$$

Here $\epsilon_o$ represents the absolute electric permittivity of vacuum,

Table II. Comparison of electric polarisabilities from electric permittivity measurements with literature values of polarisabilities from electric birefringence data at finite dilution (see text).

| Electric Permittivity | | | |
|---|---|---|---|
| $\bar{M}^w$ (in $10^5$ g $mol^{-1}$) | 3.5 | 3.1 | 3.1 |
| $c_p$ ($10^{-3}$ monomol $l^{-1}$) | 0.3 | 0.3 | 0.75 |
| $c_{salt}$ (mM) | 0 | 0.3 | 0.3 |
| $\alpha_2$ ($10^{-32}$ F $m^2$) | 37 ± 5 | 19 ± 3 | 10 ± 1 |
| $\alpha_s$ ($10^{-32}$ F $m^2$) | 260 ± 40 | 130 ± 20 | 90 ± 20 |

| Electric birefringence | | | | |
|---|---|---|---|---|
| $\bar{M}^w$ ($10^5$ g $mol^{-1}$) | 3.7 | 3.7 | 5 | 50-100 |
| $c_p$ ($10^{-3}$ monomol $l^{-1}$) | 0.3 | 0.3 | 0.75 | 0.75 |
| $c_{salt}$ (mM) | 0 | 8 | 1 | 1 |
| $\alpha$ ($10^{-32}$ F $m^2$) | 20 ± 10 | 9 ± 3 | 10 | 20 |
| reference | (28) | (28) | (29) | (29) |

$C_M$ the number of polyions per unit volume, e the elementary charge, k the Boltzmann constant, T the temperature, $\bar{N}^n$ the number averaged number of counterions per polyion, $\bar{p}^n$ the number averaged number of subunits per polyion and $L_m$ the length of a monomeric unit (0.17 nm for the double-helical DNA), all expressed in S.I. units.

Unfortunately the electric birefringence measurements have not been performed at a set of different DNA concentrations, so that extrapolation to infinite dilution for comparison with eqs. 4 and 5 is not possible. Therefore we have corrected the polarisabilities as calculated from the extrapolated increments $(\Delta\epsilon_s)_{C_p=0}$ and $(\Delta\epsilon_2)_{C_p=0}$ according to eqs. 4 and 5 for concentration effects assuming that $R_g^2$ and $b^2$ obey the same concentration dependence as $\Delta\epsilon_s$ and $\Delta\epsilon_2$ respectively. These values together with the experimental values of the polarisability derived from electric birefringence are collected in table II. The comparison suggests that the polarisability from electric birefringence is much smaller than the static polarisability from dielectric data and should rather be related to $\alpha_2$. In this respect, the finding that the polarisability from birefringence data does not show a strong molecular weight dependence is also striking. However we cannot forward at the moment a satisfactory explanation for this observation.

## REFERENCES


1 Mandel M and van der Touw F, in"Polyelectrolytes", Sélegny E, Ed., Reidel, Dordrecht, p.285 (1977).
2 van der Touw F and Mandel M, Biophys. Chem., 2, (1974) 231.
3 Müller G, van der Touw F, Zwolle S and Mandel M, Biophys. Chem., 2, (1974) 242.
4 van Beek W M, Odijk T, van der Touw F and Mandel M, J. Polymer Sci., Polymer Phys. Ed., 14, (1976) 773.
5 Paoletti S, van der Touw F and Mandel M, J. Polymer Sci., Polymer Phys. Ed., 16 (1978) 641.
6 van der Touw F, de Goede J, van Beek W M and Mandel M, H. Phys. E., 8, (1975) 840.
7 van der Touw F, Selier G and Mandel M, J. Phys. E., 8, (1975) 884.
8 van Beek W M, van der Touw F and Mandel M, J. Phys. E., 9, (1976) 385.
9 Junger G, Junger I and Allgén L-G, Nature, 163, (1949) 849.
10 Junger I, Acta Phys. Scand., 20 (suppl.) (1950) 69.
11 Allgén L-G, Bioch. Bioph. Acta., 13, (1954) 446.
12 Jacobson B, J. Amer. Chem. Soc., 77, (1955) 2919.
13 Jerrard H G and Simmons B A W, Nature, 184, (1959) 1715.
14 Johnson G A and Neale S M, J. Pol. Sci., 54, (1961) 241.
15 Neale S M and Weyl D A, Proc. Roy. Soc. (London), A 291, (1966) 368.
16 Stellwagen N C, Shirai M and O'Konski C T, Abstr. 9th Bioph. Soc. Meeting, San Francisco (1965).
17 Takashima S, J. Mol. Biol., 7, (1963) 455.

18 Takashima S, J. Phys. Chem., 70, (1966) 1372.
19 Takashima S, Adsv. Chem. Ser., 63, (1967) 232.
20 Takashima S, Biopolymers, 5, (1967) 899 and 12, (1973) 145.
21 Pollak M, J. Chem. Phys., 43, (1965) 908.
22 Hanss M, Biopolymers, 4, (1966) 1035.
23 Hanss M and Bernengo J C, Biopolymers, 12, (1973) 2151.
24 Goswami D N and Das Gupta N N, Biopolymers, 13, (1974) 1549.
25 Sakamoto M, Kanda H, Hayakawa R and Wada Y, Biopolymers, 15, (1976) 879.
26 van der Touw F and Mandel M, Biophys. Chem., 2, (1974) 218.
27 Wyllie G, in "Dielectric and related molecular processes", Vol. I, ch.2, ed. Davies M, The Chem. Soc. (1972)
28 Hornick C and Weill G, Biopolymers, 10, (1971) 2345.
29 Houssier C, Bontemeijer J, Edmonds-Alt X and Fredericq E, Roc. Symp. on Electrical Properties of Biopolymers and Biomembranes (Am. New York Acad. Science).

# ELECTRO-OPTIC MEASUREMENT OF $\gamma$-RAY INDUCED DAMAGE IN DNA

C. T. O'Konski and R.S. Farinato

Department of Chemistry, University of California,
Berkeley, California 94720

The effects on aqueous superhelical PM2 and linear calf thymus DNA's due to $\gamma$-radiation from a $^{60}Co$ source have been studied by pulsed electric birefringence relaxation techniques. In a nitrogen atmosphere the average field-free relaxation time, $\bar{\tau}$, for calf thymus DNA first increased, then decreased as a function of dose in the range 0-13 kRad. A much more rapid decrease in $\bar{\tau}$ as a function of dose was determined in the presence of ambient oxygen. Significant differences were easily measurable at the lowest dose tried (~ 0.3 kRad). Solutions of superhelical PM2 DNA in the presence of ambient oxygen showed an asymptotic increase in $\bar{\tau}$ by a factor of about 2. This was predicted for supercoil relaxation from a previous electro-optic study of PM2 DNA with ethidium bromide. Significant changes in $\bar{\tau}$ were easily discernable at the lowest doses measured (~ 0.3 kRad), with a 50% effect at 0.5 kRad.

## INTRODUCTION

Single strand breaks in DNA have been shown to be mutagenic in phage systems with double-stranded DNA (1,2). This was attributed to frameshift mutations and missense reading. The loss of transforming activity in DNA irradiated in dilute aqueous solution has been attributed as mainly due to the effects of single strand breakage (3). Double strand breaks have been correlated with lethality in phage systems (4,5). Transfection is inactivated more by double strand breaks than single strand breaks. Base damage, while not necessarily lethal, appears to play a predominant role in the inactivation of infectious DNA (3). In general, the extent of biological damage depends on the efficiency of the enzymatic repair systems operating in vivo, which varies greatly with species (3,6).

Double strand breaks are largely irreparable while single strand breaks are repaired to varying degrees in different organisms (1,4,7-9). Summers and Szybalski (5) showed that irradiation of B subtilis phage or phage DNA in dilute solution inactivates infectivity. Their results are consistent with the ideas that double strand breaks are lethal events and that single strand breaks reduce DNA transforming activity. Freifelder (4), working with T-7 phage, reported that double-strand breaks are lethal. However, Hawkins and Ginsberg (10) have shown that there are additional factors contributing to lethality. Furthermore, inactivation is not correlated with double strand breaks for the replicative form (RF) of phage ϕX174 irradiated in frozen solution (11). In dilute solution it was essential in several experiments that the relative contributions of double strand, single strand, and base damage to inactivation of ϕX174 RF DNA are approximately 3, 0-25, and 75-97% (7). For many systems with duplex DNA and active repair systems, base damage and single strand breaks are usually not lethal.

Several physical methods have been used to determine chain breaks (3) including alkaline sucrose gradients (12,13) band centrifugation (14,15) ultracentrifugation (4,14,16,17), light scattering (18-20) and viscometry (18-21). Setlow and Setlow (7) have discussed the limitations of some of these methods. We add to their list the amount of time which all of these methods require. Specifically, Christensen et al. (22) point out serious overdeterminations of single strand breaks using alkaline sucrose gradients. Alkaline treatment after irradiation of a neutral solution induces additional strand breaks in the irradiated DNA. This effect shows that there is damage to the sugars which does not produce strand breakage at neutral pH (3,6). This type of damage amounts to 20-30% of the total single strand breaks. Better in vitro assays for single strand breaks clearly are needed. The results in this study permit hope that some of the above mentioned controversies about radiation damage mechanisms might be resolved with the aid of the electro-optic relaxation technique as illustrated here.

In our original investigation of the relaxation behaviour of superhelical DNA it was found that binding to DNA of ethidium bromide, an intercalating agent, first increased and then again decreased the mean relaxation time $\bar{\tau}$ of the DNA (23). Going from the supercoil to the relaxed circles gave about a 2-fold increase in $\bar{\tau}$ under our conditions ($\sim$10 $\mu$g DNA/ml, 0.5 mM Tris, 0.1 mM $Mg^{++}$). Thus we anticipated up to a roughly 2-fold increase in $\bar{\tau}$ when supercoiled DNA is "nicked" by low level $\gamma$-irradiation. Further nicking would tend to increase the internal flexibility of the macromolecule and decrease $\bar{\tau}$; sufficiently high dosages would eventually produce opening of the circles by double strand breaks when nicks on the two chains are within about 3 base pairs of each other (24) and this would increase $\bar{\tau}$ (25).

## MATERIALS AND METHODS

The calf thymus DNA was Sigma Chemical Co.'s highly polymerized

Na salt. Aliquots referred to as sample preparation I were diluted from a stock of 0.5 mg/ml DNA in 0.1 mM Tris to 18 $\mu$g/ml in a final buffer of 0.5 mM Tris (pH 7.1). Several 1.7 ml portions in glass vials were bubbled with nitrogen for 3 minutes each, sealed with parafilm and capped. Sample preparation II was diluted from a 5 mg/ml stock solution in 10 mM Na phosphate buffer (pH 7.4) to 35 $\mu$g/ml in 0.6 mM Na phosphate (pH 7.4). Ambient oxygen remained in this second preparation. Following irradiation, electro-optic measurements were made on preparation II after the solutions were diluted to 18 $\mu$g/ml in 0.3 mM Na phosphate. Concentrations were measured spectrophotometrically using a molar absorptivity at 260 nm of 6.60 x $10^3$ $cm^2$/mol of DNA phosphate (23). The base denatured hyperchromicity of this calf thymus DNA in 0.1 mM Na phosphate was 0.32.

The PM2 DNA sample was prepared in Professor J. Wang's laboratory according to a modified procedure of LePecq (26). The original stock was 28 $\mu$g/ml in 0.5 mM Tris (pH 8.0), 0.1 mM $Mg^{++}$. This was diluted to 14 $\mu$g/ml in 0.5 mM Tris, 0.1 mM $MgCl_2$ (pH 8.0). In one run ambient oxygen was not excluded, and in a second run the samples were purged with nitrogen for three minutes and capped in glass vials until measurement.

The 3000 Curie $^{60}$Co source at the Lawrence Berkeley Laboratory was used for $\gamma$-irradiation of the samples. Dose rates and depletion factors were obtained from R. M. Lemmon and G. Wigel, and were based on Fricke dosimetry measurements. The 3000 Curie source allows insertion of the sample in a shielded cavity containing 16 $^{60}$Co slugs positioned on a circular perimeter. Doses are accurate to $\pm$ 1% except for the lowest dose (0.3 kR) which was accomplished below the central part of the radiation chamber. This dose was accurate to about $\pm$ 30%.

Pulsed electro-optic birefringence signals were measured with a computerized electro-optic spectrometer (27). The birefringence in a solution of DNA produced by an electric field pulse (Cober 605P High Voltage pulser triggered by a Tektronix 162 wave form generator) was measured as a light signal on an EMI 9856B phototube through an optical system consisting of a tungsten light source, Glan-Thompson polarizer, sample cell with electrodes (2mm spacing), $\lambda/4$ retardation plate, and Glan-Thompson analyzer. The transient signal was captured in a Biomation 805 transient recorder and read into a PDP 11/10 on-line computer equipped with an AED 2500 flexible triple disk storage device. The electric field pulses used were 500 $\mu$sec at around 6 kV/cm for calf thymus DNA preparation I; 300 $\mu$sec at about 5 kV/cm for calf thymus DNA preparation II and PM2 DNA samples. A simultaneous recording of the applied field pulse was made using ICE PTR 9200 transient recorder. The voltage divider for measuring the applied field pulse was calibrated using a Tektronix 1:1000 capacitive probe and a Hewlett Packard 738 BR voltmeter calibrator as an amplitude source. Computer programs developed here (27,28) were used to analyze EO signal amplitudes and to fit the decay portion of the curve, S(t), by a nonlinear least squares procedure to a sum of exponentials.

$$S(t) = \sum_i s_i e^{-t/\tau_i}.$$

The S/N ratio of around 100 limited us to three term fits with the spread of relaxation times characteristic of these DNA samples.

## RESULTS

The two electro-optic parameters used as a basis of comparing samples were $\bar{\tau}$ (average fitted relaxation time) and the optical retardation at the "pulse off" point ($\delta$); the latter was essentially the steady state level. Whether the DNA was linear or superhelical, the value of $\bar{\tau}$ generally increased with pulse number (amounting to ~30% change after 4-6 pulses at these field strengths). We found that repeated pulsing at 5-6 kV produced observable increases in relaxation times. Hence, records were made of pulsing history for each sample and allowances were made for this before interpreting small changes in $\bar{\tau}$ as due to other variables.

The $\bar{\tau}$ values are a birefringence weighted property of the system. They showed a 10% sample-to-sample variation; values of $\delta$ showed about a 5% sample-to-sample variation. As radiation dosage increased, $|\delta|$ decreased for linear and increased for superhelical DNA; $\bar{\tau}$ decreased for linear and increased for superhelical DNA. These trends are consistent with production of single strand scissions of the superhelical DNA, and with single and double strand scissions in the linear DNA. In this work, no physical interpretation is given to the individual terms in the relaxation time spectra.

### Calf Thymus DNA:

For preparation I ($N_2$ flush) the fitted average relaxation times first increased slightly then decreased greatly over the range 0-13 kRad (see fig. 1). This initial increase appeared to reflect mainly changes occurring in the longest relaxation component ($\tau_3$). The subsequent decrease in $\bar{\tau}$ at higher doses was more evident in the longer $\tau$ components also; and only at the highest dose was there a discernable difference in $\tau_1$ (the shortest component). Pulse history was also an important parameter. Additional pulsing tended to increase the value of $\bar{\tau}$ and may be indicative of a physical change in the DNA mediated by the electric field. The effect of the presence of ambient oxygen on the relaxation time changes in calf thymus DNA preparation II was dramatic (see fig. 1). However, retardation values ($\delta$) for calf thymus DNA preparation I showed no specific trend as a function of dosage. The quantities $\delta$ and $\bar{\tau}$ rapidly decreased as a function of dose. For 1.3 kRad, $\bar{\tau}$ averaged about 16 $\mu$sec.

We can estimate a range of chain contour lengths using a worm-like chain polymer model (29). Asymptotic formulas for the rotational diffusion constant have been given by Yamakawa (30) for the limits $\lambda\ell \ll 1$ and $\lambda\ell \gg 1$, where $\lambda$ is the reciprocal of twice the persis-

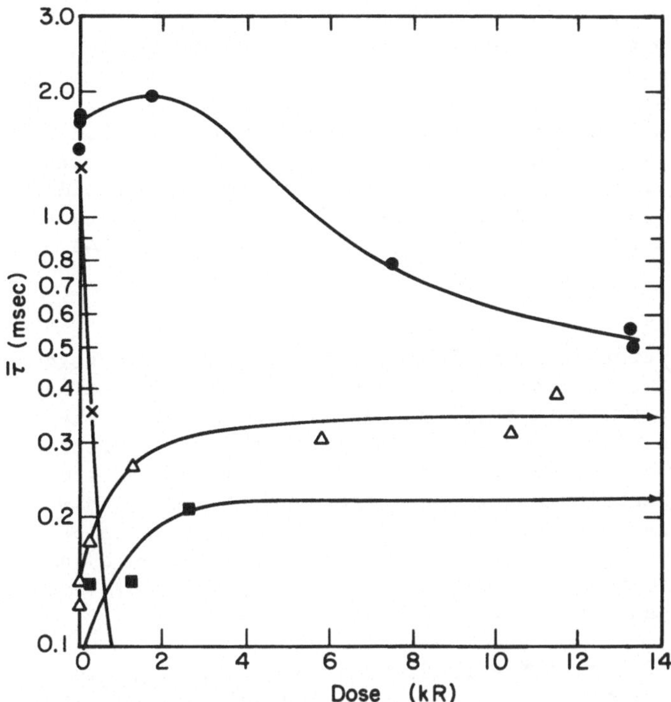


Fig. 1. Fitted average electro-optic time, $\bar{\tau}$, vs. $\gamma$ radiation dose of dilute aqueous DNA solutions. DNA samples were: circles - 18 $\mu$g/ml calf thymus DNA in 0.5 mM Tris, pH 7.1 with 3 min. $N_2$ flush; crosses - 18 $\mu$g/ml calf thymus DNA in 0.3 mM Na phosphate, pH 7.4 with ambient oxygen; triangles - 14 $\mu$g/ml PM2 DNA in 0.5 mM Tris, 0.1 mM $Mg^{++}$, pH 8.0 with 3 min. $N_2$ flush; squares - the same, with ambient oxygen. For the ambient oxygen irradiation of calf thymus DNA, there also is one point off the curve at (0.016 msec, 1.3 kR) due to the log scale for $\bar{\tau}$.

tence length of the polymer chain of contour length $\ell$. For our case $\lambda \approx \ell$ and no closed analytic expression has been derived. Hence we have made calculations with both of the available formulas and the two values bracket the actual contour lengths. Using a persistence length of 60 nm (31), a chain diameter of 2 nm, a segment length of 0.34 nm (32) and a temperature of 295° K, we computed for calf thymus DNA preparation II the contour lengths 100 and 138 nm (294 and 406 base pairs respectively). These values happen to be in the range of fragment sizes obtained by shear or sonic degredation of whole calf thymus DNA, as measured electro-optically. Measurements on calf thymus DNA preparation II were made at two different post-irradiation times: ~3 hrs. and ~54 hrs. after irradiation; there were no discernible changes in either $\delta$ or $\bar{\tau}$.

The oxygen enhancement ratio, OER, for single-strand breaks in DNA within living mammalian cells is around 4, where the OER is the ratio of the slopes of number of breaks vs. dosage with and without oxygen being present (12). The ratio of the doses at which $\bar{\tau}$ has been reduced to 1/2 of its original value without and with ambient oxygen in the calf thymus DNA samples is 47 in this work. We expect $\bar{\tau}$ to be a complicated function of the number of single-strand breaks for this flexible macromolecule. An increase in OER for pure DNA in solution is reasonable in view of the well-known fact that nuclear proteins protect DNA from radiation damage (33,34).

PM2 DNA

For the superhelical PM2 DNA the fitted average rotational relaxation time, $\bar{\tau}$ , increased as a function of $\gamma$-ray dose as shown in fig. 1. Some of the data involved second irradiations, the total accumulated dose being reported. Within experimental uncertainty the effects of $\gamma$-ray damage were cumulative. There was an increase in $\bar{\tau}$ values as a function of electric field pulse number as was seen with the calf thymus DNA results. This effect seemed to be greater when the sample was stored for several days (in the cold) and was pulsed again. Increases in $\bar{\tau}$ due to pulse number would amount to as much as a 30% change after 4-6 pulses.

The major effect of $\gamma$-irradiation in this dose range (0-20 kRad) was an increase in $\bar{\tau}$ with an approximate doubling of $\bar{\tau}$ at the asymptotic limit. This was consistent with relaxation of the superhelical turns in the DNA due to single-strand scission of all macromolecules at the highest dosages used. The plot of $\bar{\tau}$ vs. dose for the $N_2$ flushed samples was displaced vertically with respect to the samples with ambient oxygen (see fig. 1). Both curves reached an asymptote in the same dose range (3-4 kRad), indicating no oxygen enhancement of the single chain scission process.

The unextrapolated values of the specific Kerr constant, $\Delta n/(nC_vE^2)$, where $\Delta n$ is the birefringence, n the solution refractive index, C the volume fraction of solute, and E the field strength, were less for the superhelical PM2 DNA than the linear calf thymus DNA by about 20% (these numbers are not shown here). Although the molecular weights were approximately the same for PM2 DNA (35) and the native calf thymus DNA (36)$\sim$6 x $10^6$ daltons, there were dramatic differences in relaxation spectra. The average $\bar{\tau}$ for native calf thymus DNA (in 0.3 mM Na phosphate) was $\sim$ 1.26 ms whereas for PM2 DNA (in 0.5 mM Tris, 0.1 mM $Mg^{++}$) the average $\bar{\tau}$ was $\sim$ 0.10 ms. This demonstrated the large change of rotational diffusion between linear vs. superhelical structures. Three terms were necessary to fit the relaxation spectra for the PM2 DNA as well as the calf thymus DNA, which indicates flexibility in the supercoil, as well as in the linear macromolecules.

## DISCUSSION

An early study using pulsed electric birefringence was made on DNA damaged by X-ray irradiation. This was carried out by Norman and Field with DNA from E.coli, both in solution and in living cells (33). They observed changes in both the magnitude and the relaxation times as a function of radiation exposure from about 2 to 32 kRads. The linear DNA from E.coli showed a marked decrease both in the steady state birefringence and in relaxation time as a function of X-ray dose in this range. They found preferential degradation of longer molecules in the broad sample distribution based on changes in the relaxation curve with X-ray dose. An apparent single relaxation time characterizes the DNA solutions at larger doses (e.g. 0.15 msec at 8 kRad for 0.005% DNA solution). They attribute this to degradation of the macromolecule to rod-like segments and the relatively low sensitivity of their E-O experiments for very short fragments.

We have investigated similar effects in linear calf thymus DNA with sensitivity down to lower doses ($\sim$0.3 kRad) of $\gamma$ radiation. Our solutions were about 0.004% by weight and showed reduction to a single exponential term for the relaxation curve at 1.3 kRad in the presence of oxygen. Norman and Field reported a degraded segment length of 300 nm, whereas we computed a segment length closer to 100 nm. This is possibly due to greater sensitivity of the E-O spectrometer used in this work.

The effect of reducing the relaxation curve to a single component is a function of DNA concentration (33) and we expect it to depend upon molecular weight also. Thus, our results generally corroborate those of Norman and Field.

When $N_2$ was bubbled through the sample, the initial value of $\bar{\tau}$ at 0 kRad dose was larger than for the untreated PM2 DNA. This is ascribed to mechanical breakage due to shear effects.

The 13-fold difference in $\bar{\tau}$ before irradiation between the calf thymus and PM2 DNAs shows dramatically the sensitivity of the relaxation time to the compactness of the superhelical form. Under similar ionic strength conditions (0.3 mM Na phosphate for calf thymus DNA and 0.5 mM Tris, 0.1 mM $Mg^{++}$ for PM2 DNA) the average $\bar{\tau}$ values for unirradiated samples of calf thymus and PM2 DNAs were 1.26 and 0.10 msec respectively. The molecular weights are $\sim$ 6 and 6.3 x $10^6$ daltons respectively (35,36).

Using the same buffer conditions (0.5 mM Tris, 0.1 mM $Mg^{++}$, pH 8.0), Pritchard and O'Konski (23) showed that $\bar{\tau}$ for superhelical PM2 DNA as a function of bound ethidium bromide to phosphate ratio, r, goes through a maximum equivalent to nicked PM2 DNA whose $\bar{\tau}$ is about twice that of the native form. This was consistent with unwinding of negative superhelical turns in the PM2 DNA as a consequence of ethidium ion intercalation until a relaxed circular form is achieved (at the maximum in $\bar{\tau}$ vs. r). Further addition of ethidium creates positive superhelical turns which decreases $\bar{\tau}$ values. The $\bar{\tau}$ for the native superhelical PM2 DNA was 95 $\mu$s and for the relaxed circular form 190 $\mu$s. These values agreed with the $\gamma$-ray induced damage results in this work where the average $\bar{\tau}$ values were 97 $\mu$s for native PM2 DNA and 213 $\mu$s for the irradiated samples. This is consistent with the products of the radiation damage at these doses being relaxed circular structures. The levelling off of $\bar{\tau}$ values indicates that a substantial amount of double strand breakage is not occurring at these doses ($\leq$ 28 kRad), as this kind of breakage would increase $\bar{\tau}$ well beyond twice the unirradiated value and eventually decrease it again. This higher resistance of the PM2 DNA to double strand breakage compared with the calf thymus DNA appears to be due to lack of single strand nicks in the PM2 DNA.

The very large effect of $O_2$ in enhancing radiation damage on calf thymus DNA contrasts strikingly with no visible effect on PM2 DNA. This we attribute to pre-existing damage in the commercial calf thymus DNA samples. The existence of strand breaks in such preparations is well known from early studies. Oxygen may react with solvated electrons which are produced by radiolysis of the water to form the superoxide ion which in turn may undergo various reactions which have been studied kinetically (37). Prior studies of radiation damage of DNA in solution and in cells indicate that the predominant mechanism of strand breakage in dilute solutions is attack by free radicals (3,6,37-39). When such attack occurs sufficiently near a nick in the complementary strand, the duplex will be broken due to disruption of the intervening base pair hydrogen bonds and this will cut the linear DNA into smaller pieces.

The mechanisms of damage to the DNA are themselves complex, and the situation is further complicated in the living cell experiments by the effects of repair enzymes (40). Time-resolved experiments with bacteria and mammalian cells show that the oxygen effect can be resolved into at least two components, and possible mechanisms were discussed (38). Superoxide radicals appear to be agents of oxygen toxicity, e.g. in defence against disease by phagocytes; superoxide dismutases, catalases and peroxidases may therefore be involved with the oxygen effect (41).

Clearly, the biochemical and physical advantages of the radiation damage assay described here can facilitate the acquisition of new data needed for the elucidation of basic mechanisms of radiation damage and repair.

## ACKNOWLEDGEMENTS

The authors acknowledge the gift of purified superhelical PM2 DNA from Jim Wang's group at Harvard. We thank Dick Lemmon and George Wigel for arrangements and information on the use of the $^{60}Co$ source. This work was accomplished under a grant from the National Institutes of Health (Grant No. CA-12, 540-06).

REFERENCES

1 Bridges B A, Ann. Rev. Nucl. Sci., 19, (1969) 139.
2 Munson R J and Bridges B A, Biophysik, 6, (1969) 1.
3 Hüttermann J, Köhnlein W and Teouk R (eds.), Molecular Biology, Biochemistry and Biophysics, 27, "Effects of Ionizing Radiation on DNA", Springer-Verlag, Sec. III, Ch. 2 (1978).
4 Freifelder D, PNAS US, 54, (1965) 128.
5 Summers W C and Szybalski W, J. Mol. Biol., 26, (1967) 227.
6 Ward J F, Adv. Rad. Biol., 5, (1975) 181.
7 Setlow R B and Setlow K, Ann. Rev. Biophys. and Bioeng., 1, (1972) 293.
8 Munson R J, Neary G J, Bridges B A and Preston R J, Int. J. Rad. Biol., 13, (1968) 205.
9 Freifelder D, Virology, 36, (1968) 613.
10 Hawkins R B and Ginsberg D M, Biophys. J., 11,(1971) 398.
11 Taylor W D and Ginoza W, PNAS, 58, (1967) 1753.
12 Painter R B, in Proc. 5th Int. Cong. Rad. Res. (Nygaard, Alder, Sinclair, eds., Academic Press) p. 735 (1975).
13 Elkind M M and Kamper C, Biophys. J., 10, (1970) 237.
14 Ginoza W, Ann. Rev. Nuclear Sci., 17, (1967) 469.
15 Vinograd J, Bruner R, Ket R and Weigle J, PNAS US, 49, (1963) 902.
16 Studier F W, J. Mol. Biol., 11, (1965) 373.
17 McGrath R A and Williams R W, Nature, 212, (1966) 534.
18 Peacocke A R and Preston B N, J. Poly. Sci., 31, (1958) 1.
19 Peacocke A R and Wilson S, Proc. Roy. Soc. (London), B149, (1958) 511.
20 Cox R A, Overend W G, Peacocke A R and Wilson S, Proc. Roy. Soc. (London), B149, (1958) 54.
21 Lett J T, Stacey K A and Alexander P, Rad. Res., 14, (1961) 349.
22 Christensen R C, Tobias C A and Taylor W D, Int. J. Rad. Biol., 22, (1972) 457.
23 Pritchard A E and O'Konski C T, Ann. N.Y. Acad. Sci., 303, (1977) 159.
24 Dertinger H and Jung H, "Molecular Radiation Biology", Springer-Verlag, N.Y. (1970).
25 Bloomfield V and Zimm B H, J. Chem. Phys., 44, (1966) 315.
26 LePecq J-B, in "Methods of Biochemical Analysis", 20 (Glick D, ed.), Interscience, N.Y. (1971).
27 Jost J W and O'Konski C T in "Molecular Electro-Optics", 2 (C T O'Konski, ed.) Ch.15, Dekker, N.Y., (1978).
28 Kwan M, M.S. Thesis, Univ. of Calif., Berkeley (1975).
29 Kratky O and Porod G, Rec. Trav. Chim., 68, (1949) 1106.
30 Yamakawa H, "Modern Theory of Polymer Solutions", Harper and Row, N.Y. (1971).
31 Godfrey J E and Eisenberg H, Biophys. Chem., 5, (1976) 301.
32 Bloomfield V A, Crothers D M and Tinoco I Jr., "Physical Chemistry of Nucleic Acids", Harper and Row, N.Y., (1974).
33 Norman A and Fields J A, Arch. Biochem. Biophys., 71, (1957) 170.
34 Adams G E in "Advances in Radiation Chemistry", 3, (Burton M and Magee J L, eds., Wiley-Interscience)(1972).

35 Camerino-Otero R D and Franklin R M, Eur. J. Biochem., 53, (1975) 343.
36 Jordan D O in "The Nucleic Acids", 1 (Chargaff E and Davidson J N, eds.) (1955); Hornick C, Weill G and Benoit H, Abstr. Am. Chem. Soc., 152nd Meeting, N.Y., p.291 (1966).
37 Bors W, Saran M, Lengfelder E, Spottl R and Michel C, Curr. Top. Radiat. Res. Q., 9 (1974) 247.
38 Shenoy M A, Asquith J C, Adams G E, Michael B D and Watts M E, Radiation Research, 62, (1975) 498.
39 Adams G E and Wardman P in "Free Radicals in Biology", 3, Ch.2, Academic Press, N.Y. (1977).
40 Setlow R B, Nature, 271, (1978) 713.
41 Fridovich I, Science, 201, (1978) 875.

# PHYSICAL STUDIES OF Hg(II) AND Ag(I) DNA COMPLEXES - THEORY

Dawen Ding and Fritz S. Allen

Department of Chemistry, University of New Mexico
Albuquerque, NM 87131, U.S.A.

The binding of Hg(II) and Ag(I) to the bases of DNA and polynucleotides induces strong absorption and CD bands. From electric dichroism studies, the transition moment of the difference absorption band at 293 nm is in the plane of the bases and has a huge negative CD band at 285 nm; while the corresponding 293 nm difference absorption band of DNA-Ag(I) complex is relatively CD inactive. A second band at 270 nm in the difference absorption spectrum of the DNA-Ag(I) complex has a large associated CD band at 270 nm. This transition is out of the plane of the bases. The increase of the sedimentation coefficients of the rigid rod sonicated DNA by the heavy metal ions can be accounted for by the increase of molecular weight and the decrease of the partial specific volume due to the high density of the heavy metal ions. Either a charge transfer band between the bases and the heavy metal ions or a perturbation theory of the bases by the heavy metal ions would explain satisfactorily the experimental optical results of DNA-Hg(II) and DNA-Ag(I) complexes.

## RESULTS AND DISCUSSIONS

Electric dichroism is a linear dichroism induced in a solution of electrically and optically anisotropic molecules by a pulsed electric field (1,2,3). The difference in extinction coefficients with the light polarized parallel and perpendicular to the applied electric field is divided by the total unperturbed extinction coefficient, $\varepsilon$, to get the reduced dichroism ( $\varepsilon_{\parallel}$ - $\varepsilon_{\perp}$ )/ $\varepsilon$. The reduced dichroism is unitless and the total extinction coefficient of the macromolecule or macromolecular complex is used for the value of $\varepsilon$ at that particular wavelength. From electric dichroism the following information can be obtained: 1) orientation of chromophoric transition moment,

2) dipole moment and/or anisotropy of the polarizability, and 3) relaxation time - flexibility of the macromolecules.

Reduced dichroism: $$\frac{\Delta\varepsilon}{\varepsilon} = \frac{1}{3}\left(\frac{\varepsilon_{\parallel} - \varepsilon_{\perp}}{\varepsilon_{\parallel} + 2\varepsilon_{\perp}}\right) = \frac{3}{2}(3\cos^2\alpha - 1)\Phi \qquad (1)$$

$\alpha$ = angle between dipole and transition dipole moment, $\Phi$ = orientation function.

For the case of sonicated B form DNA, the $\pi \rightarrow \pi^*$ transition dipoles are polarized in the plane of the bases; and the molecules can be treated as a rigid rod with the dipole axis coaxial with the helix axis with an $\alpha$ value approaching $90^o$. From the previous results (2), it was shown that there is no significant out-of-plane transition for DNA between 250 nm and 290 nm. Any out-of-plane transition, which has a component with an $\alpha$ value of $0^o$, will make a substantial change in the reduced dichroism. A sonicated DNA, with a segment length on the order of the persistence length, was used for the ED experiments. This causes less tendency for intermolecular interaction in the complexation by heavy metal ions. Sonicated DNA can be treated hydrodynamically as a rigid rod even though there is considerable length heterogeneity. The results of reduced dichroism vs. $r_{Hg(II)}$ of sonicated CT DNA at different wavelengths with the same voltage across the electrodes are shown in fig. 1

For a permanent dipole orientation mechanism (1,4)

$$\Phi = 1 - \frac{(3\coth\beta - 3/\beta)}{\beta} \qquad (2)$$

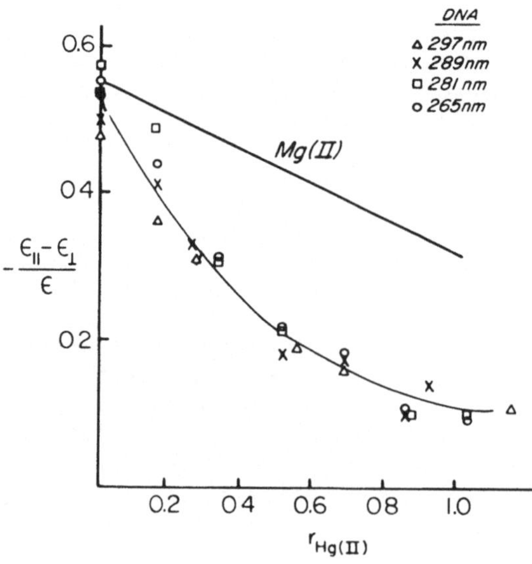


Fig. 1. Reduced Dichroism at Different Wavelengths of Sonicated CT DNA-Hg(II). The Effect of Mg(II) is also Shown for Comparison.

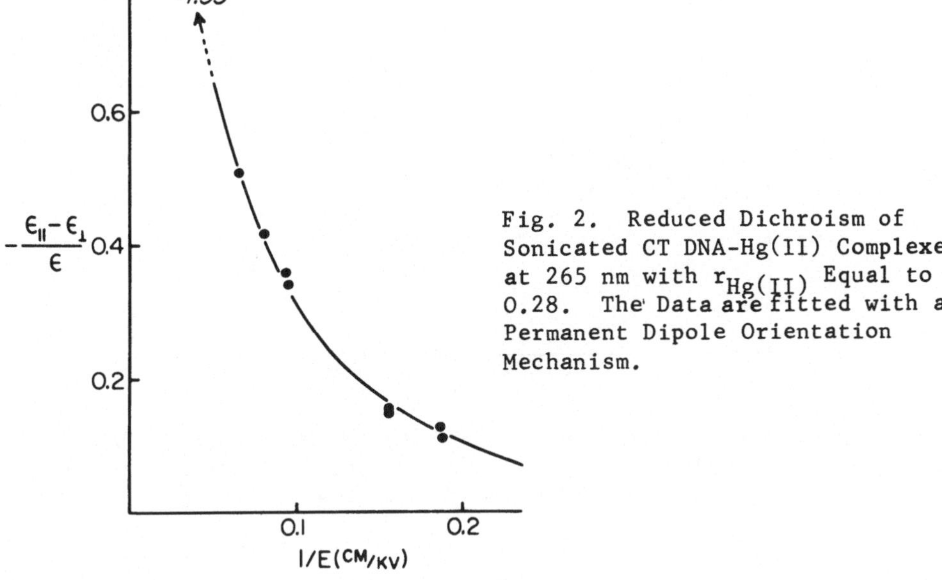


Fig. 2. Reduced Dichroism of Sonicated CT DNA-Hg(II) Complexes at 265 nm with $r_{Hg(II)}$ Equal to 0.28. The Data are fitted with a Permanent Dipole Orientation Mechanism.

where $\beta = \frac{\mu E}{kT}$, $\mu$ = permanent dipole, E = electric field, k = Boltzmann constant and T = absolute temperature.

For an induced dipole orientation (3)

$$\Phi = \frac{3}{4}\left\{\frac{\exp(\gamma)}{(\gamma)^{\frac{1}{2}}F(\gamma^{\frac{1}{2}})} - \frac{1}{\gamma}\right\} - \frac{1}{2} \qquad (3)$$

where

$$\gamma = \frac{\Delta\alpha_e E^2}{2kT}$$ $\Delta\alpha_e$ anisotropy of electric polarizability.

$$F(\gamma^{\frac{1}{2}}) = \int_0^{\gamma} e^{t^2}\,dt$$

For a permanent dipole orientation machanism, the $\Phi$ is a function of $\beta$ where $\beta$ is proportional to the electric field. On the other hand, in an induced dipole orientation mechanism, the $\Phi$ is a function of $\gamma$ where $\gamma$ is proportional to the square of the electric field. For 100 percent orientation, it is necessary to extrapolate to infinite field strength to get the limiting dichroism.

Keeping the $r_{Hg(II)}$ value constant while varying the voltage, the limiting value of reduced dichroism at infinite field can be obtained by fitting with a "permanent dipole" orientation mechanism (2). A plot of $(\varepsilon_{\parallel} - \varepsilon_{\perp})/\varepsilon$ at 265 nm vs. $\frac{1}{E}$, at r equal to 0.28, is shown

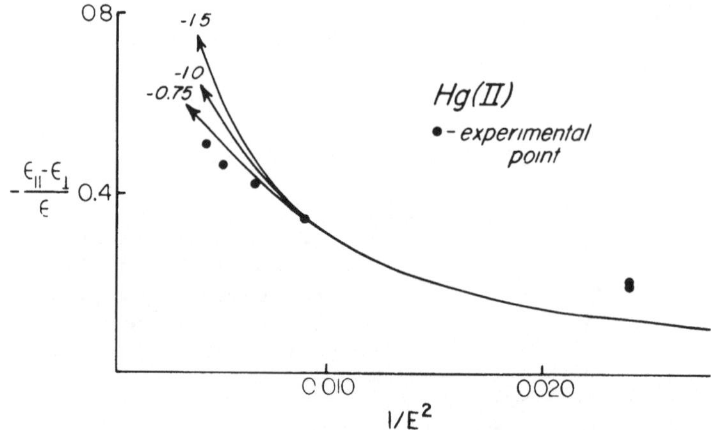


Fig. 3. An Attempt to fit Reduced Dichroism of Sonicated CT DNA-Hg(II) Complexes with an Induced Dipole Orientation Mechanism, with Different Values of Limiting Dichroism.

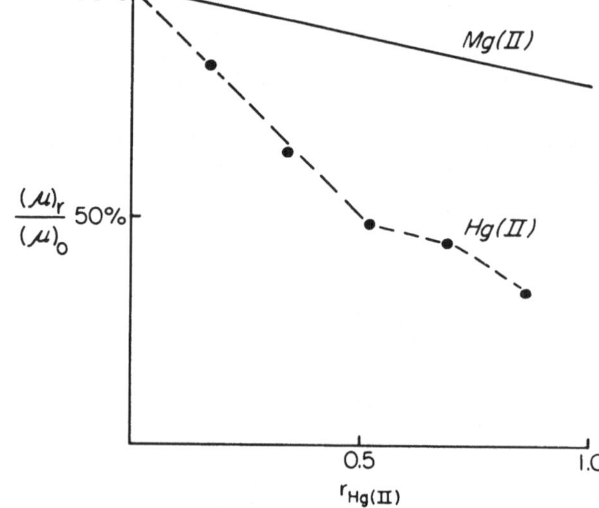


Fig. 4. The Ratio of Apparent Permanent Dipole of DNA-Hg(II) Complexes to DNA alone as a Function of $r_{Hg(II)}$. The Effect of Mg(II) is also shown for Comparison.

in fig. 2. The limiting dichroism obtained from the fitting is equal to -1.35, which is the same as that of native DNA (2). A plot of $(\varepsilon_{\parallel} - \varepsilon_{\perp})/\varepsilon$ vs. $E^{-2}$ (fig. 3) using different values for the limiting dichroism could not be fitted with an induced dipole orientation mechanism. Using the limiting reduced dichroism of -1.35, the orientation function can be obtained. Solving for $\Phi$ from eq. 2, the apparent permanent dipole at different $r_{Hg(II)}$, $(\mu)_r$, can be obtained from $(\mu)_r = \frac{\beta kt}{E}$. By plotting the ratio of $(\mu)_r$ to $(\mu)_r$ at r = 0, (i.e., $(\mu)_o$) against r in fig. 4, it can be seen that the apparent dipole of a DNA Hg(II) complex drops to approximately half when r is equal to 0.5. A further increase in r results in the reduction of the apparent dipole to one third of the original value. By comparison, introduction of Mg(II) to the solution will drop the apparent dipole 20% when $r_{Mg(II)}$ is equal to 1.0 (ref. 2).

Manning's polyelectrolyte theory (5-7) makes a number of assumptions. Principal among these are the assumption of the absence of intermolecular interactions, condensed counterion interactions, and that the polyion dielectric constant, $\varepsilon$ , is the same as that of the bulk solution. This theory has a dimensionless charge density parameter, $\xi$.

$$\xi = \frac{e^2}{\varepsilon k T b} \tag{4}$$

where e is the monovalent charge, and b is the average spacing of the charged sites in the linear array of the polyion. For monovalent counterions, the value of $\xi$ will approach 1 by condensation of the counterion on the polyion. If divalent counterions are also present these should condense in preference to monovalent ions and $\xi$ fall between 1 and 0.5.

In studies of the ED of double stranded poly A and double stranded poly C as a function of pH, Charney introduced the dielectric constant of the double strand base region, which is at least an order of magnitude smaller than the bulk dielectric constant, to account for the microstructural difference of the two double strands in the pH dependence of counterion condensation (private communication). In the binding of heavy metal ions, such as Hg(II) and Ag(I) to DNAs the quantitative binding is on the bases, and it is quite different from the binding of Mg(II) on the phosphate. These two varieties of binding should offer a check on Manning's polyelectrolyte theory. Work in this area is in progress.

We are grateful to Elliot Charney for calling our attention to this aspect of the problem.

## REFERENCES

1 Allen F S and Van Holde K E, Biopolymers, 10, (1971) 865.
2 Ding D, Rill R and Van Holde K E, Biopolymers, 11 (1972) 2109.
3 Fredericq E and Houssier C, Electric Dichroism and Electric Birefringence, Clarendon Press, Oxford (1973).
4 O'Konski C T, Yoshioka K and Orttung W H, J. Phys. Chem., 63 (1959) 1558.
5 Manning G S, J. Chem. Phys., 51 (1969) 924.
6 Manning G S, Biopolymers, 11 (1972) 937.
7 Manning G S, Biopolymers, 11 (1972) 951.

# ELECTRIC-FIELD INDUCED ORIENTATION OF POLYNUCLEOTIDES

Elliot Charney

Laboratory of Chemical Physics, National Institute of Arthritis, Metabolism and Digestive Diseases, National Institutes of Health, Bethesda, Maryland, U.S.A.

Transient dichroism produced by the orientation of a linear polyelectrolyte is shown to be related to the polarization of condensed counter-ions by demonstrating that the changes in alkali counter-ion concentration with degree of protonation of poly(A) predicted by current models of such polyelectrolytes is proportional to the field-induced orientation.

The orientation of macromolecules in strong electric fields has been discussed qualitatively in many places and treated analytically by Liptay and Czekalla (1), O'Konski and Yoshioka and Orttung (2), Shah (3), Yamaoka and Charney (4), O'Konski and Krause (5) and Kikuchi and Yoshioka (6). All except the last two of these treatments confine the interactions between the applied field and the molecule to the molecular dielectric properties associated with the neutral valence charge distribution to which permanent dipole moments and the field induced moments produced by the polarization of the valence electronic structure are attributed. The treatment by O'Konski and Krause (5) depends on the ion transport properties in polyelectrolyte solutions. A recent paper by Hogan, Dattagupta and Crothers appears to suggest a similar mechanism but no analysis is given (7). A number of authors have suggested from time to time that for macromolecules which are polyelectrolytes, the polarization of the counter-charges must also be taken into account, but only the model of Kikuchi and Yoshioka has thus far done this explicitly. As these authors point out, the effect of repulsion between the counter-ion charges is not accounted for in their treatment. Nor did they attempt to relate their theoretical treatment to the predictions of recently advanced theories (8,9) of the extent of counter-ion charge neutralization of the polyelectrolyte. In a recent paper (10), we have applied data on the electric-field induced dichroism of polyriboadenylic acid, poly(A), to the Kikuchi and Yoshioka model with the object of testing that model for its qualitative prediction that the electric-field induced orientation of a polyelectrolyte be a linear function of the number, n, of bound counter-ions per polyion at low field strengths. The boundary of low field strengths is not precisely fixed, but we have taken it to be when the first term of the expression which defines the degree of orientation of a prolate polyion,

$$\Phi(K) = \frac{nK^2}{45} + \left( \frac{n^2}{2385} - \frac{n}{1575} \right) K^4 + \text{smaller terms},$$

is smaller than the remaining terms. In this expression for the degree (0 → 1) of orientation, $\Phi$, the field strength enters in the parameter K = ZeLE/2kT, where L is the length of the polyion (in cm) and E is the applied field strength (in e.s.u./cm). In its application to the helical nucleic acids and polynucleotides, the first term is dominant only to field strengths up to about 3 or 4 kV/cm if the molecule is about 600A long, i.e. of the order of a persistence length for DNA. At higher field strengths or for higher molecular weight DNA, the second term, which becomes almost quadratic in the extent to which the counter-ion polarization contributes, becomes an important contribution to the orientation according to this model. Experimentally we observed (10) that for poly(A), the dependence is approximately linear at field strengths up to 7 or 8 kV/cm and have interpreted this difference between the observation and the model as at least partly due to the failure of the model to include the effect of counter-ion, counter-ion repulsion. Kikuchi and Yoshioka have, in fact, pointed out that this effect should reduce the effective value of n for a linear polyion of given charge density.

The difference between two recent models for the dielectric polarization of linear polyelectrolytes (11,12) which differ in the same way, that is in their accounting of the effect of counter-ion, counter-ion repulsion, also lends support to this interpretation. For example the polarization of DNA predicted by Mandel (11) is almost two orders of magnitude higher than that predicted by Manning (12) whose treatment takes into account the repulsion between counter-ions condensed on the polyion, as well as the polyion-counter-ion attraction. No completely satisfactory treatment of the orientation of a charged polyelectrolyte by an electric field yet exists. In this contribution, we describe some preliminary experiments designed to improve our understanding of the behaviour of linear polyelectrolytes in strong electric fields. DNA and poly(A) in their double helical forms at or near pH 6.5 differ in their degree of protonation; DNA is substantially unprotonated, whereas approximately one proton is associated with each base pair in the poly(A) structure at this pH. The theoretical treatments of Manning (8) and of Record (9) predict that the extent of alkali counter-ion binding is therefore very much different. The expressions obtained from their treatments is that the fraction, i. of polyion charges neutralized by bound counter-ions is:

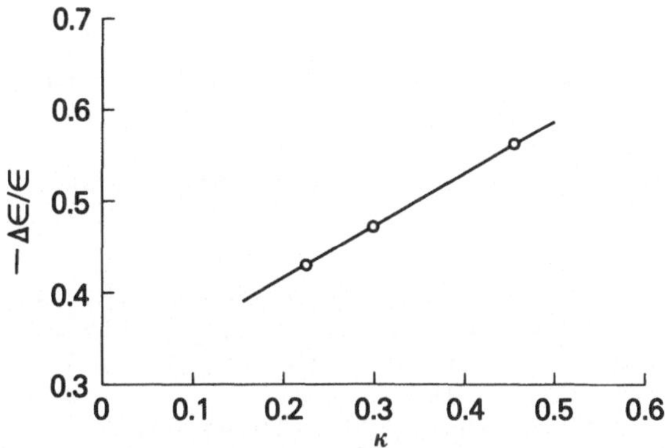


Fig. 1. The Electric Dichroism of Double-Stranded Poly(A) as a Function of the Degree of Protonation. Values of $\kappa$ were obtained by extrapolating the titration data of D.N. Holcomb and S.N. Timasheff (Biopolymers 6, (1968) 513) to an ionic strength of $0.8 \times 10^{-4}$ $MK^+$ from data at 0.15, 0.01 and 0.001 M CKl. The electric dichroism data are for Poly(A) in $2 \times 10^{-4}$ buffer at a field strength of 9.28 kV/cm and at $\lambda$ = 2520 A.

$$i = (1 - \xi^{-1})(1 - \kappa)$$

Where $\kappa$ is the degree of protonation and $\xi$, the charge density parameter for the polyion is given by $e^2/\epsilon kTb$ where $\epsilon$ is the bulk dielectric constant of the solvent, and b is the average spacing in angstroms of the projection of the charged groups on the axis of the fully extended polyion; kT is the Boltzmann thermal energy at $T^oK$. Since the number, n, of condensed cations per polyion is related to i by the relation $n = iN$ where N is total number negative charges on the polyion, it follows that for two polyions of the same N and same value of $\xi$, the degree of orientation at low field strengths should also be related linearly to i and, in turn, to the degree of protonation, $\kappa$. The electric field induced dichroism of poly(A) is plotted in fig. 1 as a function of $\kappa$ over the pH range from 4.6 to 6.5 in which it is a double stranded helix under the conditions of NaCl concentration used in the measurement. The linear relationship is observed. Taken in concord with the earlier demonstration (10) that the difference in the electric-field induced orientation of DNA and poly(A) of the same polyelectrolyte length can be accounted for by the relative concentration of condensed alkali counter-ions, and experiments on the behaviour of the orientation of DNA in buffered and unbuffered solutions (13), this result clearly demonstrates that the polarization of condensed counter-ions on a polyelectrolyte of high charge density provides one source of the induced moment necessary for electric-field orientation. Additional experiments are under way to delineate microstructural variants which are not described by current theories.

## REFERENCES


1 Liptay W and Czekalla Z, Naturforsch A., 15, (1960) 1072.
2 O'Konski C T, Yoshioka K and Orttung W H, J. Phys. Chem., 63, (1959) 1558.
3 Shah M J, IMB Technical Report TB-02-250; J. Phys. Chem., 67, (1963) 2215.
4 Yamaoka K and Charney E, J. Amer. Chem. Soc., 94, (1972) 8963.
5 O'Konski C T and Krause S, J. Phys. Chem., 74, (1970) 3243.
6 Kikuchi K and Yoshioka K, Biopolymers, 15, (1976) 583.
7 Hogan M, Dattagupta N and Crothers D M, Proc. Natl. Acad. Sci. U.S.A., 75, (1978) 195.
8 Manning G S, J. Chem. Phys., 51, (1969) 924.
9 Record M T Jr., Woodbury C P and Lohman T M, Biopolymers, 15, (1976) 893.
10 Charney E and Milstien J, Biopolymers, 17, (1978) 1629.
11 Mandel M, Mol. Phys., 4, (1961) 489.
12 Manning G S, personal communication.
13 Charney E and Yamaoka K. Manuscript in preparation.

# THE STABILIZED INDUCED DIPOLE IN SOLUTIONS OF POLYNUCLEOTIDES

Michio Shirai

Department of Chemistry, College of General Education
University of Tokyo, Komaba, Meguro, Tokyo, Japan

It has been found from electric birefringence measurements that in solutions of NaDNA and Na poly(A) orientation of solute molecules in an electric field is due to an induced dipole at low ionic strength, but a contribution of the permanent dipole to orientation becomes significant with increasing ionic strength. In solutions of intermediate ionic strength, the birefringence signals were of permanent dipole type under reversing pulses of low field intensities, but they were of induced dipole type at higher field intensities. These experimental results can be explained by supposing that the induced dipoles of the polyelectrolytes are stabilized by the dipole-ion interaction in the solutions of high ionic strength.

## INTRODUCTION

Solutions of DNA or other similar polyelectrolytes show low-frequency dielectric dispersion in addition to the regular high-frequency one. The origin of this low-frequency dispersion has been discussed by several authors (1-5). It was interpreted in terms of orientation polarization either due to a permanent dipole or due to an induced polarization from the transport of counter ions on the surface of polyions. Orientation of the induced dipole is a quadratic effect, and does not occur in such weak fields as used in dielectric measurements. Regarding the mechanism of polarization, it is interesting to detect the transient birefringence upon reversal of the applied field. If orientation of permanent dipoles occurs the birefringence signal will drop upon field reversal (6).

This paper describes experimental electric birefringence measurements for DNA and poly(A) solutions in a reversing field as a function of concentrations of DNA, added salt and field strength. The results suggest an interesting electric polarization behaviour of DNA or poly(A) molecules.

## EXPERIMENTAL

The DNA was of calf thymus origin. NaDNA was purchased from Sigma Chemical Co., as lot no. 126C-9570. Na poly(A) was purchased from Yamasa Co., as lot no. 618. The solutions were prepared by dissolving the samples in aqueous NaCl of neutral pH.

The apparatus used for birefringence measurement was that described elsewhere (7).

## RESULTS

All solutions of DNA and poly(A) exhibit negative electric birefringence signals even in the weak field down to 200 V/cm. This result shows that the DNA molecules orient with their long axes parallel to the electric field. In solutions of high concentration of NaCl, the electric birefringence signal of DNA in a reversing pulse has the shape shown in fig. 1. The dip in fig. 1 is characteristic of a permanent dipole moment.

On the other hand, in solutions of low concentration of NaCl the signal of DNA has the shape shown in fig. 2. In fig. 2 a spike appears on reversing the applied field. The spike can be explained by the presence of a slow induced transverse dipole. If we take the following values, the plot of Tinoco and Yamaoka's equation (6) can fit the experimental curve of DNA shown in fig. 2.

$$\theta = 500\ s^{-1}, \qquad \tau_1 = 3 \times 10^{-4}\ s, \qquad \tau_3 = 10^{-4}\ s,$$

$$\alpha_1 = 1.5 \times 10^{-14} cm^3, \qquad \alpha_3 = 2.5 \times 10^{-14} cm^3.$$

where $\theta$ is the rotational diffusion constant for rotation about the transverse axis, $\alpha_1$ and $\alpha_3$ are the electric polarizabilities perpendicular and parallel to the symmetry axis, and $\tau_1$ and $\tau_3$ are the relaxation times for $\alpha_1$ and $\alpha_3$ respectively.
The value of $\theta$ mentioned above corresponds to a relaxation time of 0.33 ms, whereas the longest relaxation time obtained from the decay curve is 5 ms, which corresponds to the molecular length of 1000 nm. A wide distribution of relaxation times was also found from the decay curve, and the components of shorter relaxation times

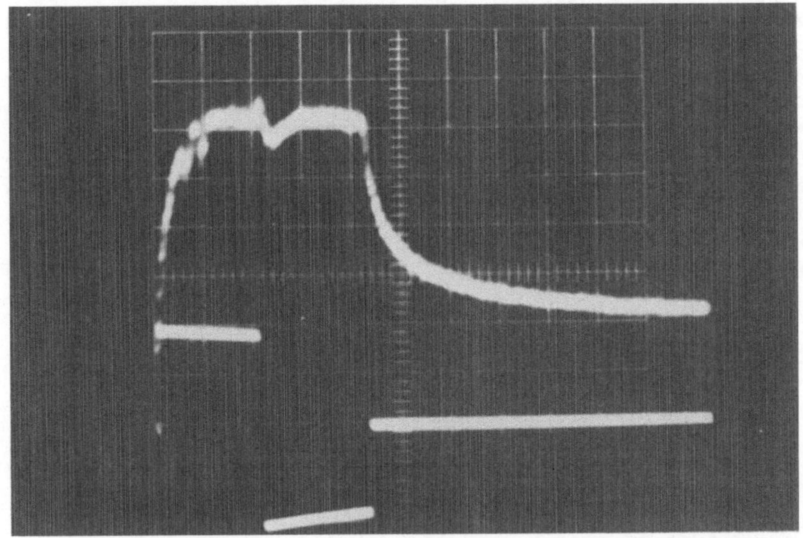

Fig. 1. Birefringence signal of DNA in a reversing pulse. Concentration of NaDNA 0.1 g/l in $10^{-3}$ mol/l NaCl. E = ± 900 V/cm. 1 ms/div.

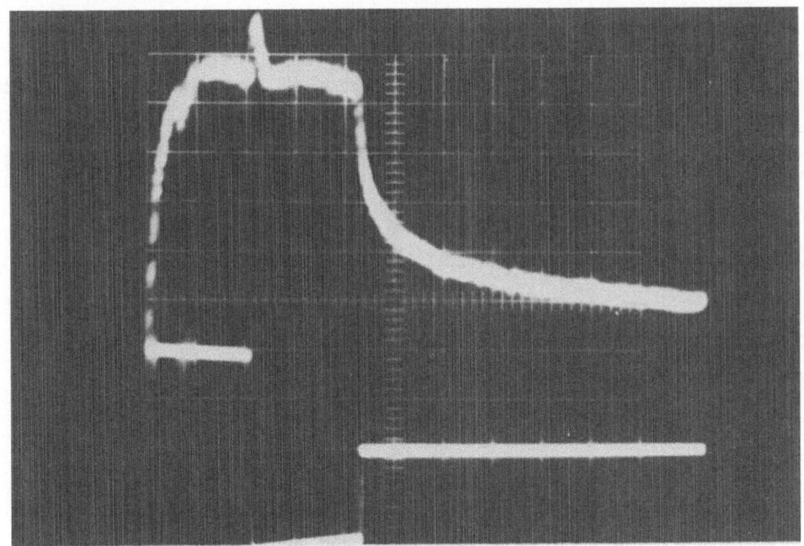

Fig. 2. Birefringence signal of DNA in a reversing pulse. Concentration of NaDNA 0.1 g/l in 2 x $10^{-4}$ mol/l NaCl. E = ± 1000 V/cm. 1 ms/div.

are effective in a rapidly reversing field, so that the value of $\theta$ obtained seems reasonable. Since $(\alpha_3 - \alpha_1) > 0$, the DNA molecules orient with their long axes parallel to the electric field. If adsorbed mobile ions move along the helix of the DNA molecules, an induced transverse polarizability is possible. Thus, the signal with the spike shown in Fig. 2 can be explained by the induced dipole.

Generally, the electric birefringence signal arises from both the permanent and the induced dipole. The ratio of contributions of the above-mentioned two origins is denoted by $\beta$ defined by Tinoco and Yamaoka (6) as follows,

$$\beta = \frac{\mu^2/k^2T^2}{(\alpha_3 - \alpha_1)/kT}$$

where $\mu$, k, and T are the dipole moment, Boltzmann constant, and the absolute temperature respectively. The value of $\beta$ can be obtained by analysing the electric birefringence signal in a reversing pulse.

Table I shows the values of $\beta$ obtained for DNA as a function of concentrations of DNA and NaCl. At low ionic strength $\beta = 0$. This means that the orientation is due to the induced dipole. However, at higher ionic strength the contribution of a permanent dipole to the orientation becomes significant.

In solutions of $3 \times 10^{-4}$ mol/l NaCl $\beta$ is 0 to 3, depending upon the applied sield strength. Table II shows the value of $\beta$ for NaDNA solutions as a function of the field strength. In low fields, contribution of the permanent dipole is relatively large, but in fields higher than 500 V/cm orientation is due to the induced dipole.

Table I. $\beta$ for various concentrations of NaDNA and NaCl

| NaDNA concn. (g/l) / NaCl concn. ($10^{-4}$mol/l) | 0.5 | 0.2 | 0.1 | 0.05 |
|---|---|---|---|---|
| 10 | 0.8 | 0.8 | 0.5 | 0.5 |
| 5 | 0.5 | 0.5 | 0.3 | 0.3 |
| 3 | 0.5 | 0.3 | 0-0.3 | 0-0.3 |
| 2 | 0.3 | 0 | 0 | 0 |
| 1 | 0.3 | 0 | 0 | 0 |

Table II. $\beta$ of NaDNA solution, 0.1 g/l in 3 x $10^{-4}$ mol/l NaCl

| Field strength (V/cm) | 200 | 300 | 400 | 500 | 600 | 700 | 1,000 |
|---|---|---|---|---|---|---|---|
| $\beta$ | 0.3 | 0.3 | 0.2 | 0 | 0 | 0 | 0 |

The experimental results of poly(A) are similar to those of DNA. At low ionic strength the birefringence signal of poly(A) in a reversing pulse is of the induced dipole type, whereas at high ionic strength it is indicative of a permanent dipole. In the poly(A) solution of 0.38 g/l in 2 x $10^{-4}$ mol/l NaCl, the value of $\beta$ depends on the field strength as shown in Table III.

Table III. $\beta$ of Na poly(A) solution, 0.38 g/l in 2 x $10^{-4}$ mol/l NaCl

| Field strength (V/cm) | 100 | 160 | 180 | 500 | 750 |
|---|---|---|---|---|---|
| $\beta$ | 0.7 | 0.3 | 0 | 0 | 0 |

## DISCUSSION

In polyelectrolyte solutions there are two kinds of counterions (8). Some counterions are adsorbed on the polyions, and are mobile along the surface of a polyion. These ions belong to the same kinetic unit as the polyion, and it is the shift of the center of these ions with respect to the center of the polyionic charge which gives rise to the dipole moment. Other unadsorbed ions form a different kinetic unit. They are subject to electrostatic interaction with the polyion, and build up the ionic atmosphere. In the presence of much added salt, the electrostatic interaction between the polyion with adsorbed counterions and the ionic atmosphere is strong. In such a circumstance the dipole induced in a polyion may be stabilized by the dipole-ion interaction. Then, the polyions have the stabilized dipole induced by shift of the adsorbed mobile counterions, and are expected to behave as if they have a permanent dipole.

In the experimental results shown in Table I the birefringence signal is of induced dipole type at low ionic strength, and is of permanent dipole type at high ionic strength. These results can be accounted for by the dipole-ion interaction stated above. The results of the field dependence of $\beta$ shown in Table II can be explained as follows. A high field reverses the polarity of the dipole, and accordingly, the reversing field signal is of an induced dipole type. On the other hand, a low field cannot reverse the polarity of the dipole, so that the signal is of permanent dipole type.

If the center of adsorbed mobile ions is shifted from the centre of polyionic charge, the polyion will have a dipole moment. This dipole will interact with the ionic atmosphere, and we denote this interaction energy by $W_s$. If the coulombic attraction energy between the center of adsorbed mobile ions and that of polyionic charge is denoted by $W_c$, the condition of minimum $(W_s + W_c)$ determines the dipole moment. When an external field is applied, this field interacts with the dipole. If we denote this interaction energy by $W_E$, then

$$W_E = -\mu E |\cos\theta|_{av}$$

where $\mu$ is the dipole moment, E the intensity of the applied field, and $|\cos\theta|_{av}$ is the average of the absolute value of cos $\theta$, $\theta$ being the angle between the directions of the applied field and the dipole moment. Since the induced dipole has no definite polarity, the absolute value of cos $\theta$ is taken. In weak fields $|\cos\theta|_{av}$ is about 0.7.

In equilibrium $W_E < 0$, but immediately after the field is reversed the sign of $W_E$ is changed. Then, if $|W_E| > |W_s|$, the polarity of the dipole is reversed because generally $|W_s| \gg |W_c|$. If $|W_E| < |W_s|$ the applied field cannot reverse the polarity of the dipole, and so the dipole orients. In the former case the birefringence signal is of the induced dipole type, and in the latter case it is of the permanent dipole type. As is shown in table II, $|W_E| = |W_s|$ in a field of 500 V/cm.

$W_s$ was calculated by Kirkwood (9), and the result depends on the excluded volume of a polyion, which is difficult to estimate. If the radius of the orienting polyion be 500 nm, the equation $|W_E| = |W_s|$ produces the value $\mu = 5 \times 10^4$ D using Kirkwood's formula. This result is smaller than that obtained by Sakamoto et al. using dielectric dispersion (5). The reversing pulse technique detects a part of the polarization which includes dipolar orientation. The dielectric dispersion detects also the induced polarization and the permanent dipole polarization due to residual proteins. In view of these facts the value of $5 \times 10^4$ D for $\mu$ is the dipole moment stabilized by interaction with the ionic atmosphere. The results for poly(A) are similarly explained. This is a new type of dipole peculiar to polyelectrolytes.

## REFERENCES

1 Jungner G and Jungner I, Nature, 163, (1949) 849.
2 O'Konski C T, J. Phys. Chem., 64, (1960) 605.
3 Takashima S, Biopolymers, 5, (1967) 899.
4 Hanss M and Bernengo J C, Biopolymers, 12, (1973) 2151.

5 Sakamoto M, Kanda H, Hayakawa R and Wada I, Biopolymers, 15 (1976) 879.
6 Tinoco I Jr. and Yamaoka J K, J. Phys. Chem., 63, (1959) 423.
7 Ikeda K, Watanabe H, Shirai M and Yoshioka K, Sci. Pap. Coll. Gen. Educ., 15 (1965) 139.
8 Manning G S, J. Chem. Phys., 51, (1969) 924.
9 Edsall J T and Wyman J, "Biophysical Chemistry" Vol. 1, p.296, Academic Press, New York (1957).

# Biological Systems

'Whether or not we find what we
are seeking
Is idle, biologically speaking.'

EDNA ST. VINCENT MILLAY (1892-1950)

Sonnet: I shall forget you presently

# ELECTRO-OPTICAL PROPERTIES OF NUCLEOSOMES AND NUCLEOSOMAL DNA

B. Roux, C. Marion and J. C. Bernengo*

Laboratoire de Chimie Biologique, Université C. Bernard Lyon I, 43 Bd. du onze Novembre 1918, 69621 Villeurbanne, France

Rat liver soluble chromatin is digested by micrococcal nuclease. The birefringence of the seven fractions obtained from sucrose gradient is studied in 1 mM phosphate buffer pH 7.4, 0.2 mM EDTA. Electro-optical properties and relaxation times show a sharp transition when the number n of nucleosomes in the oligomer chain is greater than 6: the negative birefringence becomes positive and no signal is observed for the hexamer. The monomer relaxation time (0.6 μs) agrees with a prolate ellipsoid model of dimensions about 140 x 140 x 70 Å. The relaxation time varies also with n: it is a maximum for the dimer (1.8 μs) and decreases slightly up to the pentamer. These results may be interpreted in terms of a superstructure of nucleosome chains.
DNA extracted from nucleosomes has been studied by the reverse pulse technique. The observed transient birefringence may be interpreted with the theory of Tinoco and Yamaoka. Monomer DNA presents a 4 μs relaxation time, corresponding to a length of 195 base pairs, which is the value also found by electrophoresis. For other DNA fractions, relaxation times show a strong field dependence: even at low field, the calculated lengths are always lower than these obtained by electrophoresis.

## INTRODUCTION

Recent studies on the chromatin structure have demonstrated that DNA is associated with five histones in a repeating subunit structure (1,2). For all tissues and organisms, each subunit ( ν - body or

---

* Present address: Laboratoire de Biophysique, Université de Nice, Parc Valrose, 06034 Nice - Cedex, France

nucleosome) contain a substructure called the "core particle" with a constant DNA length (140 base pairs) and two each of histones H2A, H2B, H3 and H4. The fifth histone H1 is bound to the remaining DNA, the length of which varies from 15 to 100 base pairs (about 60 for rat liver). So, in rat liver chromatin, the total DNA length is about 200 base pairs, approximately 680 Å (for reviews see ref. 3).

When the earlier electro-optical data on chromatin were reported (4-7), these biochemical results were not well established. Now, such studies can be carried out either to investigate the structure of the nucleosome or to specify the arrangement of the nucleosomes in the higher order chromatin structure. Recently, Klevan et al. (8) described the electric dichroism of mononucleosome and dinucleosome. In this paper, we describe experiments made on chains of nucleosomes up to 7 nucleosomes and on the extracted DNA from these particles. Indeed, these DNA samples are not appreciably polydisperse and present a number of base pair multiples of 200.

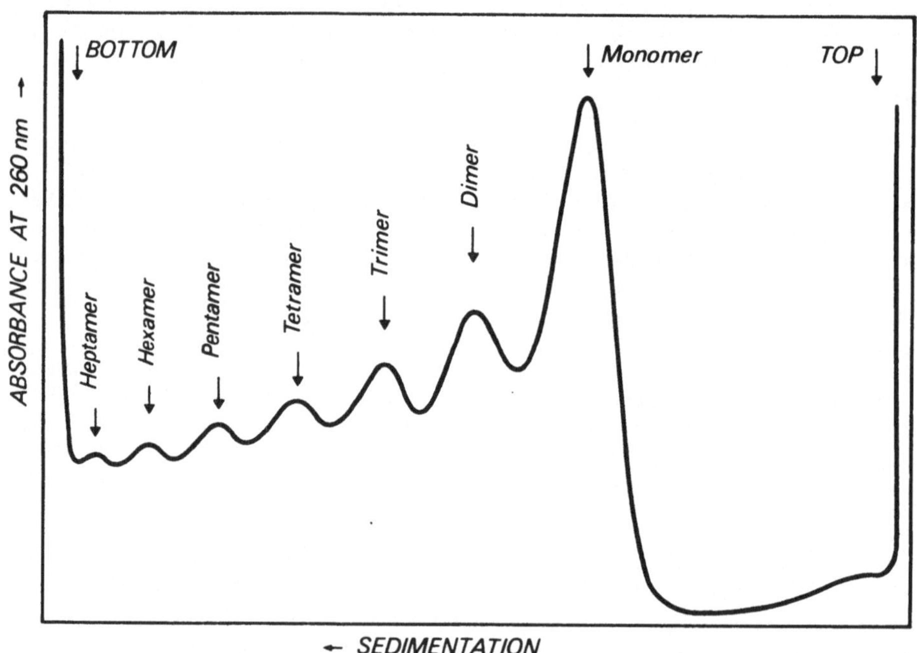


Fig. 1. Isokinetic sucrose gradient (5 - 28.2%) fractionation of chromatin digested with micrococcal nuclease (Worthington). Centrifugation was performed at 27 000 rpm for 20 hours in a SW27 rotor (Beckman). Digestion time was 2½ minutes with 300 units of nuclease per ml of nuclei suspension.

## MATERIALS

Native chromatin was obtained from rat liver nuclei prepared by the method of Hewish and Burgoyne (9) and suspended at a concentration of $2.10^8$ nuclei/ml. Chromatin extraction was performed according to Noll et al. (10) with a micrococcal nuclease digestion. Chromatin subunits were fractionated by isokinetic sucrose gradient ultracentrifugation as described by Finch et al. (11). The fractionation pattern is shown in fig. 1. For the electro-optical study of nucleosomes, the fractions corresponding to each of the seven peaks were pooled and dialysed against 1 mM phosphate buffer pH 7.4, 0.2 mM EDTA.

The full complement of histone proteins in all chromatin subunits was confirmed by electrophoretic analysis and thermal denaturation profiles.

DNA was extracted by incubation with proteinase K (Merck) in 1 M NaCl, followed by extractions with chloroform-isoamylalcohol mixtures and precipitation with ethanol. Estimation of proteins by thin layer chromatography on cellulose showed less than 0.5 %. The purity of DNA fractions is variable with the number of oligomers: it may be estimated by analytical electrophoresis on polyacrylamide gel. Up to the trimer, the population is highly homogenous in length, but for higher fractions the purity decreases: for instance, only 80% of the heptamer fraction has a length corresponding to seven times the DNA monomer length (7 x 195 base pairs).

## METHODS

The birefringence apparatus uses a sensitive optical device including a quarter-wave retarder and a stable and noise-free He-Ne laser (Spectraphysics model 120) supplied by a high performance voltage source. The low noise solid state optical detector is a photodiode followed by a linear or logarithmic amplifier. The fastest signals obtained on nucleosomes are displayed on a storage oscilloscope, and photographed. The transition times of the apparatus are about 50 ns. For the slowest phenomena (essentially for studies on DNA), a transient recorder (Datalab) is used, then the signal is plotted on a XY recorder. Single and reversed electric pulses have been used in this work, using classical, limiting low field, dynamic birefringence equations: namely (12)

$$\Delta_D = e^{-t/\tau} \qquad (1)$$

for the decay where $\Delta_D$ is the normalised birefringence. Also

$$\Delta_R = 1 - \frac{3}{2}\frac{r}{r+1} e^{-t/3\tau} + \frac{1}{2}\frac{r-2}{r+1} e^{-t/\tau} \qquad (2)$$

for the rise where $\Delta_R$ is the normalised birefringence for the rise.

In these equations $r = \mu^2 / \Delta\alpha_E \cdot kT$, where $\mu$ is the permanent moment and $\Delta\alpha_E$ is the electrical polarizability anisotropy.

The reverse pulses used for DNA are interpreted in terms of Tinocco and Yamaoka's theory (13). The values of r and $\tau$ may be calculated from the coordinates of the observed extrema according to the equations:

for the reversal

$$r = \frac{1 - \Delta_m}{0.1547 + \Delta_m} \tag{3}$$

$$\tau = 0.60\, t_m \tag{4}$$

and for the rise

$$\Delta_m = 1 - \frac{r}{r+1}\left(\frac{r}{r-2}\right)^{\frac{1}{2}} \tag{5}$$

$$\tau = \frac{2}{3} t_m / \ln\left(\frac{r-2}{r}\right) \tag{6}$$

(if the rise signal presents an extremum)

$t_m$ is the time at which the extremum occurs and $\Delta_m$ is the normalised birefringence at this time.

Dimensions of nucleosome particles are determined from the rotational diffusion constants (14-16) of revolution ellipsoids:

$$D_r = 1 / 6\tau = kT\, r(p) / 8\pi\eta ab^2 \tag{7}$$

where p is the axial ratio a/b, with a and b respectively the semi-long and the semi-short axis.

For DNA, the formula of Broersma (17) is used, relating to a cylindrical model:

$$D_r = 3\, kT \left\{ \ln 2p - 1.57 + 7\,(1/\ln 2p - 0.28)^2 \right\} / 8\pi\eta a^3 \tag{8}$$

## ELECTRO-OPTICAL PROPERTIES OF NUCLEOSOMES

The electric field is varied from 250 to 5,000 V/cm. The birefringence is proportional to the nucleosome concentration for all the studied range (50 to 400 mg/$\ell$) and $\Delta n$ could be expressed in terms of a Kerr constant B. In these experimental conditions, Kerr's law is obeyed for all the chromatin subunits. Fig. 2 shows the variation of B and relaxation time $\tau$ versus the number of multimer subunits, n. The main result is the change in sign of B and the sharp increase of $\tau$ when n is greater than 6. This transition is obtained whatever the applied electric field and the chromatin subunits concentration.

The birefringence decays show only one relaxation time for all the oligomers and the linearity is observed on more than 3 neperian units. On the other hand, the rise and decay curves are almost symmetrical

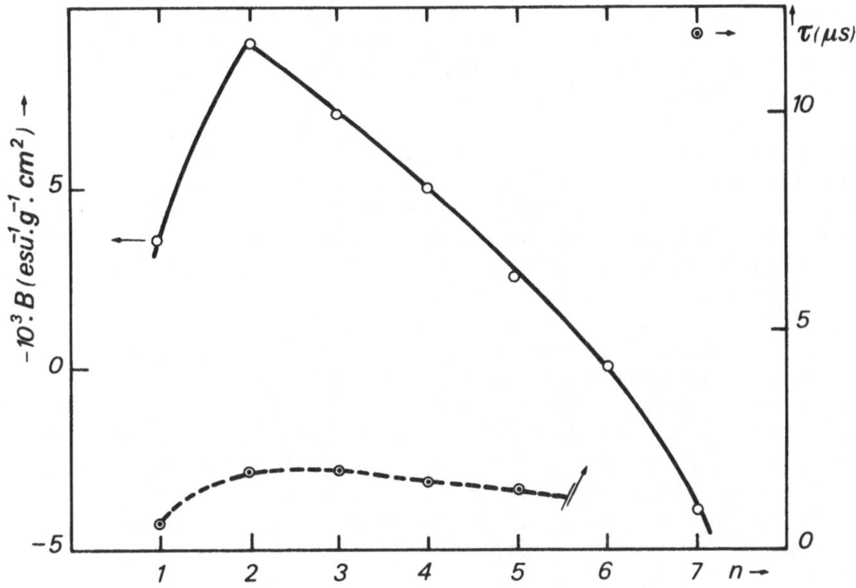


Fig. 2. Variation of Kerr constant B and relaxation time τ versus the nucleosome number n in the oligomers. E = 3000 V/cm. As the hexamer exhibits no birefringence signal, no relaxation time could be determined.

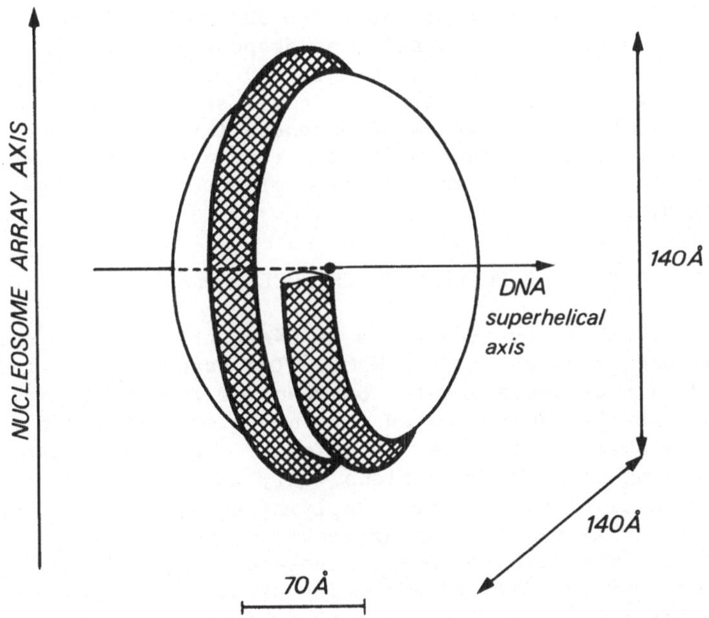


Fig. 3. Nucleosome model. Thick gray thread represents DNA.

and r values of about 0.5 can be calculated. The orientation of nucleosome oligomers seems mainly to be due to an induced dipole interaction with the field. However, the existence of a small permanent dipole cannot be disregarded.

As the mononucleosome and DNA birefringences have the same sign, it can be supposed that the axis of larger polarizability of nucleosome particles is the superhelical axis of the DNA, which is wrapped around the histone core: so, the birefringence is essentially due to the nucleic part of the complex and DNA is oriented with its long axis in the direction of the electric field.

From the relaxation time and by considering this orientation mechanism, subunit molecular dimensions can be calculated, assuming simple geometrical shapes. The nucleosome model which fits correctly with our results ( $\tau$ = 0.6 $\mu$s for the monomer) is an oblate ellipsoid or disc-shaped particle of dimensions about 140 x 140 x 70 Å. This is in quite good agreement with other reported values (8,18-22). In this model, the DNA superhelical axis is parallel to the short axis of the equivalent ellipsoid (fig. 3), as recently proposed (23).

As, on one hand, the relaxation times are approximately constant for oligomers up to the pentamer and, on the other hand, the hexamer structure must be symmetric because it presents no birefringence, the proposed model for this hexanucleosome structure is an helical array of nucleosomes in tight contact by their longer side. This structure is approximately spherical with a diameter about 250 Å as suggested by other authors (24-27). In this arrangement, the DNA superhelical axis is perpendicular to the nucleosomes axis, as proposed by Rill and Van Holde (4), Houssier et al. (7) and Klevan et al. (8) when the birefringence is negative. To explain this negative sign, a second hypothesis retained by Houssier et al. is the presence of extended internucleosome fragments of free DNA. Consequently, the positive birefringence obtained in our experiments for the heptamer can only be explained by a change in the orientation axis.

## ELECTRO-OPTICAL PROPERTIES OF NUCLEOSOMAL DNA

Birefringence and relaxation times of DNA extracted from nucleosome oligomers have been studied in 1 mM NaCl. Two ranges of electric field have been used in these experiments: between 300 and 500 V/cm and between 1 and 3 kV/cm. Kerr's law is obeyed for the lower fields and a beginning of saturation is observed for the higher ones. Table 1 summarises the results obtained on these DNA fractions. Only monomer and dimer show a single relaxation time. For monomer, applying eq.8, a 660 Å length may be calculated and this value is in agreement with 195 base pairs as has been demonstrated from electrophoretic measurements, considering a diameter of 20 Å for rod-like DNA and a spacing of 3.4 Å between 2 base pairs. For higher fractions, at low field, small relaxation times appear. It is still possible to characterise a well-defined long relaxation time as has already been found for calf thymus DNA (molecular weight 6 x $10^6$) (28).

Table I. Electro-optical parameters of nucleosomal DNA

| Fractions | Concentrations (mg/l) | Electric field (V/cm) | Decay | | Rise | Reverse | |
|---|---|---|---|---|---|---|---|
| | | | T (μs) | a (%) | $t_{m1}$ (μs) | $t_{m2}$ (μs) | r |
| Monomer | 120 | ≤500 | 4. 3 | 1. 00 | - | - | - |
| Dimer | 130 | ≤500 | 15 | 1. 00 | - | - | - |
| Trimer | 70 | ≤500 | 30 | 0. 70 | - | - | - |
| Tetramer | 85 | ≤500 | 55 | 0. 75 | 220 | 95 | -0. 1 |
| | | 2000 | 50 | 0. 45 | 91 | - | -0. 25 |
| | | 3000 | no linear part | | 65 | - | -0. 25 |
| Pentamer | 51 | ≤500 | 140 | 0. 67 | (250-300) | 135 | -0. 1 |
| | | 2000 | no linear part | | 210 | - | -0. 25 |
| | | 3000 | no linear part | | 100 | - | -0. 25 |
| Hexamer | 17 | ≤500 | 190 | 0. 50 | - | - | - |
| Heptamer | 27 | ≤500 | 250 | 0. 45 | - | - | - |

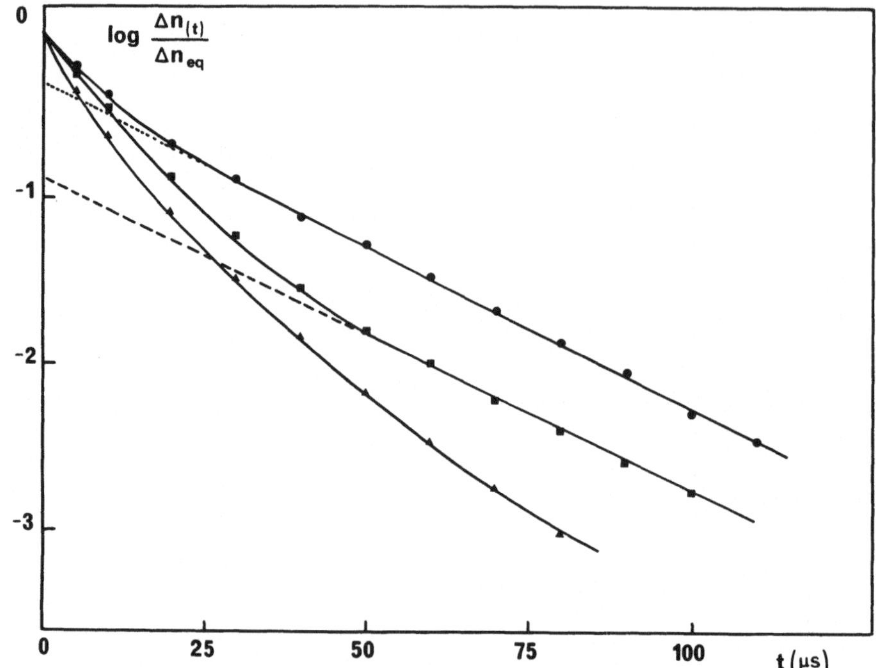


Fig. 4. Birefringence decay analysis for DNA tetramer fraction. DNA concentration of 85 mg/ℓ in 1 mM NaCl. The contribution a of the longest relaxation time decreases as the field strength increases. Log a is calculated extrapolating the linear part of the logarithmic plot. Circles; 300 V/cm, a = 75%, $\tau$ = 55 μs: squares; 2000 V/cm, a = 45%, $\tau$ = 50 μs: triangles; 3000 V/cm.

The contributions to the total birefringence a may be calculated as indicated on fig. 4 and reported in table I. These values vary with the nature of fraction and with the applied electric field. At low field it decreases with the molecular weight, indicating a more flexible conformation. Such a behaviour has been noted by Kovacic and Van Holde (29) and Godfrey and Eisenberg (30) from sedimentation and diffusion scattering experiments: only the DNA extracted from the monomer is a rigid molecule.

Fig. 5 shows the variation of the longest relaxation time and of the relaxation time which may be calculated from eq. 8, if all samples are considered as rod-like molecules.

Variation of a with electric field is shown on fig. 4 for tetramer DNA: this indicates a deformation and a decrease of the longest relaxation mechanism contribution to the total birefringence, when the

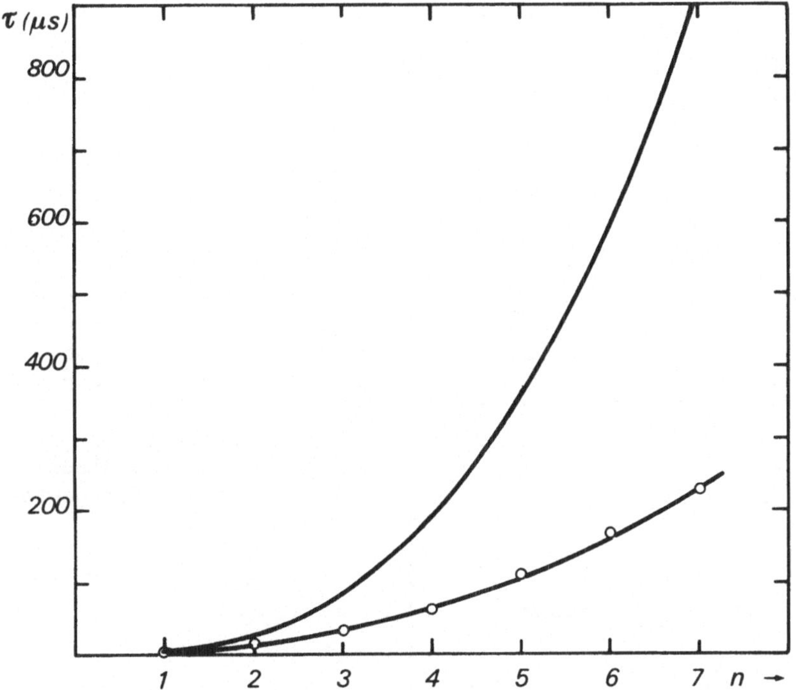


Fig. 5. Variation of the DNA longest relaxation time with the nucleosome number in the oligomer chain n. Theoretical curve from eq. 8, assuming each fraction as a 200 base pairs multiple.

applied forces increase. Hogan et al. (31) observed a variation of $\tau$ with field strength and suggested that this result was due to polydispersity of the samples.

On the other hand, with the reverse pulse technique, some preparations show anomalous signals, even at low field (fig. 6 B). Colson et al. (32) have described such signals, for sonicated DNA at a higher field.

Other preparations exhibit signals shown on fig. 6 A, which may be interpreted in terms of theory of Tinocco and Yamaoka (13). The origin of this difference is currently being studied in our laboratory. In table I, $t_{m1}$ and $t_{m2}$ are the abscissa of respectively maxima for rise and reverse birefringence. From these times and according to eqs. 4 and 6, a relaxation time may be calculated: for the tetramer, with E $\leqslant$ 500 V/cm, there is a good accordance (55 $\mu$s) with the $\tau$ value, determined from the linear part of the semi-logarithmic decay curves (fig. 4). The same agreement is not observed for other fractions: contribution of the longest time decreases and an average for all the relaxation mechanisms is measured.

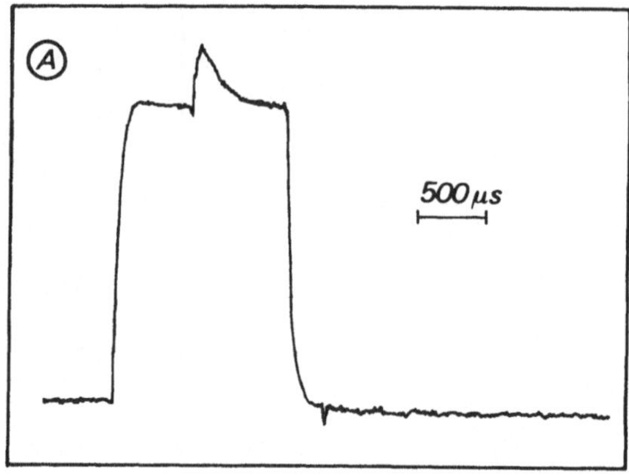


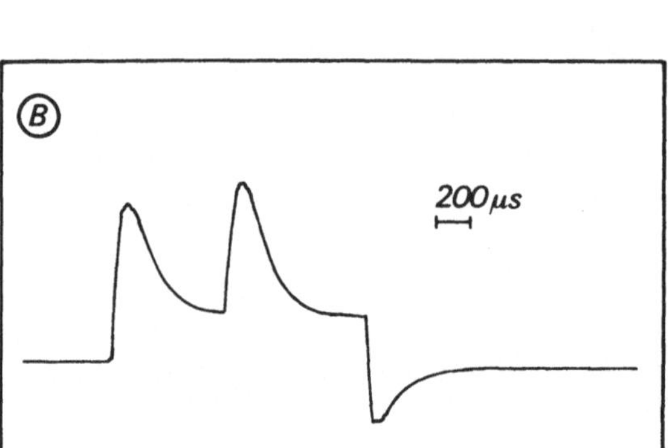


Fig. 6. DNA birefringence signals obtained by reverse electric pulse with an applied electric field 400 V/cm: (A) "Classical signal" for tetramer fraction, (B) Anomalous signal obtained for pentamer fraction.

From the classical electro-optical properties of DNA (6), a negative value of r may implicate an excess of electrical polarizability in a direction perpendicular to the molecular axis; in absolute value, r increases with electric field. No molecular explanation exists for these maxima.

## CONCLUSIONS

This study presents results obtained on a carefully characterized material and permits investigation of current problems in the biochemical field. The existence of an hexanucleosome superstructure in chromatin, which has been postulated by many authors, seems to be confirmed experimentally. Data on monodisperse DNA samples show a deformation by high electric fields, for molecular weight higher than $0.5.10^6$.

## REFERENCES


1 Olins A L and Olins D E, Science, 183, (1974) 330-332.
2 Oudet P, Gross-Bellard M and Chambon P, Cell, 4, (1975) 281-300.
3 (a) Kornberg R D, Ann. Rev. Biochem., 46, (1977) 931-954.
(b) Felsenfeld G, Nature, 271, (1978) 115-122.
4 Rill R L and Van Holde K E, J. Mol. Biol., 83, (1974) 459-471.
5 Itzhaki R F, Proc. Royal Soc., 164, (1966) 75-95.
6 Fredericq E and Houssier C, "Electric dichroism and electric birefringence", Oxford University Press, London (1973).
7 Houssier C, Bontemps J, Emonds-Alt X and Fredericq E, Ann. N.Y. Acad. Sci., 303, (1977) 170-189.
8 Klevan L, Hogan M, Dattagupta N and Crothers D M, Cold Sprg. Harb. Symp. Quant. Biol., in press.
9 Hewish D R and Burgoyne L A, Biochem. Biophys. Res. Comm., 52, (1973) 504-510.
10 Noll M, Thomas J O and Kornberg R D, Science, 187, (1975), 1203-1206.
11 Finch J T, Noll M and Kornberg R D, Proc. Natl. Acad. Sci. USA, 72, (1975) 3320-3322.
12 Benoit H, Ann. Phys., 6, (1951) 561-587.
13 Tinoco I and Yamaoka K, J. Chem. Phys., 63, (1959) 423-432.
14 Perrin F, J. Phys. Radium, 5, (1934) 497-511.
15 Daune M, Freund L and Spach G, J. Chim. Phys., 59, (1962) 485-493.
16 Small E W and Isenberg I, Biopolymers, 16, (1977) 1907-1928.
17 Broersma S, J. Chem. Phys., 32, (1960) 1626-1631.
18 Suau P, Kneale G G, Braddock G W, Baldwin J P and Bradbury E M, Nucl. Acids Res., 4, (1977) 3769-3786.
19 Finch J T, Lutter L C, Rhodes D, Brown R S, Rushton B, Levitt M and Klug A, Nature, 269, (1977) 29-36.
20 Bram S, Biochem. Biophys. Res. Comm., 81, (1978) 684-691.
21 Langmore J P and Wooley J C, Proc. Natl. Acad. Sci. USA, 72, (1975) 2691-2695.
22 Wooley J C and Langmore J P, J. Supramol. Structure Suppl.,1, Abstr. 103, (1977).
23 Trifonov E, Nucl. Acids Res., 5, (1978) 1371-1380.
24 Carpenter B G, Baldwin J P, Bradbury E M and Ibel K, Nucl. Acids Res., 3, (1976) 1739-1746.
25 Finch J T and Klug A, Proc. Natl. Acad. Sci. USA, 73, (1976) 1897-1976.
26 Renz M, Nehls P and Hozier J, Proc. Natl. Acad. Sci. USA, 74, (1977) 1879-1883.
27 Worcel A and Benyajati C, Cell, 12, (1977) 83-100.
28 Roux B, Bernengo J C, Marion C and Hanss M, J. Colloid & Interface Sci., in press.
29 Kovacic R T and Van Holde K E, Biochemistry, 16 (1977) 1490-1498.
30 Godfrey J E and Eisenberg H, Biophysical Chemistry, 5 (1976) 301-318.
31 Hogan M, Dattagupta N and Crothers D M, Proc. Natl. Acad. Sci. USA, 75,(1978) 195-199.
32 Colson P, Houssier C, Fredericq E and Bertolotto J A, Polymer, 15, (1974) 396-397.

# POLARISABILITY ANISOTROPY AS AN INDICATOR OF THE EFFECTS OF AMINOGLYCOSIDE ANTIBIOTICS ON SENSITIVE, DEPENDENT AND RESISTANT STRAINS OF E.coli

V. J. Morris*, B. R. Jennings*, N. J. Pearson+ and F. O'Grady+

*Electro-Optics Group, Physics Department, Brunel University, Uxbridge, Middlesex, U.K.

and

+Department of Microbiology, University of Nottingham, Queen's Medical Centre, University Hospital, Nottingham, U.K.

Using high frequency, low amplitude pulsed electric fields, the anisotropy of the electrical polarisability ( $\Delta\alpha$) of E.coli bacteria may be evaluated from scattered light intensity changes. $\Delta\alpha$ has been found to be very sensitive to the addition of antibiotics to aqueous E.coli suspensions. With increasing antibiotic concentration $\Delta\alpha$ decreases in a manner reminiscent of an adsorption isotherm. Illustrative data on the interaction of aminoglycoside antibiotics with a sensitive, a dependent and two resistant strains of E.coli K 12 is presented and discussed.

## INTRODUCTION

This communication is concerned with the use of electric field light scattering to observe changes in the anisotropy of the electrical polarisability ($\Delta\alpha$) induced by the interaction of various antibiotics with bacterial surfaces.

Measurements of $\Delta\alpha$ are made by observing the rotational response of the bacteria to an applied high frequency alternating electric field (1,2). For large particles such as bacteria, in ionic suspending media, there is strong evidence for the contention that the major contribution to $\Delta\alpha$ arises from interfacial polarisation (1,3,4). Further, comparative studies of $\Delta\alpha$ and the electrophoretic mobility ($u$) suggest that both parameters monitor the charge density at the shear plane of the particle-medium interface (5). Measurements of $u$ are made by noting the translational response of the particles to an applied d.c. electric field (6). This is a well established technique for studying particle-medium interfaces (6). Studies have been made of the surface structure of normal (7) and resistant (8) bacteria and of changes in the electrical properties of bacterial surfaces due to interactions with added antibiotics (9,10).

Recently it has been shown that $\Delta\alpha$ may also be used to monitor the interaction of various additives (4,11), and in particular antibiotics (12-14), with bacterial surfaces. In this paper we will describe the use of electric field light scattering for studying such interactions. The method will be illustrated with data on the interaction of aminoglycoside antibiotics with a sensitive, a dependent and two resistant strains of E.coli K 12.

## EXPERIMENTAL METHOD

Consider a dilute suspension of particles of major size dimension ($\xi$), refractive index (n) suspended in a medium of refractive index ($n_o$) and illuminated by light of wavelength ($\lambda$). Provided the conditions

$$\gamma_1 = \frac{2\pi\xi}{\lambda}\left|\frac{n-n_o}{n_o}\right| \ll 1 \quad \text{and} \quad \gamma_2 = \left|\frac{n-n_o}{n_o}\right| \ll 1 \qquad (1)$$

are upheld then the intensity of the scattered light is describable by the Rayleigh-Gans-Debye (R.G.D.) approximation (15). In this case for a suspension of randomly orientated particles the angular dependence of the scattered radiation is determined by the particle shape. For anisodiametric particles particle orientation leads to a change in the polar scattering pattern. By monitoring at a fixed scattering angle one observes this change, and hence the orientation, as a change in the scattered intensity. Orientation is induced by applying a burst of high frequency alternating electric field to the suspension. The frequency of the applied field is chosen to be sufficiently high so that (a) the permanent dipole moments of the particles are incapable of contributing to particle orientation and (b) the amplitude of the double frequency component of induced change in scattered intensity has become zero. Hence particle orientation arises solely due to a coupling between the applied electric field and the induced dipole moment of the particle. For sufficiently long pulse lengths an equilibrium situation is achieved in which the orientating influence of the electric field balances the disorientating effect of Brownian motion. For low r.m.s. field amplitudes (E) the relative change in intensity of scattered light ($\frac{\Delta I}{I}$) is given by (1,2)

$$\frac{\Delta I}{I} = F \cdot \frac{\Delta \alpha}{kT} \cdot E^2 \quad (2)$$

Here k is the Boltzmann constant, T the absolute temperature and F a constant which depends on the size and shape of the particles. Values of $\Delta\alpha$ can be obtained from a plot of $\frac{\Delta I}{I}$ versus $E^2$.

After termination of the applied pulse the particles disorientate and the change in scattered intensity $\Delta I$ decays to zero as (1)

$$\Delta I(t) = \Delta I(o) \exp(-6 D_R t) \quad (3)$$

where $D_R$ is the rotary diffusion coefficient which may be used to monitor particle size if the particle shape is known.

For bacterial suspensions the condition $\gamma_2 \ll 1$ is generally upheld but the condition $\gamma_1 \ll 1$ is often exceeded. However, measurements may be restored to the R.G.D. regime by elevating the wavelength of the incident light to near infra-red values and working at low scattering angles (16,17). Further, comparative electro-optic measurements at visible and near infra-red wavelengths suggest that qualitative studies of changes in $\Delta\alpha$ may be made at visible wavelengths (18).

## ANALYSIS OF DATA

Measurements are made on dilute suspensions ($\sim 10^7$ organisms/ml) initially in distilled water. The anisotropy of the electrical polarisability is evaluated using eq. 2 from the field dependence of the equilibrium change in $\frac{\Delta I}{I}$. Let this value be $\Delta\alpha_o$. Measurements of $\Delta\alpha$ are then made on bacterial suspensions containing increasing concentrations (c) of added antibiotic. In such measurements it is essential to check that the values of $\Delta\alpha$ do not change appreciably during the measurement period. Such time dependent changes may arise either as a result of the interaction of the antibiotic with the bacterial surface or simply due to degradation of the bacterial surface. It is also important to monitor $D_R$ (eq. 3) to ensure that aggregation or size or shape changes are absent. Finally values of $\Delta\alpha_o$, and hence $\Delta\alpha$, are sensitive to the preparative techniques, the pH and the conductivity of the suspending medium. Hence it is convenient to display the results as plots of $\frac{\Delta\alpha}{\Delta\alpha_o}$ vs. c. A schematic diagram of the type of plot obtained is shown in fig. 1. The shape of the curve obtained is reminiscent of the monolayer adsorption isotherms familiar in

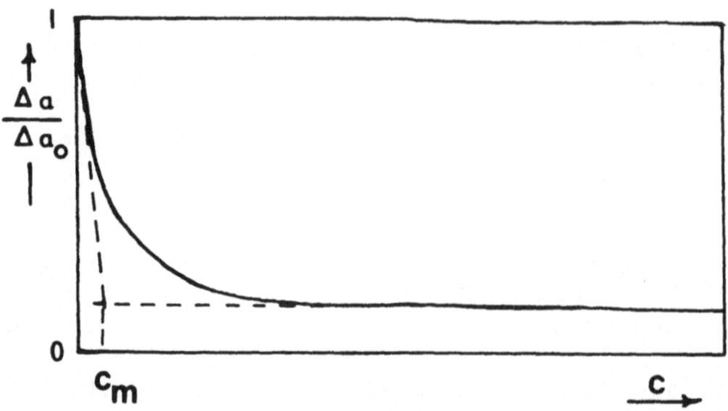


Fig. 1. Schematic diagram of the variation of the anisotropy of electrical polarisability ($\Delta\alpha$), normalised with respect to the value at zero antibiotic concentration ($\Delta\alpha_o$), with bulk concentration of added antibiotic (c). The concentration $c_m = 1/\beta$

physical chemistry. It has been demonstrated previously (12) that if the accumulation of the antibiotic at the bacterial surface may be pictured on a monolayer model and, if the changes in charge density at the shear plane are linearly proportional to the surface concentration of antibiotic, then

$$\frac{\Delta\alpha}{\Delta\alpha_o} = 1 - \frac{K_o\beta c}{1+\beta c} \tag{4}$$

with $\beta = \frac{1}{K}\exp\left(\frac{\Delta\epsilon}{kT}\right)$ (5)

Here K is an equilibrium constant, $\Delta\epsilon$ a binding energy and $\beta$ and $K_o$ characteristic parameters of the curve.

Eq. 4 may be rewritten in the form

$$\frac{1}{\left(1-\frac{\Delta\alpha}{\Delta\alpha_o}\right)} = \frac{1}{K_o\beta c} + \frac{1}{K_o} \tag{6}$$

and the parameters $\beta$ and $K_o$ evaluated by a Least Squares fit to the experimental data. A useful parameter is the quantity $c_m = 1/\beta$. It is relatively easy to show that the limiting tangent to the experimental curve (eq. 4) at low concentrations intersects the limiting tangent at high concentrations (saturation value) at $c = c_m = 1/\beta$. Hence $c_m$ may be considered as a representative concentration at which maximum coverage of the bacterial surface occurs.

## ILLUSTRATIVE DATA

Illustrative data on the interaction of aminoglycoside antibiotics with sensitive, dependent and resistant strains of E.coli K 12 are given in tables I and II. Details of the preparation of the bacteria and the experimental measurements of $\Delta\alpha$ are given elsewhere (14).

Table I. Characteristic parameters and MIC's for the interaction of aminoglycoside antibiotics with the sensitive E.coli K 12/711.

| Antibiotic | $K_o$ | $\beta$ (ml/$\mu$g) | $c_m$ ($\mu$g/ml) | MIC ($\mu$g/ml) |
|---|---|---|---|---|
| Gentamicin | 0.86 | 9.0 | 0.11 | 0.2 |
| Sisomicin | 0.80 | 5.1 | 0.20 | 0.25 |
| Framycetin | 1.0 | 12.2 | 0.08 | 0.5 |
| Neomycin | 1.0 | 5.6 | 0.18 | 0.75 |
| Paromomycin | 0.95 | 0.2 | 5.0 | 1.0 |
| Streptomycin | 0.92 | 0.16 | 6.3 | 9.0 |

Table I contains data on the interaction of several members of the aminoglycoside family of antibiotics with the sensitive bacterium E.coli K 12/711. Besides listing the parameters $\beta$, $K_o$ and $c_m$ the table also contains values of the minimum inhibitory concentration (M.I.C.). The M.I.C. is defined as the lowest concentration of antibiotic which completely inhibits bacterial growth during overnight incubation. The measurement of the M.I.C. is described in detail elsewhere (14). Table I represents an attempt to relate the physical characteristics of the 'adsorption curve' embodied in $\beta$ and $K_o$ with the biological effectiveness of the antibiotics expressed in terms of the M.I.C. We note with interest that, although there is no complete correlation between $c_m$ and the M.I.C., there is evidence that large values of $c_m$ correspond to high values of the M.I.C. The results are somewhat surprising since the variation in M.I.C. is much greater than the difference in molecular weights and the antibacterial action of the aminoglycosides is believed to involve binding of a single drug molecule per ribosome, leading to an interference with protein synthesis (19). Hence the present results may be tentatively interpreted in the following manner. The variation in the M.I.C. of these aminoglycosides may result from structural differences in the molecules altering the ease with which the molecules can be adsorbed at the surface and then enter the bacterium, rather than affecting their action at the ribosomal site once inside the microorganism.

Table II. Interaction of streptomycin sulphate with a sensitive, a dependent and two resistant variants of E.coli K 12.

| Microorganism | $K_o$ | $\beta$ (ml/ $\mu$g) |
|---|---|---|
| Sensitive | 0.92 | 0.16 |
| Dependent | 0.91 | 0.048 |
| Chromosomally mediated | 0.94 | 0.11 |
| Plasmid bearing | 0.86 | 0.047 |

Table II contains data on the interaction of the antibiotic streptomycin sulphate with a sensitive, a dependent and two resistant strains of E.coli K 12. The dependent form was obtained by incubating the sensitive parent strain K 12/711 on nutrient agar containing streptomycin. A clone was selected which required the presence of streptomycin for its growth (14). The resistant forms were of two types. A variant K 12/712 was chromosomally resistant to streptomycin (M.I.C. $>$ 1000 $\mu$g/ml) but sensitive to the remaining aminoglycosides. The resistance involves an alteration of the ribosome leading to failure to bind the drug. The second resistant form was obtained by crossing K 12/711 with E.coli strain G1056, which carries a plasmid conferring resistance to streptomycin and sulphonamide, to obtain a variant with plasmid borne resistance to streptomycin (14). The plasmid codes for a surface enzyme which degrades the antibiotic before it can reach the sensitive ribosome target. Comparison of the data for the sensitive parent and chromosomally resistant variant suggests that the uptake of antibiotic may be slightly less in the case of the resistant form although the curves are very similar. This is perhaps not surprising since the resistance involves a modification of the ribosomal site within the bacterium and the surface interaction ought to remain unchanged. The dependent and plasmid bearing micro-organisms show marked differences from the behaviour of the sensitive parent strain. In the case of the dependent form there appears to be a reduced uptake ($\beta K_o$) of antibiotic with added bulk concentration but the saturation value $K_o$ remains the same as for the sensitive form. This difference may arise because the variant requires streptomycin in its growth medium and hence there may be a finite concentration of the antibiotic at the bacterial surface at zero bulk concentration of added antibiotic. The plasmid-bearing variant yields lower values of both $\beta$ and $K_o$ in comparison with the sensitive strain. This could be indicative of reduced overall binding of the drug due to the breakdown of the antibiotic by a surface enzyme.

## ACKNOWLEDGEMENTS

V.J.M. thanks the Medical Research Council for a Junior Research Fellowship during the tenure of which this work was commenced. Messrs. ICI are thanked for a grant to B.R.J. which also provided a Fellowship for V.J.M. for the continuation of this work. In addition all authors are grateful to 'Microbios' for permission to reproduce tables I and II.

## REFERENCES

1 Stoylov S P, Advances Colloid Interface Sci., 3 (1971) 45.
2 Jennings B R and Morris V J, J. Coll. Interf. Sci., 49 (1974) 89.
3 O'Konski C T and Haltner A J, J. Amer. Chem. Soc., 79 (1957) 5634.
4 Morris V J and Jennings B R, J. Chem. Soc., Far. Trans. II, 71 (1975) 1948.
5 Brownsey G J, Jennings B R and Morris V J, J. Coll. Interf. Sci., 63 (1978) 597.
6 Shaw D J, "Electrophoresis", Academic Press, London (1969).
7 Schott H and Young C Y, J. Pharm. Sci., 61 (1972) 182.
8 Pechey D T and James A M, Microbios, 10A (1974) 111.
9 McQuillen K, Biochim. Biophys. Acta, 6 (1951) 534.
10 McQuillen K, Biochim. Biophys. Acta, 7 (1951) 54.
11 Morris V J and Jennings B R, J. Coll. Interf. Sci., 55 (1976) 143.
12 Morris V J and Jennings B R, Biochim. Biophys. Acta, 392 (1975) 328.
13 Morris V J and Jennings B R, Biochim. Biophys. Acta, 497 (1977) 253.
14 Morris V J, Jennings B R, Pearson N J and O'Grady F, Microbios, 17 (1976) 133.
15 Van de Hulst H C, "Light Scattering by Small Particles", J. Wiley, New York (1957).
16 Morris V J, Coles H J and Jennings B R, Nature, 249 (1974) 5454.
17 Coles H J, Jennings B R and Morris V J, Phys. Med. Biol., 20 (1975) 225.
18 Jennings B R and Morris V J, J. Coll. Interf. Sci. 50 (1975) 352.
19 Franklin T J and Snow G A, "Biochemistry of Antimicrobial Action", Chapman and Hall Ltd., London (1971) Chap. 5, 87.

# ROTATIONAL DIFFUSION COEFFICIENTS OF COMPLEX MACROMOLECULES

Victor A. Bloomfield, Jose Garcia de la Torre and
Robert W. Wilson
Department of Biochemistry, University of Minnesota
St. Paul, MN 55108, U.S.A.

It is necessary to have an accurate theory enabling calculation of the rotational diffusion coefficient $D^{\theta\theta}$ in terms of molecular structure, if electric birefringence relaxation times are to be interpreted. We have constructed such a theory for rigid macromolecular structures composed of arrays of nonidentical, spherical subunits. Hydrodynamic interaction between the subunits is expressed by a tensor which takes into account the finite size of the subunits. The set of simultaneous interaction equations is solved completely by numerical methods. While substantial improvements over prior theories are obtained, difficulties still remain when a large spherical subunit is located near the center of rotation. The rotation of this subunit about its axis is calculated to contribute less to rotational friction than is in fact the case. We have remedied this by replacing the large central sphere by an array (typically octahedral or cubic) of smaller spheres chosen to have the same $D^{\theta\theta}$. Agreement with the Perrin equation for prolate ellipsoids is now good over the entire range of axial ratios. For T-even phage, experimental (20°C) and theoretical values are now in excellent accord: 250-325 $sec^{-1}$ (exp) and 250-330 $sec^{-1}$ (theor) for fiberless or fiber-retracted phage; and 100-120 $sec^{-1}$ (exp) and 112 $sec^{-1}$ (theor) for fiber-extended phage. We feel that the theory is now sufficiently refined that adequate calculations of $D_R$ for any rigid structure, regardless of shape or symmetry, can be made with a moderate investment of computer time.

## INTRODUCTION

One of the major aims of electric birefringence and other electro-optics experiments is to determine the rotational diffusion coefficient $D^{\theta\theta}$ of the dissolved macromolecule. In order for $D^{\theta\theta}$ to be interpreted, it is necessary to have an accurate theory relating this transport coefficient to macromolecular structure. While exact expressions for $D^{\theta\theta}$ as a function of length and axial ratio have been derived for ellipsoids of revolution by Perrin (1), and accurate equations for cylindrical rods have been obtained by Broersma (2), a trustworthy theory for less symmetrical structures has been lacking.

Recently, we (3) have constructed such a theory for rigid macromolecular structures composed of arrays of nonidentical, spherical subunits. Although similar theories had been developed earlier (4,5), our recent version has two improvements which substantially improve its accuracy. The first is a modification of the Oseen (6) - Burgers (7) hydrodynamic interaction tensor in the manner proposed by Yamakawa (8) and by Rotne and Prager (9). This modified tensor takes into account the finite size of the subunits. The second improvement is solution of the set of simultaneous hydrodynamic interaction equations completely by numerical methods, rather than reliance on the first few terms of a perturbation series. As will be reviewed in this paper, these modifications give asymptotic agreement with the Perrin equations for prolate ellipsoids. The theory has also been applied to plane polygonal rings, lollipops, dumbbells, and T-even bacteriophage models (3). These last structures, in particular, pose a severe test of any hydrodynamic theory.

A difficulty still remains, however, when a large spherical subunit is located near the center of rotation. The rotation of this subunit about its axis is calculated to contribute less to rotational frictional resistance than is in fact the case. We (10) have remedied this defect in a simple but satisfactory way by replacing the large central sphere by an array (typically octahedral or cubic) of smaller spheres chosen to be touching and to have the same $D^{\theta\theta}$ or volume as the original sphere. Agreement with the Perrin equation is now good over the entire range of axial ratios, and accordance with experimental determinations on T-even bacteriophage is excellent.

In this paper we shall review the basic theoretical hydrodynamical ideas and the computational methods, describe how the center of frictional resistance is located, and how the replacement of a large central sphere is carried out. We shall then describe results for prolate ellipsoids, lollipops and dumbbells, and T-even phage structures. We conclude with some brief remarks on the computer requirements and reliability of this approach.

## THEORY

In this section we summarize the general approach and most important equations of the hydrodynamic theory. For details, the reader is referred to our previous work (3,11). We consider a macromolecular complex composed of N spherical subunits, the jth with radius $\sigma_j$, rotating in a solvent of viscosity $\eta_o$ with angular velocity $\Omega$. The complex is rotating about the z-axis of a right-handed Cartesian coordinate system, of which the x-axis is the principal symmetry axis of the complex, if such exists. In cylindrical coordinates $r_j$ is the distance of the jth subunit from the rotation axis, and $\theta_j$ is its angle of rotation. The vector distance between subunits i and j is $\underline{R}_{ij}$.

In the absence of particle rotation the solvent would be at rest. Macromolecular rotation causes a solvent viscosity $\underline{v}_i$ at the position of the ith subunit due to the forces $\underline{F}_j$ exerted on the solvent by all the other subunits:

$$\underline{v}_i = \sum_{j=1}^{N}{}' \underline{\underline{T}}_{ij} \cdot \underline{F}_j. \tag{1}$$

The torque P on the article is

$$P = \sum_{i=1}^{N} r_i F_i = \zeta_R^{\theta\theta} \Omega \tag{2}$$

where $\zeta_R^{\theta\theta}$, the rotational frictional coefficient, is related to the rotational diffusion coefficient $D^{\theta\theta}$ by the familiar Einstein equation

$$D^{\theta\theta} = kT/\zeta^{\theta\theta} \tag{3}$$

Combining these equations with Stoke's law for the translational frictional coefficient of the jth subunit, $\zeta_j = 6\pi\eta_o\sigma_j$, and noting that the relative velocity of the ith sphere with respect to the solvent is $\Omega r_i - v_i$, we obtain a set of N linear simultaneous equations for the N unknown forces $F_i$

$$F_i = \zeta_i(\Omega r_i - v_i) = \zeta_i \Omega r_i - \zeta_i \sum_{j=1}^{N}{}' (r_i r_j)^{-1}(T_{ij})_{\theta\theta} F_j. \tag{4}$$

Once these forces are known, $D^{\theta\theta}$ can be computed according to eqs. 2 and 3.

The hydrodynamic interaction tensor is of the Oseen form as modified by Yamakawa (8) and by Rotne and Prager (9):

$$\underline{\underline{T}}_{ij} = (8\pi\eta_o R_{ij})^{-1}\left\{(\underline{\underline{I}} + \frac{\underline{R}_{ij}\underline{R}_{ij}}{R_{ij}^2}) + \frac{(\sigma_i^2+\sigma_j^2)}{R_{ij}^2}(\frac{1}{3}\underline{\underline{I}} - \frac{\underline{R}_{ij}\underline{R}_{ij}}{R_{ij}^2})\right\} \quad (5)$$

where $\underline{\underline{I}}$ is the unit tensor of rank three.

The $\theta\theta$ component of this is readily shown to be

$$(T_{ij})_{\theta\theta} = \frac{r_i r_j}{8\pi\eta_o R_{ij}}\left\{(1+\frac{\sigma_i^2+\sigma_j^2}{3R_{ij}^2})\cos(\theta_j-\theta_i) + \frac{r_i r_j}{R_{ij}^2}(1-\frac{\sigma_i^2+\sigma_j^2}{R_{ij}^2})\sin^2(\theta_j-\theta_i)\right\}. \quad (6)$$

To solve the simultaneous hydrodynamic interaction equations it is convenient to define N scalars $H_i$ by

$$F_i = \Omega\,\zeta_i H_i \quad (7)$$

which when known enable calculation of $\zeta_R^{\theta\theta}$ according to eq. 2 by

$$(\zeta^{\theta\theta}/6\pi\eta_o) = \sum_{i=1}^{N} r_i\sigma_i H_i . \quad (8)$$

Substituting eq. 7 in eq. 4, the set of simultaneous equations becomes

$$\sum_{j=1}^{N} C_{ij}H_j = r_i \qquad i=1,\ldots,N \quad (9)$$

where the coefficients are

$$C_{ij} = \delta_{ij} + (1-\delta_{ij})\frac{3\sigma_j}{4R_{ij}}\left\{(1+\frac{\sigma_i^2+\sigma_j^2}{3R_{ij}^2})\cos(\theta_j-\theta_i) + (1-\frac{\sigma_i^2+\sigma_j^2}{R_{ij}^2})\frac{r_i r_j}{R_{ij}^2}\sin^2(\theta_j-\theta_i)\right\}. \quad (10)$$

and $\delta_{ij}$ is the Kronecker delta.

A second order perturbation solution of eqs. 9, starting with the unperturbed forces $F_i = \Omega\theta r_i \zeta_i$ and using the unmodified Oseen tensor ($\sigma_i^2 = \sigma_j^2 = 0$) in the square brackets of eq. 10 yields

$$\frac{6\pi\eta_o}{kT} D^{\theta\theta} = \frac{1}{\alpha-\beta}\left\{1 + \frac{3}{4(\alpha+\beta)} \sum_{i=1}^{N} \sum_{j=1}^{N}{}' \left(\frac{\sigma_i\sigma_j(X_iX_j+y_iy_j)}{R_{ij}} + \frac{\sigma_i\sigma_j(X_iy_j - X_jy_i)^2}{R_{ij}^3}\right)\right\} \tag{11}$$

where

$$\alpha = \sum_{i=1}^{N} \sigma_i X_i^2, \quad \beta = \sum_{i=1}^{N} \sigma_i y_i^2. \tag{12}$$

Eq. 11 is a generalization of Hearst's (4) theory for $D^{\theta\theta}$ to encompass nonidentical subunits. While this equation is convenient for rapid approximate calculations, it does not have the accuracy we desire. Therefore, we now describe the method we have used to solve the complete system of eq. 9.

## COMPUTATIONAL METHODS

It is first necessary to locate the center of frictional resistance, which is the point of intersection between the principal symmetry (x) axis, and the (z) axis about which rotation occurs. In some cases, such as ellipsoids, rods, or rings, this point will be obvious from symmetry; but in many other structures of interest, such as bacteriophage, it must be determined by calculation.

Physically, the center of frictional resistance is that point about which rotation causes the least resistance, so that $\zeta^{\theta\theta}$ is a minimum or $D^{\theta\theta}$ a maximum. Therefore, we calculate $D^{\theta\theta}$ for various choices of displacement  from the initial arbitrary origin x' and locate the position $x = x' - \delta$ for which $D^{\theta\theta}$ is a maximum. Because solution of the complete set of linear equations 9 is time-consuming and expensive, we first use eq. 12 to obtain an approximate position for the maximum. Following Filson and Bloomfield (5), we rewrite eq. 12 as

$$\frac{6\pi\eta_o}{kT} D^{\theta\theta} = \frac{1}{\alpha+\beta}\left\{1 + \frac{3}{4(\alpha+\beta)} (T_1 + T_2\delta + T_3\delta^2)\right\} \tag{13}$$

$$\alpha = U_1 + U_2 \delta + U_3 \delta^2 \tag{14}$$

where

$$U_1 = \sum_{i=1}^{N} \sigma_i X_i^2, \quad U_2 = 2\sum_{i=1}^{N} \sigma_i X_i, \quad U_3 = \sum_{i=1}^{N} \sigma_i \tag{15}$$

and

$$T_1 = \sum_{i=1}^{N} \sum_{j=1}^{N}{}' \sigma_i \sigma_j \left\{ \frac{X_i'X_j'+y_iy_j}{R_{ij}} + \frac{(X_i'y_j-X_j'y_i)^2}{R_{ij}^3} \right\} \tag{16}$$

$$T_2 = \sum_{i=1}^{N} \sum_{j=1}^{N}{}' \sigma_i \sigma_j \left\{ \frac{X_i'+X_j'}{R_{ij}} + 2\frac{(y_j-y_i)(X_i'y_j-X_j'y_i)}{R_{ij}^3} \right\} \tag{17}$$

$$T_3 = \sum_{i=1}^{N} \sum_{j=1}^{N}{}' \sigma_i \sigma_j \left\{ \frac{1}{R_{ij}} + \frac{(y_i-y_j)^2}{R_{ij}^3} \right\} \tag{18}$$

Thus the double summations need be evaluated only once, and the maximum in the approximate value of $D^{\theta\theta}$ located by varying $\delta$. Once this approximate maximum is located, the system of linear equations 9 is solved to the desired accuracy by Gauss-Seidel iteration (11) at four points around $\delta^*$. Then the true maximum value of $D^{\theta\theta}$ is obtained by a third-degree interpolating polynomial. For small structures, the simultaneous equations 9 may be solved by matrix inversion rather than iteration.

## THE PROBLEM OF A LARGE CENTRAL SUBUNIT

The procedure outlined above gives excellent results save when a dominant frictional element is located close to the center of frictional resistance. Among such structures are short prolate ellipsoids of revolution, and bacteriophage with tail fibers retracted. The source of the difficulty is evident from eq. 6. In theories of the Oseen type, the frictional resistance of each subunit is concentrated at its center, a distance $r_i$ from the center of resistance, instead of being distributed over its surface. This idealization is unimportant when $r_i$ is large or when $\sigma_i$ is small; but when the situation is reversed, eq. 6 severely underestimates the contribution of the subunit to the rotational frictional resistance of the complex.

Table I. Relative Subunit Radii and Rotational Diffusion Coefficients of Arrays Equivalent to a Single, Large Sphere

| Array | $\sigma/\sigma_1$ | $D^{\theta\theta}/8\pi\eta_o\sigma_1^3$ |
|---|---|---|
| Trigonal bipyramid | 0.585 (V), 0.535 (R) | 0.765 (V), 1 (R) |
| Octahedron | 0.550 (V), 0.532 (R) | 0.905 (V), 1 (R) |
| Cube | 0.500 (V), 0.465 (R) | 0.804 (V), 1 (R) |

V = Volume-equalized array; R = Rotation-equalized array

The most fundamental way to overcome this problem would be to coat the surface of the large subunit with an array of smaller spheres, thus moving the resistance from the center to the surface in the spirit of the original shell model (12). However, this would be unduly expensive of computer time, so we have devised a simpler but still satisfactory expedient (10). We replace the original sphere by a trigonal bipyramidal, octahedral, or cubic array of smaller spheres. These are constrained to touch and either to occupy the same volume as the original, or to have the same $D^{\theta\theta}$ about the axis of interest. In the first case, if n is the number of replacement subunits, the ratio of the small subunit radius $\sigma$ to the original sphere radius $\sigma_1$ is $n^{-1/3}$. In the second case, the radii $\sigma$ are adjusted so that $D^{\theta\theta}$, calculated according to the method detailed above is $8\pi\eta_o\sigma_1^3$. The resulting sizes and rotational diffusion coefficients for various arrays are given in table I.

## RESULTS

### Prolate Ellipsoids of Revolution

In a previous paper (3) we computed $D^{\theta\theta}$ for subunit models of prolate ellipsoids of revolution and compared them with their exact values according to the Perrin (1) equations. At large axial ratios agreement was excellent, but became very poor as the axial ratio approached unity. This is shown by the dotted line in fig. 1. As discussed in the previous section, this difficulty at low axial ratios is associated with the underestimation of the contribution of the large central sphere, shown in the inset of fig. 1. We have therefore recalculated $D^{\theta\theta}$ after replacing the central sphere with a cubic array of smaller spheres, shown schematically in the inset. Both volume-equalized and rotation-equalized arrays were used. As expected, the divergence in $D^{\theta\theta}$ as axial ratio approaches unity disappears. In this limit the rotation-equalized array is

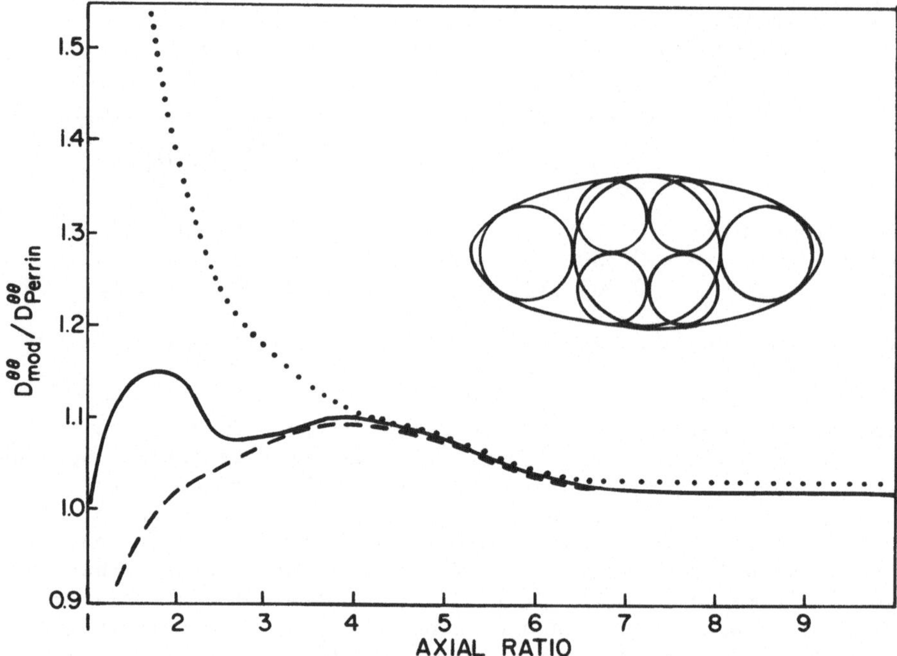


Fig. 1. Ratios of the rotational diffusion coefficients of volume-corrected subunit models for prolate ellipsoids of revolution to the exact Perrin values, as functions of axial ratio. Central element: (···) sphere; (---) volume equalized cubic array; (——) rotation equalized cubic array.

better (by definition) but there is little significant difference over much of the range. The oscillations in the curves are unavoidable residual artifacts of the representation of the smooth ellipsoid by a small number of spheres.

## Lollipops

The inset fig. 2 shows a structure denoted as a "lollipop" which should serve as a useful structural model for tailed bacteriophage and for some protein and nucleoprotein complexes. It is constructed of a single large sphere of radius $\sigma_1$ attached to a rod of (N-1) smaller spheres of radius $\sigma_2$. The length of the rod is $L = 2(N-1)\sigma_2$. In fig. 2 we show the reduced rotational diffusion coefficient $6\pi\eta_0\sigma_1^3 D_R/kT$ calculated as a function of the reduced rod length $L/\sigma_1$ for various values of the relative rod thickness $\sigma_1/\sigma_2$. As for ellipsoids, proper convergence to the single sphere limit $L/\sigma_1 \rightarrow 0$ or

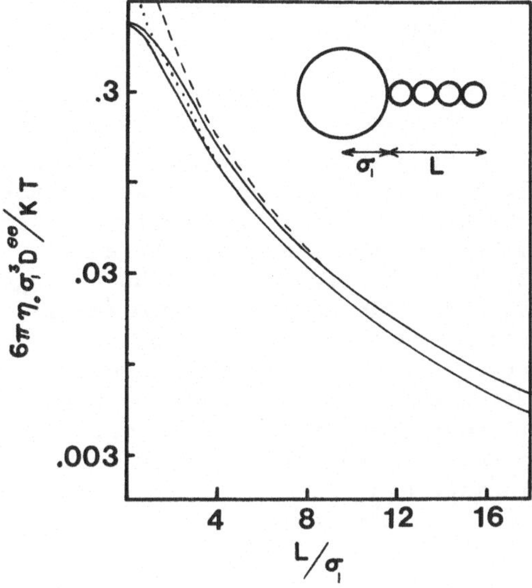


Fig. 2. Reduced rotational diffusion coefficient of a lollipop with head radius $\sigma_1$, tail length L, and tail radius $\sigma_2$. Dotted and dashed lines are for a single large sphere, solid lines for equivalent cubic arrays. These merge for large $L/\sigma_1$.

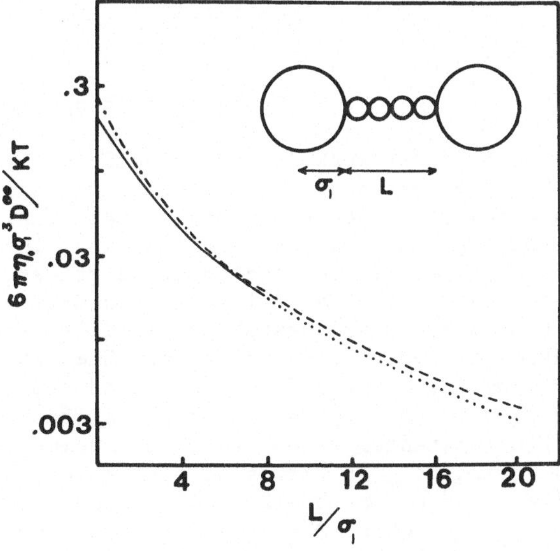


Fig. 3. Reduced rotational diffusion coefficient of a dumbbell with two large spheres of radius $\sigma_1$ separated by a rod of length L and radius $\sigma_2$. Curves merge for $L/\sigma_1 < 7$, as represented by dot-dash line (single sphere) or solid line (equivalent cube).

$\sigma_2/\sigma_1 \to 0$ is obtained only when the large subunit is replaced by an array of smaller ones, chosen to give the same $D_R$. The case $\sigma_1/\sigma_2=1$ corresponds to a cylindrical rod of axial ratio N. Fig. 2 shows that $D^{\theta\theta}$ is very sensitive to $\sigma_1$ and L, but less so to $\sigma_2$.

## Dumbbells

The inset fig. 3 shows a structure denoted as a "dumbbell" which may serve as a useful model for nucleoprotein complexes such as nucleosomes and polyribosomes, and for proteins consisting of globular domains connected by a thin spacer. The dumbbell consists of two spheres of radius $\sigma_1$, connected by a rod of N-2 spheres of radius $\sigma_2$, whose length is $L = 2(N-2)\sigma_2$. The reduced diffusion coefficient $6\pi\eta_o\sigma_1^3 D_R/kT$ is plotted in fig. 3 as a function of $L/\sigma_1$ and $\sigma_1/\sigma_2$, using both the large spheres and their rotation-corrected cubic array equivalents. Not much difference between these two models is found, since even in the limit L = 0 both spheres are well away from the center of rotation. The limit $\sigma_1/\sigma_2 = \infty$ corresponds to two spheres held by an infinitely thin spacer at a center-to-center distance $R_{12} = L + 2\sigma_1$. Eqs. 9 and 10 may be solved exactly for this case, yielding (3)

$$\frac{8\pi\eta_o\sigma_1^3 D_R}{kT} = \frac{4}{3}\;\frac{\left\{1 - \frac{1}{16}\left(\frac{R_{12}}{\sigma_1}\right)^{-2}\left(1 + \frac{2\sigma_1^2}{2R_{12}}\right)^2\right\}}{\left\{\frac{1}{2}\left(\frac{R_{12}}{\sigma_1}\right)^2 + \frac{3}{8}\frac{R_{12}}{\sigma_1}\left(1 + \frac{2\sigma_1^2}{3R_{12}}\right)\right\}^{-1}} \tag{19}$$

## T-even bacteriophage

Much of our work in hydrodynamics, dating back to 1967 (5,13,14), has been motivated by the desire to reconcile theoretical calculations with experimental observations on bacterial viruses. Particularly challenging has been the understanding, in hydrodynamic terms, of bacteriophages T2, T4, and their relatives. Electron microscopy shows these to consist of a prolate icosahedral head, a cylindrical tail built on a hexagonal baseplate distal to the head, and six long, kinked fibers attached to the baseplate. The fibers may be either extended away from the phage body or retracted near the head and tail, giving rise to virus particles with quite different hydrodynamic properties. The fiber-extended (slow) form has sedimentation coefficient $S^o_{20,w}$= 700S, translational diffusion coefficient $D^o_{20,w} = 2.5 \times 10^{-8}$ $cm^2/sec$, and $D^{\theta\theta} \approx 100$-$120\ sec^{-1}$. The fiber-retracted (fast) form has $S^o_{20,w}$ = 1000S, $D^o_{20,w} = 3.6 \times 10^{-8}\ cm^2/sec$, and $D^{\theta\theta}$ = 250-325 $sec^{-1}$. Fiberless phage can also be prepared; they have hydrodynamic properties similar to those of the fiber-retracted form. For citations of the original literature see ref. 15.

Our earliest calculations (5,13) severely underestimated the frictional resistance of the tail fibers in determining the hydrodynamic properties of the fiber-extended form of T2L phage (14). More recent calculations (15), using the modified interaction tensor and computer solution of the simultaneous hydrodynamic interaction equations, accounted adequately for the effect of the fibers, but gave $D^{\theta\theta}$ too high for the fiber-retracted form. This arises because, as discussed above, the large sphere representing the head lies too close to the center of rotation. Therefore, we have redone these calculations replacing the head subunit by one or another equivalent array of smaller subunits (10). This simple stratagem gives much improved agreement with experiment.

The simplified structures used to model phage hydrodynamically are shown in fig. 4. The head has a Stokes radius $\sigma_H$ = 55 nm. There are five co-linear spheres for the tail, with $\sigma_T$ = 8.0 nm; and a larger sphere for the baseplate, with $\sigma_B$ = 20 nm. Each of the six fibers is represented by a kinked linear array of 69 spheres with $\sigma_F$ = 1 nm. Thus 421 subunits in all model the phage, before modification of the head is made. $D^{\theta\theta}$ calculated for the fiber-extended form is 119 $sec^{-1}$, in excellent agreement with experiment; the center of rotation is 108 nm from the center of the head, well down the tail toward the baseplate. For the fiberless phage model, however, $D^{\theta\theta}$ is 511 $sec^{-1}$, considerably greater than the experimental value. The center of rotation is calculated to be only 36 nm from the center of the head, substantially inside the head itself.

When the large head subunit is replaced with a cubic array of smaller spheres having the same $D^{\theta\theta}$ (the long axis of the tail passes through diagonally opposite vertices of the head), the rotational diffusion coefficient for the fiberless phage drops to 327 $sec^{-1}$, within the range of experimental values. The center of rotation is 45.3 nm from the center of the head, still within the head itself. Other arrays - trigonal bipyramids, octahedra, and differently oriented cubes - equivalent either with respect to volume or rotational diffusion coefficient, gave similar results (10). This indicates that the

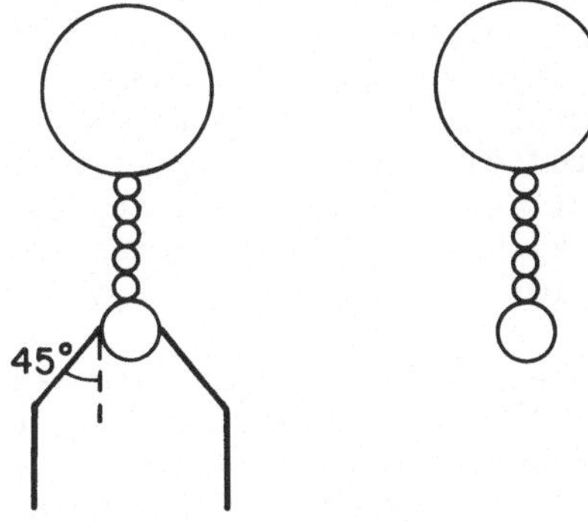


Fig. 4. Schematic subunit structures of fiberless (left) and fibered (right) phage. For simplicity only two of the up to six fibers are shown.

Table II. Rotational Diffusion Coefficient and Position of Center of Rotation for Phage as Functions of Number of Extended Fibers

| Fibers | $D^{\theta\theta}_{20,w}$, $sec^{-1}$ | $\delta$, nm |
|---|---|---|
| 0 | 327 | 45.3 |
| 1 | 202 | 63.5 |
| 2 | 160 | 86.4 |
| 3 | 140 | 98.0 |
| 4 | 128 | 104.6 |
| 5 | 120 | 113.8 |
| 6 | 114 | 113.3 |

details of the modelling are not crucial, but that replacement of the head sphere by a small equivalent array is sufficient to bring theory and experiment into good agreement.

We (10) have also calculated the effect of varying the number of extended fibers on $D^{\theta\theta}$. These results, shown in Table II, will be of interest for studies of fiber attachment and cooperativity of the slow-fast transition. For phage with 2 to 6 fibers, the geometry in fig. 2 was used. For phage with a single fiber, the fiber extended straight down the tail axis 98 nm, which is the projection of the kinked fiber on the tail axis. As might be expected, the first few fibers make a large difference; but it would be difficult to distinguish between phage with 4, 5 or 6 fibers extended.

## CONCLUSION

It now appears possible to compute, with acceptable accuracy, rotational diffusion coefficients and other hydrodynamic parameters for macromolecular complexes of almost any conceivable structure. In assessing "acceptable accuracy" it should be recalled that $D^{\theta\theta}$ is roughly proportional to the cube of the greatest linear dimension of the particle. Thus a 6 - 9% error in calculation reduces to only a 2 - 3% error in dimension, an uncertainty usually acceptable in structural studies. Indeed, experimental determinations of $D^{\theta\theta}$ are not often reliable, or at least reproducible between different laboratories working with different preparations of the same material, to better than 5 - 10%.

These computations are not very expensive. For example, calculation of $D^{\theta\theta}$ for T2 bacteriophage with six fibers, containing 421 subunits, required 68 seconds of CPU execution time on a CDC Cyber 74 computer. At current prices, this is about $10. Smaller structures, or those with more symmetry, would cost less.

## ACKNOWLEDGEMENTS

This research was supported in part by grants from NIH (GM 17855) and NSF (PCM 75-22728). J.G.T. was supported by a postdoctoral fellowship from the committee of Cultural Exchange between the U.S.A. and Spain.

## REFERENCES

1 Perrin F, J. Phys. Rad., 5, (1934) 497.
2 Broersma S, J. Chem. Phys., 32, (1960) 1626.
3 De la Torre J G and Bloomfield V A, Biopolymers, 16, (1977) 1765.
4 Hearst J E, J. Chem. Phys., 38, (1963) 1062.
5 Filson D P and Bloomfield V A, Biochemistry, 6, (1967) 1650.
6 Oseen C W, "Hydrodynamik" Akademisches Verlag, Leipzig (1927).
7 Burgers J M, "Second Report on Viscosity and Plasticity, Amsterdam Acad. Sci.", Nordemann, Amsterdam (1938), Ch. 3.
8 Yamakawa H, J. Chem. Phys., 53, (1970) 436.
9 Rotne J and Prager S, J. Chem. Phys., 50, (1969) 4381.
10 Wilson R W and Bloomfield V A, submitted for publication.
11 De la Torre J G and Bloomfield V A, Biopolymers, 16, (1977) 1747.
12 Bloomfield V A, Dalton W O and Van Holde K E, Biopolymers, 5, (1967) 135.
13 Bloomfield V A, Van Holde K E and Dalton W O, Biopolymers, 5, (1967) 149.
14 Douthart R J and Bloomfield V A, Biochemistry, 7, (1968) 3912.
15 De la Torre J G and Bloomfield V A, Biopolymers, 16, (1977) 1779.

# DETERMINATION OF ROTATIONAL DIFFUSION COEFFICIENTS OF THE BACTERIOPHAGES T4B AND T7 BY DEPOLARIZED DYNAMIC LIGHT SCATTERING. THE INFLUENCE OF DOUBLE SCATTERING

P. C. Hopman, G. Koopmans and J. Greve

Physics Laboratory of the Vrije Universiteit, Biophysics Department, De Boelelaan 1081, 1081 HV Amsterdam, The Netherlands.

Methods to correct for double scattering effects in the determination of rotational diffusion coefficients by depolarized dynamic light scattering are discussed. Rotational diffusion coefficients obtained by this technique for the bacteriophages T4B and T7 are compared to the ones measured by electro optical methods. The results obtained by both methods agree. This indicates that for both T4B and T7 the orienting electric field used in the electric birefringence technique does not influence the conformation of the bacteriophages during the field-free decay.

## INTRODUCTION

Rotational diffusion coefficients of optically anisotropic macromolecules in solution can be determined by using transient electric birefringence. A major drawback of this technique is the need for an orienting electric field. This limits the ionic strength of the buffer in which the macromolecules are dissolved. Moreover, the orienting force may distort the macromolecules under study.

A recently developed alternative technique is the method of depolarized dynamic light scattering (1). Here the autocorrelation function of the depolarized intensity scattered in the forward direction is determined, and no electric field is needed.

In this contribution we present some results of measurements on the bacteriophages T4B and T7. Special attention will be given to the problem of double scattering.

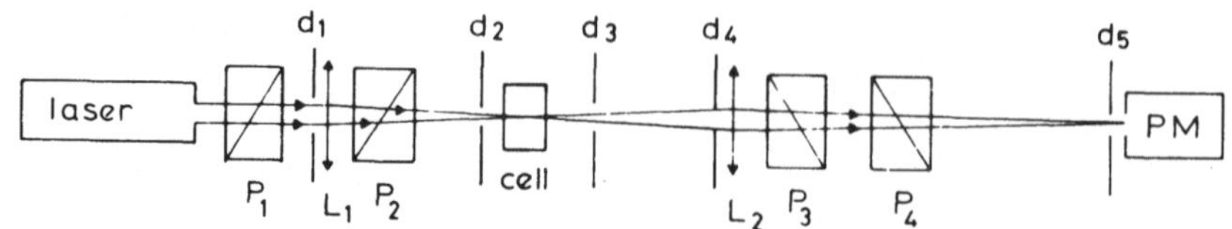


Fig. 1. Schematic diagram of the depolarized dynamic light scattering apparatus. Lens L1 focusses the light (514 nm) into the scattering cell. Lens L2 depicts the scattered light on to pinhole d5, in front of the photomultiplier PM. The polarizers ($P_1$, $P_2$) and analyzers ($A_1$, $A_2$) are crossed. Diaphragms d1, d2 and d3 are inserted to minimize stray light, d4 determines the aperture of the system.

## MATERIALS AND METHODS

Details of the dynamic light scattering apparatus are given in fig. 1. The transmission of the polarizer-analyzer system with a scattering cell in position was smaller than $5.10^{-7}$.

According to reference 4 the scattering volume and the volume which is seen by the detector must coincide. Moreover, the radii of these volumes must be small, when averaged over the length of the cell. This can be taken care of by a proper choice of lenses L1 and L2 and diaphragm d5. We used lenses with focal distances $f_1 = 220$ mm and $f_2 = 100$mm and a diaphragm d5 with a diameter of 0.1 mm. This resulted in a diameter for the scattering volume and the volume as seen by the detector of $\approx 0.1$ mm.

Translational diffusion coefficients were determined by standard dynamic light scattering techniques (1). For T4B (retracted fibers) and T7 we found resp. $D_{T,25,w} = 3.8 \times 10^{-8} cm^2 sec^{-1}$ and $D_{T,25,w} = 6.6 \times 10^{-8} cm^2 sec^{-1}$. Rotational diffusion coefficients were also determined by electric birefringence measurements, using the apparatus described elsewhere (2).

Special care was taken to minimize the amount of dust in the cell. The dust removal techniques (3) used were filtering and centrifugation.

## THEORY

The correlation function of the depolarized electric field scattered in the forward direction by freely rotating, identical particles is given by (1,4)

$$C_1^s(\tau) = A \, \ell \, \beta^2 \, C_N \, f_1^s (D_R, \tau) \qquad (1)$$

$$f_1^{s}(D_R,\tau) = \exp(-6 D_R \tau) \tag{2}$$

where $\beta$ is the anisotropy in the optical polarizability and $C_N$ is the number concentration and $D_R$ is the rotational diffusion coefficient of the macromolecules. $\ell$ is the length of the scattering cell and A is a geometrical factor determined by the experimental set-up used.

The correlation function of the depolarized electric field resulting from double scattering can be written as

$$C_1^{d}(\tau) = B\,\ell\,\alpha^4 C_N^{\,2} f_1^{d}(D_T,\tau) \tag{3}$$

with $$f_1^{d}(D_T,\tau = 0) = 1 \tag{4}$$

B is a factor determined by the geometry of the set-up, whereas $D_T$ and $\alpha$ are the translational diffusion coefficient and the polarizability of the particles. In reference 4 an analytical expression for $f_1^{d}(D_T,\tau)$ is given, while in figure 2 its shape is shown. If $\beta \ll \alpha$ the correlation function of the total depolarized field scattered by a solution of macromolecules will be

$$C_1^{m}(\tau) = C_1^{s}(\tau) + C_1^{d}(\tau) \tag{5}$$

After normalization this can be written as

$$f_1^{m}(\tau) = d\, f_1^{s}(D_R,\tau) + (1 - d)\, f_1^{d}(D_T,\tau) \tag{6}$$

with

$$f_1^{m}(\tau) = C_1^{m}(\tau) / C_1^{m}(0) \tag{7}$$

and $0 \leqslant d \leqslant 1$.

Here d is a measure for the relative amount of double scattering present at $\tau = 0$. If d = 1 no double scattering is present.

## RESULTS AND DISCUSSION

Eq. 6 was fitted (3) to the measured correlation function to determine $D_R$ and d. Correlation functions, obtained in a set-up with $\ell = 1$ cm, for T4B phages are shown in fig. 2. For a concentration of $23 \times 10^{11}$ particles/ml we found $D_{R,25,w} = (357 \pm 18)\ \mathrm{sec}^{-1}$ and $d = 0.55 \pm 0.15$, whereas for a concentration of $3.6 \times 10^{11}$ particles/ml. $D_{R,25,w} = (278 \pm 46)\ \mathrm{sec}^{-1}$ and $d = 0.47 \pm 0.2$ was found. Clearly a considerable amount of double scattering is present at these concentrations. As can be seen from the scatter in the measured data at the lower concentration, we could not decrease the concentration further, due to the finite sensitivity of the set-up. This makes it impossible to measure with this set-up for T4B a correlation function of the electric field depolarized by single scattering only.

Correlation functions for T7 bacteriophages were measured in the same set-up. In this case we could decrease the concentration of the phages so far that double scattering effects could be neglected ($d>0.97$). Apparently $\beta^2/\alpha^4$ is much larger for T7 than it is for T4B. At a concentration of $6.5 \times 10^{12}$ phages/ml we obtained $D_{R,25,w} = (5200 \pm 150)\ sec^{-1}$.

In order to decrease the influence of double scattering further for the T4B case, we used a longer scattering cell. Then the ampli-

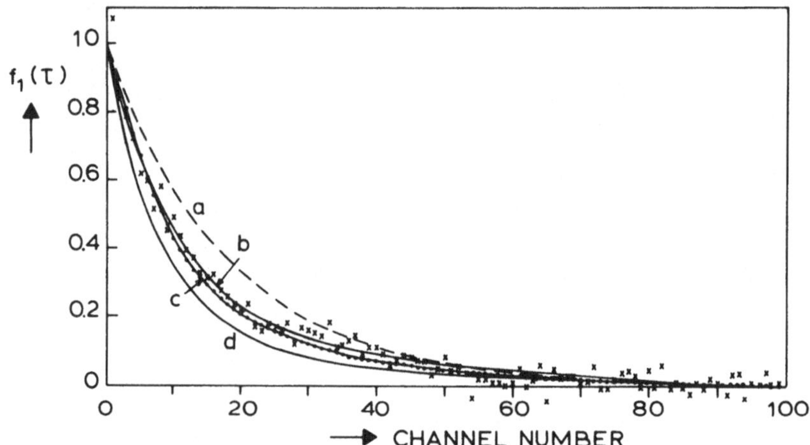


Fig. 2. Normalized correlation functions for the depolarized intensity scattered by solutions of T4B (retracted fibers). $\tau$ = channel number x 33.3 $\mu$sec. Curve a was calculated assuming the depolarized intensity is due to single scattering only. Curve d gives the expected correlation function if only double scattering contributes to the depolarized intensity. In the calculation of curves a and d we used $D_{R,25,w} = 280\ sec^{-1}$ and $D_{T,25,w} = 3.8 \times 10^{-8} cm^2/sec$. Correlation functions were measured at $23 \times 10^{11}$ particles/ml (circles) and $3.6 \times 10^{11}$ particles/ml (crosses). Curve b and c give the result of fitting eq. 6 to these data. At $C_N = 3.6 \times 10^{11}$ phages/ml we obtained $D_{R,25,w} = (278 \pm 46)\ sec^{-1}$ and $d = 0.47 \pm 0.2$. At $C_N = 23 \times 10^{11}$ phages/ml the results were $D_{R,25,w} = (357 \pm 18)\ sec^{-1}$ and $d = 0.55 \pm 0.15$. In both fitting procedures $D_{T,25,w} = 3.8 \times 10^{-8} cm^2\ sec^{-1}$ was used.

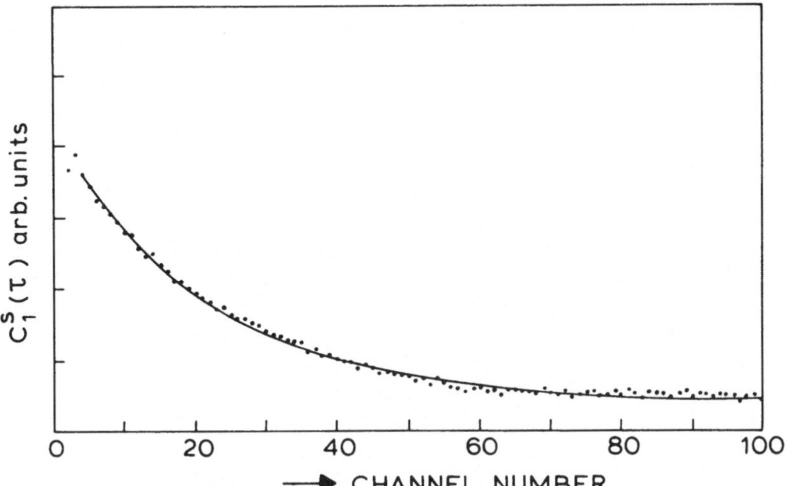


Fig. 3. Correlation function for the depolarized scattered intensity in the forward direction for T4B (retracted fibers). Concentration of $1.4 \times 10^{11}$ particles/ml. $\tau$ = channel number x 33.3 $\mu$sec. Full line gives the result of fitting eq. 1 to the measured data. The fitting procedure yields $D_{R,25,w} = (295 \pm 14)s^{-1}$ ($d > 0.97$).

tudes of $C_1{}^s(\tau)$ and $C_1{}^d(\tau)$ are increased while their ratio is not influenced (eq. 1 and 3). This means that we increased the sensitivity of the apparatus for depolarized scattering due to single scattering, and that we are able to use lower concentrations. In fig. 3 we show a typical correlation function, obtained in a set-up with $\ell$ = 10 cm for a sample of T4B ($C_N = 1.4 \times 10^{11}$ phages/ml.). From fitting eq. 6 to this function we obtained $D_{R,25,w} = (295 \pm 14)$ sec$^{-1}$ and $d > 0.97$. This means that the influence of double scattering can be neglected.

The rotational diffusion coefficients as determined from electric birefringence experiments were $D_{R,25,w} = (5100 \pm 100)$ sec$^{-1}$ and $D_{R,25,w} = (280 \pm 10)$ sec$^{-1}$ for T7 and T4B respectively. Comparison of these values with the results of depolarized dynamic light scattering leads to the conclusion that the electric field used in transient electric birefringence does not influence the conformation of the bacteriophages T4B (retracted fibers) and T7 during the field free decay.

From the results obtained in this study we conclude that rotational diffusion coefficients of particles like the T-phages, which have a small optical anisotropy, can be obtained by depolarized dynamic light scattering, provided double scattering effects are taken into account.

We thank Prof.Dr. Joh. Blok for the many stimulating discussions and suggestions during this work.

## REFERENCES

1 Berne B J and Pecora R, "Dynamic light scattering", John Wiley & Sons, New York (1976) and references herein.
2 De Groot G, Greve J and Blok Joh., Biopolymers, 16 (1977) 639.
3 Hopman P C, Koopmans G and Greve J, manuscript in preparation.
4 Koopmans G, Hopman P C and Greve J, accepted for publication J. of Phys. A.
5 Kelly H C, J. of Phys. A 6 (1973) 373.
6 Sorensen C M, Mockler R C and O'Sullivan W J, Phys. Rev. A, 14 (1976) 1520.

# EFFECTS OF CALCIUM AND ATP ON THE CONFORMATION OF F-ACTIN IN VIVO AND IN VITRO

M. Taniguchi

Department of Physics, Faculty of Science
Nagoya University, Nagoya, Japan

The dichroism spectrum of I-band in a glycerinated single fiber of muscle was very similar to that of an oriented F-actin filament in vitro. The negative dichroism of F-actin in the thin filament in vivo or in vitro was found to change with the addition of Ca ions, suggesting a conformational change of F-actin. In the absence of ATP, a small amount of bound H-meromyosin induced a remarkable decrease of the dichroism of F-actin. Upon the addition of Mg-ATP, the negative dichroism of the F-actin-H-meromyosin complex became larger than that of pure F-actin, suggesting that F-actin interacting with H-meromyosin assumes a different conformation from pure F-actin. A similar increase was also observed in the F-actin-tropomyosin-troponin-H-meromyosin complex. The maximum increase was greater under the condition that super-precipitation was delayed. The possible role of F-actin in muscle contraction is discussed.

## INTRODUCTION

Electron microscopic observations have shown that the contractile apparatus of striated muscle consists of two filaments, the thick and the thin (1). Thick filament mainly consists of myosin and the thin filament of actin, TM and TN. According to the sliding theory (2), the elementary process of muscle contraction is a cyclic reaction between thick and thin filaments, releasing the chemical energy of ATP by dephosphorylation.

---

Abbreviation used: HMM, heavy meromyosin; TM, tropomyosin; TN, troponin

The purpose of this study is to investigate by ultraviolet linear dichroism measurements and microspectrophotometry whether or not any change takes place in the structure of F-actin in the thin filament during excitation and contraction of muscle.

## Linear dichroism of F-actin or the thin filament in vivo and in vitro

The linear dichroism of F-actin from rabbit skeletal muscle when oriented by flow was found by Higashi et al. (3); the dichroism spectrum has a negative peak at 260 nm which is due to ADP tightly bound to actin, another negative peak at 295 nm which is due to a specific tryptophan residue and a positive peak at 285 nm which is due to tyrosine and tryptophan residues in actin.

In order to compare F-actin or the thin filament between in vivo and in vitro, the ultraviolet absorption and dichroism of the muscle fiber and F-actin gel were measured by a microspectrophotometer (4). I-band (isotropic band which is made up mainly of thin filaments and Z-line) in a glycerinated single fiber of crab leg muscle was used. This fiber was very convenient for measurements of the dichroism of I-band and A-band (anisotropic band which is made up of both thick filaments and interdigitating thin filaments) respectively, because of its extremely long sarcomere. Different spectra were obtained for light polarized parallel and perpendicular to the fiber axis. The scattering makes a contribution to apparent dichroism. In the range of wavelengths from 320 nm to 400 nm where the sample had no absorption, the scattering was found to be inversely proportional to the approx. third power of the wavelength. Therefore, by extrapolating this relation to shorter wave length, the correction of the scattering was subtracted. The dichroism of I-band after correction of scattering had two negative peaks, as shown in fig. 1. One appeared at about 260 nm, which is due to ADP bound to actin, and the other at about 295 nm, which is probably due to specific tryptophan residue in actin. To examine the contribution of Z-line, the dichroism in this range of wavelengths was measured by using a smaller beam with the center at a Z-line and the negative peak at 295 nm became smaller with the same shape. This suggests that the Z-line itself had no dichroism but only the thin filament had. Thus, the average orientation of the transition moments of ADP and tryptophan in the thin filament is perpendicular rather than parallel to the fiber axis, and these dichroism peaks are a good indicator of the arrangement of actin molecules in the thin filament in vivo.

In addition, a remarkable difference was found in the dichroism around 280 nm between F-actins from rabbit and from crab, suggesting a difference in the structure of actin monomers. On the other hand, the dichroism spectrum of H-band (the non-overlapping region of A-band) was different from that of I-band and gave only a large positive peak at about 280 nm (ref. 4).

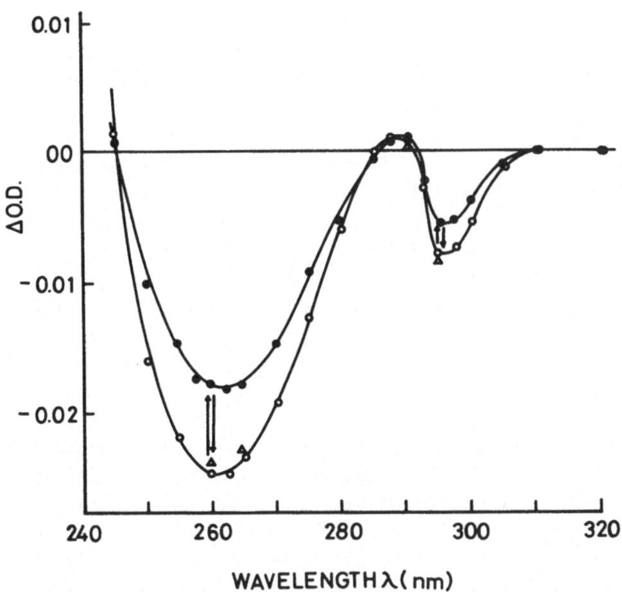


Fig. 1. Ultraviolet dichroism spectra of the I-band of a glycerinated crab leg muscle fiber in Ca-free solution (o, Δ ) and in a Ca containing solution (•). The length of the I-band was about 10 μm and the diameter of the light beam was about 8 μm.

The dichroism spectrum of I-band was very similar to that of an orientated gel of pure F-actin from the crab muscle; that is, it has two negative peaks at 260 nm and at 295 nm (ref. 4).

The following facts also suggest that F-actin in the thin filament has nearly the same structure in vivo and in vitro. The birefringence of I-band from crab muscle measured by the polarizing microscope with an optical rectifier was found to be of the same order as the flow birefringence of F-actin or the thin filament in solution at the same concentration (5-8). The measurements of quasi-elastic light scattering, optical diffraction and analysis of electron-micrographs showed that F-actin or the thin filament have the same flexibility in vivo and in vitro (9-12).

<u>Effect of calcium ions on F-actin in the thin filament in vivo and in vitro</u>

Activation of striated muscle is believed to be caused by calcium ions released from the sarcoplasmic reticulum. In vertebrate muscle, and some other muscles, the receptor for binding Ca ions in TN in the thin filament and the information of binding is transferred via TM to

F-actin which interacts with myosin to develop the contractile force (13). Therefore, it is important to examine if any change takes place in the state of F-actin in the thin filament, depending on whether TN binds Ca ions or not.

The response of the thin filament in vivo to Ca ions was investigated by the measurement of ultraviolet dichroism of I-band (4). Two negative peaks in the dichroism spectrum due to actin both decreased by about 30 percent with the addition of Ca ions, as shown in fig. 1. This is the first clear evidence that a conformational change occurs in F-actin in the thin filament, associated with the binding of Ca ions. The dichroism spectrum of H-band was not changed by Ca ions (4). So, in the crab or rabbit muscle the response to Ca ions was found only in the thin filament.

A similar response of the dichroism to Ca ions was found in the case of the F-actin-TM-TN complex reconstituted in vitro from rabbit skeletal muscle when oriented by flow (4). The negative dichroism of the complex at about 260 nm and 295 nm became smaller in the presence of $Ca^{2+}$ than in its absence, both by about 20%. Two negative peaks of dichroism at 260 nm and 295 nm changed in parallel to each other with the Ca ion concentration. The dichroism change in the crab muscle which has long thin filaments was larger than in the rabbit muscle which has shorter ones.

The magnitude of dichroism of F-actin in the complex in the presence of Ca ions was the same as that of F-actin alone. This means that the complex is more rigid in the absence of Ca ions than in their presence. In other words, Ca ions make the complex more flexible. The study of laser light scattering (14) supported that the complex of F-actin, TM and TN reconstituted in vitro at molar ratio 6 : 1 : 1 is more rigid in the absence of Ca ions than in their presence.

Therefore, it is very probable that the binding of Ca ions changed not only the overall flexibility of the thin filament but also the intramolecular conformation and/or the intermolecular bond structure in F-actin.

## Effect of HMM on F-actin or the thin filament in the absence of ATP

Now, it is important to examine whether there are any clues as to how the required conformational change of F-actin in the thin filament might be brought about during interaction with crossbridges after activation. Here, we have undertaken the in vitro experiment on the reconstituted F-actin-HMM complex or the F-actin-TM-TN-HMM complex by flow dichroic method. In the absence of ATP, the dichroism spectrum of the F-actin-HMM complex from rabbit skeletal muscle has a negative peak at about 260 nm and a large positive peak at about 285 nm, and a small negative peak at 295 nm. Each of these peaks is nearly equal in the position to the dichroism spectrum of pure F-actin previously obtained. However, the magnitude of the dichroic ratio of the negative peak at 260 nm is much smaller in the complex than that of pure F-actin. At the molar ratio of HMM to actin of 1 : 10, the negative dichroism was decreased to about 40% of pure F-actin. A similar decrease by HMM was found in the F-actin-TM-TN complex.

Therefore, a small amount of HMM bound to F-actin induces a remarkable decrease of the dichroism at 260 nm. There was no difference in the dichroism spectrum of the F-actin-HMM complex in the absence of ATP with and without the Ca ion.
Tawada (15) showed that in the absence of ATP, the flow birefringence of F-actin is decreased by the binding of HMM. At the molar ratio of 1 : 10, the decrease of birefringence was about 30% of that of pure F-actin. The decrease was considered to be due to the disorientation of F-actin and also to some intrafilamentous conformational change. The quasielastic light scattering (16) and the electron-microscopic observation (12) suggested the flexibility increase of F-actin by binding of HMM. The electron micrographs of the F-actin-HMM complex showed that at different molar ratios of HMM to F-actin there are different configurations of actin-HMM attachment. Therefore, the dichroism decrease must be also due to the disorientation and the conformational change of the complex.

The dichroism of the F-actin-TM-TN-HMM complex was smaller than that in pure F-actin of F-actin-TM-TN complex. There was a difference in the magnitude of the negative peak at 260 nm in the absence and presence of Ca ions. As previously mentioned, the negative peak at 260 nm of the F-actin-TM-TN complex was smaller in the presence of Ca ions than in their absence. After binding of a small amount of HMM, this difference became more remarkable. This means that the conformational change of F-actin in the thin filament by Ca ions was amplified by the interaction of F-actin with HMM.

## ATP induced conformational changes of the F-actin-HMM complex and of F-actin-TM-TN-HMM complex

When Mg-ATP of 300 $\mu$M was added to the F-actin-HMM complex in the absence of Ca ions, the negative dichroism at 260 nm increased markedly. A few minutes later the negative dichroism became about six times larger than that before the addition. This value was about two times larger than that of pure F-actin. The dichroism gradually decreased, although it did not completely recover to the initial level. In the presence of Ca ions, the addition of Mg-ATP increased the negative dichroism at 260 nm, but the observed increase was smaller than in the absence of Ca ions. The positive dichroism at 280 nm of the complex showed decrease upon the addition of Mg-ATP and then recovered to the initial level. It is likely that the orientation of HMM bound to F-actin changed. The negative dichroism at 295 nm did not change in the presence or absence of Ca.

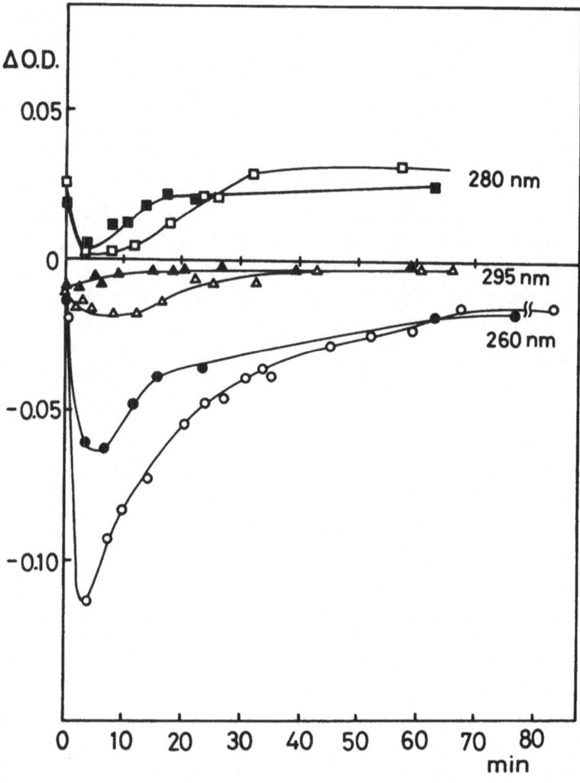


Fig. 2. Time course of the dichroism of the F-actin-TM-TN-HMM complex in Ca-free solution (o, □ and △) and Ca containing solution (●, ■ and ▲ ). The abscissa is the time after addition of 300 μM Mg-ATP. o,● ; 260 nm: □ , ■ ; 280 nm.

In the case of the F-actin-TM-TN-HMM complex (see fig. 2) also, the addition of a small amount of Mg-ATP induced a large increase of the negative dichroism at 260 nm. In the absence of Ca ions the dichroism change was larger than in the presence of Ca ions, and the recovery was slower. In this case there was no difference between the initial level and the final level. The negative dichroism at 295 nm which is due to tryptophan residues changed, only in the case of the complex in the absence of Ca.

If myosin was used instead of HMM, superprecipitation took place after the addition of Mg-ATP. The superprecipitation finished within five minutes after the addition of Mg-ATP of 300 μM. Only in the case of the F-actin-TM-TN-HMM complex in the absence of Ca ions it took about ten minutes. The inverse relation between the rates of

ATPase and the rates of superprecipitation was shown. The viscosity of the F-actin-HMM complex dropped by the addition of Mg-ATP of 300 μm recovered to the initial value within three minutes. Therefore, the first measurement of dichroism after the addition of ATP could be done only after almost all ATP was split. Simple dissociation of HMM from F-actin by ATP must give the same dichroism as pure F-actin. The large negative dichroism suggested the specific interaction of HMM with F-actin. Two possibilities may be suggested as the origin of large negative dichroism at 260 nm. It may come from ADP bound to F-actin. ADP bound to F-actin is oriented with its base plane nearly perpendicular to the long axis of F-actin (17). Hence, the increase of negative dichroism, if it happened, must be due to better orientation of F-actin. Under the same condition, HMM was found to bind ADP. If HMM were bound to F-actin with ADP oriented perpendicular to F-actin axis, these ADP have contributions to negative dichroism at 260 nm. Although the amount of ADP bound to HMM was too small to explain the total increase of the negative dichroism, it is very probable that the large negative dichroism is, at least partially, due to ADP bound to HMM. Finally, the dichroism at 260 nm recovered nearly to the initial level, suggesting the loss of ADP bound to HMM.

The positive dichroism at 280 nm of the complex showed decrease upon the addition of Mg-ATP. The value of the dichroism after decrease was quite different from that of pure F-actin. This dichroism at 280 nm recovered also to the initial level. It is likely that the orientation of HMM bound to F-actin changed. The negative dichroism at 295 nm changed, upon the addition of Mg-ATP, only in the case of the F-actin-TM-TN-HMM complex in the absence of Ca.

In summary, on the basis of the findings described in this paper, it is very probable that conformational changes of F-actin are involved in the mechanism of muscle contraction and its regulation (18-21).

## REFERENCES


1 Huxley H E, J. Biophys. Biochem. Cytology, 3, (1957) 631.
2 Huxley A F, Progress in Biophysics, 7, (1957) 255.
3 Higashi S, Kasai M, Oosawa F and Wada A, J. Mol. Biol., 1, (1963) 421.
4 Yanagida T, Taniguchi M and Oosawa F, J. Mol. Biol., 90, (1974) 509.
5 Oosawa F, Maeda Y, Fujime S, Ishiwata S, Yanagida T and Taniguchi M, J. Mechanochemistry and Cell Motility, 4, (1977) 63.
6 Kasai M, Kawashima H and Oosawa F, J. Polym. Sci., 44, (1960) 51.
7 Tanaka K and Oosawa F, Biochim. Biophys. Acta, 180, (1969) 199.
8 Maeda U, Eur. J. Biochem., in press.
9 Fujime S, J. Phys. Soc. Japan, 29, (1970) 751.
10 Oosawa F, Fujime S, Ishiwata S and Mihashi K, Cold Spring Harbor Symp. Quant. Biol., 37, (1973) 277.

11 Umazume Y and Fujime S, Biophys. J., 15, (1975) 161.
12 Takebayashi T, Morita Y and Oosawa F, Biochim. Biophys. Acta, 492, (1977) 357.
13 Ebashi S and Endo M, in Prog. Biophys. and Mol. Biol., 18, (1968) 123.
14 Ishiwata S and Fujime S, J. Mol. Biol., 68, (1972) 511.
15 Tawada K, Biochim. Biophys. Acta, 172, (1969) 311.
16 Fujime S and Ishiwata S, J. Mol. Biol., 62, (1971) 251.
17 Miki M and Mihashi K, Biophys. Chem., 6, (1977) 101.
18 Oosawa F, Asakura S and Ooi T, Prog. Theor. Phys. Sup., 17 (1961) 14.
19 Asakura S, Taniguchi M and Oosawa F, J. Mol. Biol., 7, (1963) 55.
20 Ishiwata S and Oosawa F, J. Mechanochem. Cell Motility, 3, (1974) 9.
21 Taniguchi M, Biochim. Biophys. Acta, 427, (1976) 126.

# ELECTRIC BIREFRINGENCE OF CARTILAGE PROTEOGLYCAN AND ITS ASSOCIATION WITH HYALURONIC ACID

E.Y. Hawkins*, M. Isles*, A.R. Foweraker*, B.R. Jennings*, T. Hardingham** and H. Muir**

*Electro-Optics Group, Physics Department, Brunel University, Uxbridge, Middlesex, U.K.

**Kennedy Institute of Rheumatology, Bute Gardens, Hammersmith, London, U.K.

Electric birefringence measurements are reported for dilute aqueous solutions of proteoglycans. Complicated birefringence transients obtained over a range of applied electric fields of up to 6 kV $cm^{-1}$ amplitude are shown to be consistent with the presence of three discrete relaxation phenomena, involving different rate processes. Analysis of the relaxation times suggests that self-association of the proteoglycan molecules occurs in the form of dimers which appear to be of head-to-head form. Single proteoglycan molecules and dimers co-exist in solution and contribute independent components to the birefringence at relatively low field strengths. With high applied fields, a third birefringence component can be observed which is attributed to motion of the chondroitin sulphate side chains in the field. In the presence of hyaluronic acid the proteoglycan molecules form complexes which are consistent with the model in which proteoglycans attach radially to extended hyaluronic acid chains. Proteoglycan samples for which the cystine residues have been reduced and alkylated show no evidence of this association with hyaluronic acid, indicating the importance of the conformation of the proteoglycan molecule in its ability to bind to hyaluronic acid.

## INTRODUCTION

Electric birefringence offers a rapid and sensitive method of elucidating macromolecular dimensions, associations and conformational changes. The method is being increasingly applied in the study of molecular biology. In the present work this technique has been used to observe the structure and aggregation properties of cartilage proteoglycans.

These molecules, in the form of large aggregates, bind to the collagen matrix of cartilage, giving the structure compressional strength by impeding the free flow of water through the tissue. A detailed knowledge of the structure, aggregation and binding of these molecules is essential in understanding how the cartilage functions and what happens when it degenerates.

Cartilage proteoglycans consist of a central protein core, about 350 nm long, to which two types of carbohydrate side-chain are attached radially (fig. 1a). The predominant side-chains are of chondroitin-4-sulphate, each of which is about 40 nm long. There are some 100 such chains to each proteoglycan molecule. A smaller number of shorter keratan sulphate chains also occur (1). The chondroitin sulphate chains carry a large number of negatively charged carboxyl and sulphate groups causing the molecules to behave as expanded polyelectrolytes in aqueous solution. Proteoglycans are found in cartilage in the form of large molecular complexes involving hyaluronic acid. Each proteoglycan molecule has a globular protein head which contains a specific binding site with a high affinity for hyaluronic acid. From measurements using other physico-chemical methods (2,3), the complex is thought to consist of an extended hyaluronic acid chain to which the proteoglycans are regularly attached in a radial manner (fig. 1b). The minimum length of hyaluronate occupied by each proteoglycan molecule has been estimated to be about 24 nm (4). Thus, as the hyaluronic acid chains can have an extended length of the order of a few microns, a large number of proteoglycan molecules can bind to form very large aggregates.

## THE STUDY

The electric birefringence apparatus was of conventional design. A circulating cell system was used to enable incremental additions of hyaluronic acid to be made to the proteoglycan solutions without disturbing the sample cell in the optical assemblage.

The study is presented below in two parts. The first concerns solutions of proteoglycan molecules alone. The second describes the proteoglycan-to-hyaluronic acid association.

### Proteoglycan Solutions

All measurements were made on samples of pig laryngeal proteoglycan (5) in deionised water at a pH of approximately 5.5 and a concentration of 0.7 mg $ml^{-1}$. Pulsed electric fields of up to 6 ms duration and 6 kV $cm^{-1}$ were applied to these solutions and the resulting birefringence detected in the linear mode (6). Changes in the birefringence of the

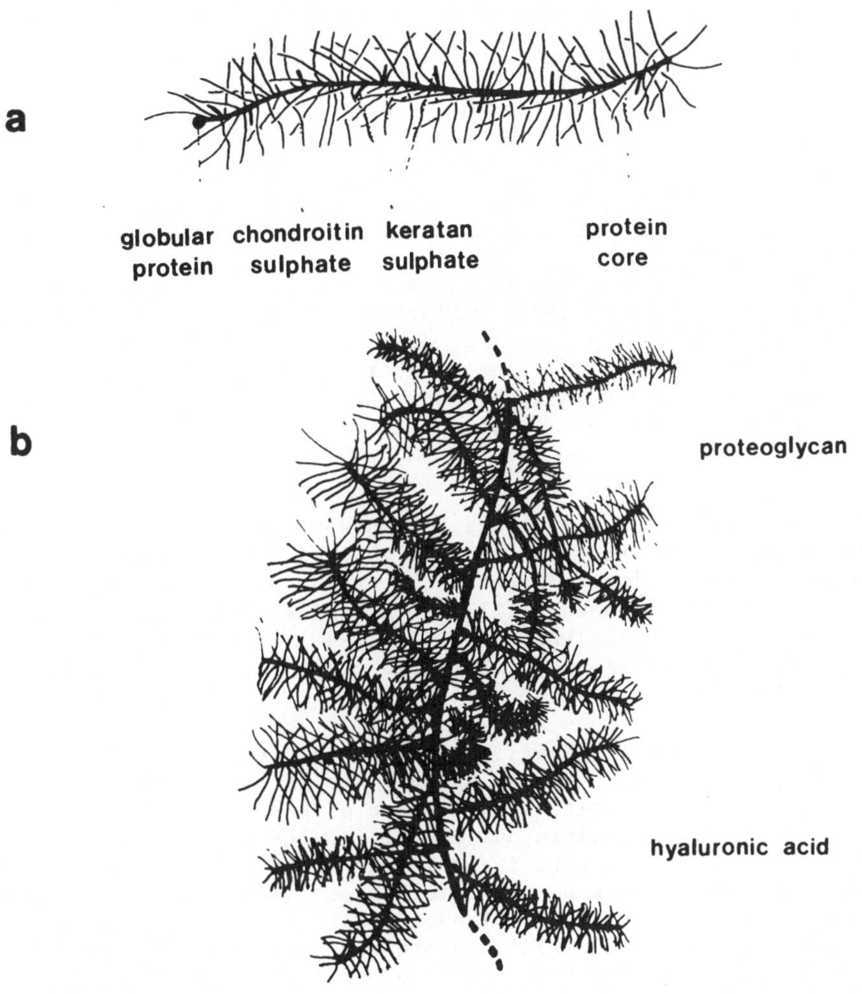


Fig. 1. Schematic diagrams of the structure of a proteoglycan molecule (a) and its aggregate with hyaluronic acid (b).

sample with increasing field strength were monitored. A sequence of responses of increasing complexity was obtained. An example is shown in fig. 2. Whereas at low field amplitude, regular, single component birefringence transients were obtained, the complete sequence showed that at higher fields more than one relaxation process was evidenced. It was possible, by the method described elsewhere (7) to deconvolute these responses into three distinct birefringent components each of which was characterised by a discrete relaxation time. These have been designated as component I, whose effect is seen alone in frame a of fig. 2, with a characteristic decay time of about 3 ms. It

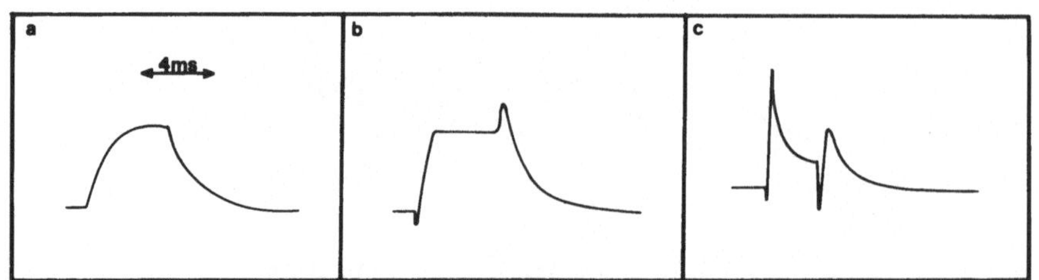


Fig. 2. Transient electric birefringence traces for a proteoglycan solution with increasing electric field amplitudes. Frames a to c are for pulsed d.c. fields of 200, 680 and 2400 V $cm^{-1}$ respectively. Time runs from left to right, with the same time scale in each case. Frames b and c have the same ordinate magnitude, whilst a is half this value.

exhibited a positive birefringence. Component II had a shorter decay time (between 400 $\mu$s and 800 $\mu$s) and gave rise to a negative birefringence. Frame b of fig. 2 shows the presence of both components I and II. The third component, III, which dominated at very high fields, had a short decay time of the order of a few $\mu$s, and manifested a large positive birefringence. All three components are in evidence in fig. 2c.

The fast component III was easily assigned to the flexing of the chondroitin sulphate side chains as very similar responses in relaxation time, sign and field amplitude dependence of the birefringence had been recorded for free chondroitin sulphate molecules in an independent study. The decay times of components I and II were used to obtain size estimates for the entities responsible for them. Assuming molecular shapes to be approximated by prolate ellipsoids with semi-minor axes of 40 nm (the length of the chondroitin sulphate side chains), Perrin's (8) equations gave an estimate of the semi-major axis dimensions. The sizes obtained led to the interpretation of component II as arising from single proteoglycan molecules and component I from particles of twice the length of these molecules. End-to-end proteoglycan dimers are a probable interpretation.

Two further experiments were conducted to demonstrate the presence of such aggregates. The first involved the addition of dithiothreitol, which is known to denature the globular protein of the proteoglycan molecules (9). In the presence of this additive, the birefringence component I disappeared. The second experiment involved the addition of oligomers of hyaluronic acid which are known to have a high affinity with this globular head of the proteoglycan molecules. Oligomers were used because of their negligible size compared with the proteoglycan molecules. Thus, they should not affect noticeably the observed relaxation time of the molecules. Here again, the presence of the hyaluronic acid oligomers resulted in the disappearance of component I and an increase in the contribution from component II. It appears that component I is probably due to a proteoglycan dimer formed by head-to-head associations. Dimerisation has been reported in certain other studies of this system but under different conditions of pH and ionic strength (10,11).

The aforementioned experiments involving short chain oligomers of hyaluronic acid should not be confused with the following studies using high molecular weight material.

Proteoglycan - Hyaluronic Acid Association

In this study, the electric birefringence apparatus was operated in the quadratic detection mode (6). Fields of 360 V $cm^{-1}$, corresponding to a value in the $E^2$ region of component I, were applied. High molecular weight, hyaluronic acid, with a relative molecular mass of 6.7 x $10^5$ daltons, corresponding to an extended length of approximately 1.6 $\mu$m, was increasingly added to a proteoglycan solution and the birefringence transients recorded. The amplitude of these decreased in a continuous manner with increasing hyaluronic acid concentration (fig. 3) although the final decay rate remained constant. Full details have been given elsewhere (12). These experiments were performed before the study reported in the first sub-section. The field strength was kept low so as to obtain regular transients which were easier to analyse. It would thus appear that the birefringence changes reported in ref.12 and reproduced in fig. 3, primarily (i.e. component I in the previous sub-section) monitor the behaviour of the dimer proteoglycan molecules.

When much longer field pulses (of about 160 ms duration) were used to study the proteoglycan/hyaluronic acid solutions, and for hyaluronic acid concentrations in excess of 3 $\mu$g $ml^{-1}$, a component of very much longer decay time (about 650 ms) became apparent. The birefringence amplitude of this component increased as the hyaluronic acid concentration increased, whilst that of component I diminished and ultimately disappeared at a total hyaluronic acid content of 4.5 $\mu$g $ml^{-1}$. This long decay process is attributed to the large proteoglycan-hyaluronic acid complex (fig. 1b) which became more abundant as the available hyaluronic acid content was increased. This simultaneously caused a decrease to zero of free proteoglycan and a corresponding decline in the birefringence amplitude of component I. Again, taking the shape of the molecular complex to be approximated by a prolate ellipsoid, but this time with a semi-minor axis equal to the length of the proteoglycan molecule, Perrin's (8) equations gave a major axis of the order of 3 $\mu$m for the extended complex.

If one takes the zero point of the birefringence amplitude of component I as indicating that concentration of hyaluronic acid (4.5 $\mu$g $ml^{-1}$ from fig. 3) for which no free proteoglycan occurs, an estimate can be made of the density of packing of the proteoglycan molecules along the hyaluronic acid chains within the molecular

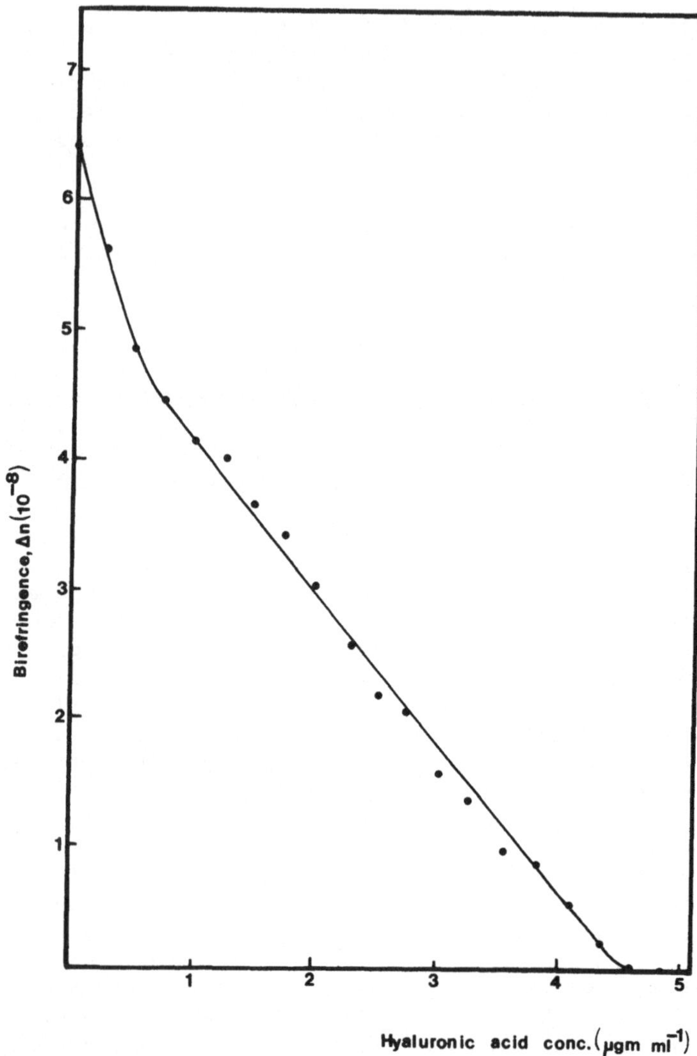


Fig. 3. Effect of hyaluronic acid concentration on the birefringence of a proteoglycan solution.

complex. From the relative concentrations of the two species, a value of 25 nm is obtained for the distance between proteoglycan molecules along the hyaluronate molecule. This is in concord with previous estimates of 24 nm (4), 24 to 28 nm (13) and 20 to 30 nm (14).

It has been mentioned in the previous sub-section that dithiothreitol denatures the globular protein of proteoglycan molecules.

The denatured state can be stabilised by alkylating with iodacetimide (9). When hyaluronic acid was added to such reduced and alkylated proteoglycan solutions, and the birefringence transients recorded using the linear detection mode, two noteworthy factors were evidenced. Firstly, the traces were not significantly influenced by the addition of the hyaluronic acid, and nothing corresponding to the proteoglycan/ hyaluronic acid was observed. The protein head appears to be the active binding centre as predicted elsewhere (9). Secondly, the birefringence trace was regular, but of negative sign and of faster decay than that for the regular proteoglycan molecule. It was, in fact, very similar to that of component II of the study reported in the first sub-section of this paper. The reduced and alkylated material does not appear to contain any proteoglycan dimer. This change in sign of the birefringence was not apparent in the studies reported in ref. 12 owing to the inability of the quadratic birefringence display system to differentiate the sign of the birefringence. An important advantage of the linear detection principle is illustrated hereby.

## CONCLUSIONS

Electric birefringence transients on dilute solutions of cartilage proteoglycans at low ionic strength and pH 5.5 indicate the presence of three electrically excited relaxing species. These are interpreted as (a) the fast, flexing motion of the chondroitin sulphate side chains, (b) the rotation of the proteoglycan molecules as single entities and (c) the rotation and deformation of aggregates which are probably head-to-head dimers.

Upon addition of hyaluronic acid, a very much slower relaxing species is observed in increasing amplitude with increasing hyaluronic acid addition. The characteristics of this species are consistent with those of the molecular complex in which the proteoglycan molecules attach radially to extended hyaluronic acid chains. The globular protein of the proteoglycan molecules appears to contain the active binding site and these bind to the hyaluronic acid at 25 nm intervals.

## ACKNOWLEDGEMENTS

The Arthritis and Rheumatism Council is acknowledged for general support to T.E.H. and H.M. and for a grant to B.R.J. under which E.Y.H. was funded. The Science Research Council is thanked for a studentship to M.I. and Messrs. Ciba-Geigy Ltd. for a grant under which A.R.F. held a fellowship.

## REFERENCES


1 Muir H and Hardingham T E, in MTP International Review of Science Series I, Vol.5 (1975).

2 Hardingham T E and Muir H, Biochem. Biophys. Acta., 279, (1972) 401.

3 Atkins E D T, Hardingham T E, Isaac D H and Muir H, Biochem. J. 141, (1974) 919.

4 Hardingham T E and Muir H, Biochem. J., 135, (1973) 905.

5 Hardingham T E and Muir H, Biochem. J., 139, (1974) 565.

6 Riddiford C L and Jerrard H G, J. Phys. D: Appl. Phys., 3, (1970) 1314.

7 Hawkins E Y, Foweraker A R and Jennings B R, Polymer, 19, (1978) 1233.

8 Perrin F J, J. Phys. Radium (Paris), 5, (1934) 497.

9 Hardingham T E, Ewins R J F and Muir H, Biochem. J., 157, (1976) 127.

10 Wells P J and Serafini-Fracassini A, Nature, New Biology, 243, (1975) 266.

11 Sheehan J K, Nieduszynski I A, Phelps C F, Muir H and Hardingham T E, Biochem. J., 171, (1978) 109.

12 Isles M, Foweraker A R, Jennings B R, Hardingham T and Muir H, Biochem. J., 173, (1978) 237.

13 Hascall V C and Heinegard D, J. Biol. Chem., 249, (1974) 4242.

14 Rosenberg L, Hellman W and Kleinschmidt A K, J. Biol. Chem., 250, (1975) 1877.

# AN ELECTRICAL BIREFRINGENCE STUDY OF THE CONTRIBUTION OF PERMANENT AND INDUCED DIPOLE MOMENTS TO ACID-SOLUBLE COLLAGEN POLARIZATION

J. C. Bernengo*, B. Roux and D. Herbage**

*Laboratoire de Biophysique, Université de Nice
Parc Valrose, 06034 Nice-Cedex, France

**Laboratoire de Biochimie et de Chimie Macromoléculaire
Université Lyon I, 43 Bd du 11 Novembre 1918
69621 Villeurbanne, France

A method is described using decay, reverse and sine wave birefringence experiments to measure the electrical parameters of acid soluble collagen molecules in solutions containing both monomeric and aggregated forms. Mathematical analysis of decay and reverse signals is performed according to classical limiting low fields equations, and values of permanent and induced dipole moments can be obtained for both molecular species. A confirmation of these results and an evaluation of molecular flexibilities is given by comparison with sine wave birefringence measurements at very low applied fields.
First results show that collagen molecules present a permanent dipole moment around 10 000 D which is not influenced by side chains or telopeptides, while the induced polarizability and aggregation mechanism are highly dependent on the external charges along the molecule, as observed on enzyme treated or chemically substituted collagen.

## INTRODUCTION

The hydrodynamical properties of acid soluble collagen molecules have been studied for many years to determine the molecular shape, size and aggregation properties. For this kind of experiment, the electrical birefringence has the advantage over pure hydrodynamical techniques in that it gives information about molecular electrical parameters. Nevertheless, the number of published works reporting such electrical results is not large: Yoshioka and O'Konski (1) in

1966 and later Kahn and Witnauer (2) in 1971, determined a permanent and an induced dipole moment through mathematical calculations involving extrapolation at low fields. In 1974, we measured directly the permanent dipole moment of acid soluble collagen in $10^{-1}$M acetic acid and found a value well in accordance with previous results (3). But the reverse pulse technique showed a non negligible induced moment with some preparations. A further study carried out to charaterize aggregates in native and enzyme treated collagen solutions at various pH (4) confirmed the existence of an induced dipole moment, strongly dependent upon the experimental conditions, though this was not the prime purpose of the work.

Considering the importance of the polarizability of collagen molecules in the early steps of fibrogenesis (5), it was decided to measure the relative contribution of permanent and induced dipole moment to the molecular polarization in aggregating conditions, i.e. at weakly acid pH. Under these conditions, the reverse pulse response is not straightforward, since it corresponds to the reorientation of both monomer and aggregates. An experimental procedure combining the analysis of decay, reverse and sine wave birefringence responses has been set up and used to study native, enzyme treated and chemically substituted acid soluble collagen molecules.

## THEORY

The equations used for the determination of molecular parameters are rather classical, since two basic considerations are taken into account:

- the experiments are carried out at limiting low fields for all the molecular entities involved in the birefringence process (for further details, refer to experimental section).

- the solutions are considered to contain "monomeric" tropocollagen (rigid rod with a length of 300 nm and a diameter of 1.4 nm, with same optical, electrical and hydrodynamical axis) and aggregates of well defined characteristics, as described in a previous work (4).

In the following formulas, the index m refers to the monomer and a to the aggregates,

$\mu$ is the permanent dipole along the main axis,
$\Delta\alpha_E$ the difference between longitudinal and transverse polarizability,
D the rotational diffusion constant around a transverse axis and
$\tau$ the corresponding relaxation time ( $\tau = 1/6$ D).
r the ratio $\mu^2 / \Delta\alpha_E.kT$ which indicates the relative contribution of permanent and induced moment to the total polarizability.

As Kerr law is followed for both monomer and aggregates, the steady state birefringence $\Delta n$ is proportional to the square of the electric field E:

$$\Delta n = \Delta n_m + \Delta n_a = (K_m + K_a)\, E^2$$

A specific Kerr constant $K_{sp}$ is normally defined as:

$K_{sp} = \dfrac{K}{C_v . n}$ where $C_v$ is the volume fraction of the solute and n the refractive index of the solution.

$K_{sp}$ is related to the molecular parameters by the classical first order equation (6):

$$K_{sp} = \frac{2\pi\, \Delta\alpha_o}{15\, n^2 k^2 T^2} (\mu^2 + \Delta\alpha_E\, kT) \quad (1)$$

in which $\Delta\alpha_o$ is the optical anisotropy factor.

The normalized birefringence decay is given by (7):

$$\Delta D(t) = \alpha_m\, e^{-t/\tau_m} + \alpha_a\, e^{-t/\tau_a} \quad (2)$$

with $\alpha_m = \Delta n_m / \Delta n$ and $\alpha_a = \Delta n_a / \Delta n$

For the birefringence in a rapidly reversed field we get (8):

$$\Delta R(t) = 1 + C_m (e^{-t/\tau_m} - e^{-t/3\tau_m}) \quad (3)$$

$$+ C_a (e^{-t/\tau_a} - e^{-t/3\tau_a})$$

with $C_m = \dfrac{3 r_m}{1 + r_m} \cdot \alpha_m$ and $C_a = \dfrac{3 r_a}{1 + r_a} \cdot \alpha_a$

When the applied field is a sine wave alternate field $E = E_o e^{j\omega t}$, the birefringence response (at a frequency $2\omega$) is given by the Thurston and Bowling equations (9):

$$\Delta n' = \frac{\Delta n_o}{1+r} \left\{ \frac{r(1 - 6\omega^2 \tau^2)}{(1+9\omega^2 \tau^2)(1+4\omega^2 \tau^2)} + \frac{1}{1+4\omega^2 \tau^2} \right\} \quad (4)$$

$$\Delta n'' = \frac{\Delta n_o}{1+r}\left\{\frac{5\omega\tau r}{(1+9\omega^2\tau^2)(1+4\omega^2\tau^2)} + \frac{2\omega\tau}{(1+4\omega^2\tau^2)}\right\} \quad (4)$$

$\Delta n_o$ being the peak value of the sine birefringence at low frequency below any relaxation mechanism.

In fact, only the modulus $|\Delta n| = (\Delta n'^2 + \Delta n''^2)^{\frac{1}{2}}$ of the sine birefringence can be measured with precision by a lock-in detector used in our experiments. In the case of two relaxing molecular species, it is the sum of $|\Delta n_m|$ related to monomer and $|\Delta n_a|$ for the aggregates.

## EXPERIMENTAL

### Material

Acid soluble collagen was extracted from calf skin by the method of Piez et al. (10), and pepsin treated collagen prepared according to Drake et al. (11). After extraction and enzyme treatment, collagen was freeze-dried and stored frozen until used. Solutions at a concentration around 150 mg/l were obtained by dialysis at 4°C against the final solvent. Immediately before use, dialyzed solutions were centrifuged (80,000 g for 2h) to eliminate precipitate, and the final concentration measured by optical rotation.

(a) Chemically substituted collagen. Three types of substitution have been carried out:

- Methylation of aspartic and glutamic acid carboxyle groups

$$-\underset{\underset{\displaystyle O}{\|}}{C} - OH \rightarrow -\underset{\underset{\displaystyle O}{\|}}{C} - O - CH_3$$

(Through treatment during 24 hours in methanol + 0.2M HCl at 20°C).

- Acetylation of Lysines

$$-NH_3^+ \rightarrow -NH-\underset{\underset{\displaystyle O}{\|}}{C}-CH_3$$

(24 hours at 10°C in a sodium acetate + 10% acetic anhydride solution maintained at pH 8 (12) according to Green et al.).

- Desamidation of asparagine and glutamine

$$-\underset{\underset{O}{\|}}{C}-NH_2 \rightarrow -\underset{\underset{O}{\|}}{C}-OH$$

(72 hours at 20°C in a sodium sulfate saturated solution with 5% NaOH added)

(b) Electrical birefringence measurements. The major parts of the birefringence apparatus have already been described (13,14,15) and only the optical features required for the present study have to be recalled here.

A laboratory made, linear wide band amplifier has been designed to apply voltages of any shape up to ± 300 V to conducting solutions (transition time = 1 μs, band pass for sine waves around 150 kHz). A He - Ne laser used as light source combined with a silicon photodetector followed by an integrated current amplifier (linear or logarithmic) provide a high sensitivity and signal over noise ratio: pulsed birefringence as low as $10^{-10}$ and sine wave birefringence down to $10^{-12}$ could be detected (the latter using a lock-in detector PAR model 186).

(c) Signal processing. Birefringence signals are stored in a transient recorder (DATALAB model 901) and averaged through a microprocessor interface. Then, the experimental data are sent to a computer (CII Iris 80) via a transmission line. A calculation programme has been set up in our laboratory to evaluate the decay, reverse, and sine wave responses parameters, for up to 3 relaxation mechanisms. It is based on several iterative fitting methods (subtractive, least squares, steepest slope) and does not require any parameter value to be entered initially.

(d) Experimental procedure. For a particular solution, the measurements sequence is as follows: Firstly, it has to be verified that the applied field is in the Kerr law validity domain, even for the biggest aggregates. In other words, the normalized birefringence decay has to be completely field independent in the selected field range. For all the solutions studied in this work, the experimental Kerr law validity domain extended from a few V/cm (sine wave measurements) up to 300 V/cm. Secondly, decay curves are recorded using the logarithmic detector, and analyzed to get $\alpha$ and $\tau$ values for the monomer and the aggregates. For instance, fig. 1a shows a typical decay presented by a 150 mg/1 acetylated collagen solution in 0.1 M acetic acid. Thirdly, from reversing pulse signals obtained under the same experimental conditions, curve fitting according to eq. 3 gives the closest values of C and $\tau$ for both molecular species. For this last step, great care has to be taken to apply strictly symmetrical reversed pulses, since the mathematical analysis could be totally wrong on the end of the transient signal in the presence of a significant dissymmetry. In

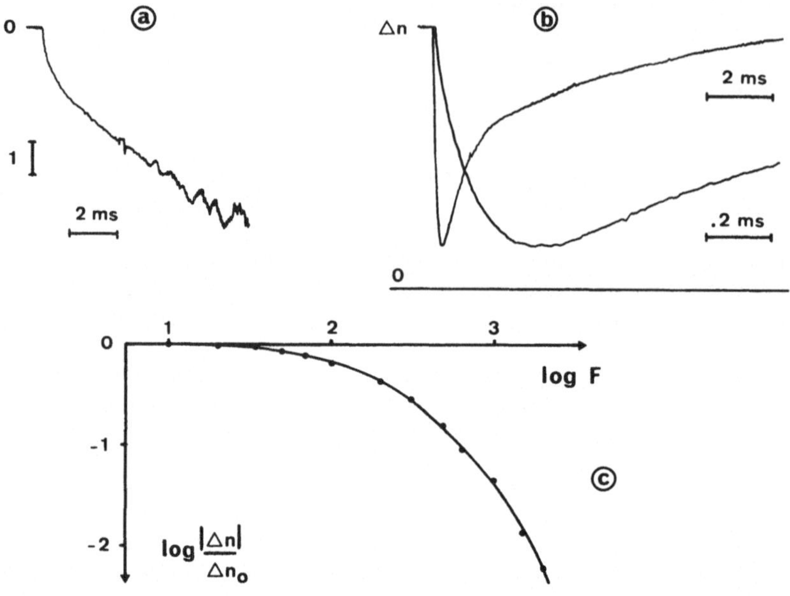


Fig. 1. Birefringence signals, obtained on acetylated acid soluble collagen dissolved at a concentration of 150 mg/l.
Frame a: Decay; applied field 300 V/cm, pulse duration 10 ms. Vertical axis is the natural logarithm of the normalized birefringence.
Frame b: Response to a reversed pulse; applied field ± 300 V, pulse duration 10 ms for each polarity.
Frame c: Dispersion curve at an applied field of 10 V/cm. Dots are experimental data, solid line is mathematically built from table 1 values according to eq. 4.

Table 1

| | Decay | | Reverse | | | Sine waves | | |
|---|---|---|---|---|---|---|---|---|
| | $-\alpha$ | $\tau$ ms | C | $\tau$ | r | $\alpha$ | r | $\tau$ ms |
| monomer | 0.8 | 0.32 | 2.7 | 0.33 | -10 | 0.8 | -12 | 0.3 |
| aggregates | 0.2 | 2.0 | 0.4 | 2.3 | 1.4 | 0.2 | 1.2 | 2.0 |

Sine wave values correspond to the best fit of dispersion curve of fig. 1 c according to eq. 4.

fig. 1b, a reverse pulse response is represented for the same solution as that of fig. 1a with the same applied field.

When the relaxation times obtained from both single pulse and reverse pulse experiments are similar (taking into account the measurement errors), values of $\alpha$ (eq. 2) and C (eq. 3) are combined to get r for monomer and aggregates, as $r = \frac{\alpha C}{3 - \alpha C}$. It has to be pointed out here that this procedure is applicable for the aggregates only if they contribute significantly to the total birefringence signal. A contribution of less than 1% ($\alpha_a = 0.01$) is easily detectable on the decay with the logarithmic amplifier, but cannot be separated on the less sensitive reverse pulse signal.

A final confirmation is carried out by feeding the $\alpha$, r and $\tau$ values obtained through this procedure into the corresponding program to compute a sine wave dispersion curve according to eq. 4. The experimental data measured at very low fields (3 to 10 V/cm) should fit correctly on this curve, thereby showing that the behaviour of the solution is not field dependent. As an example, fig 1c presents such a verification, and the results of the three analyses according to the described procedure are summarized in table 1.

Referring to static birefringence measurements, ($\Delta n_m$ and $\Delta n_a$ on monomer and aggregates), eq. 1 and a previously determined r value enable us to calculate $\mu$ and $\Delta\alpha_E$ for both molecular species. Unfortunately, these calculations require the determination of the optical anisotropy factor $\Delta\alpha_o$ which could be achieved by saturation or by laser induced birefringence measurements (16). The value of reference 1 has been considered for calculations on both monomer and aggregates, though it is not obviously valid for the aggregated form of collagen.

## RESULTS

### Non Treated Acid Soluble Collagen

As we have already found (3), normal skin extracted acid soluble collagen in 0.1 M acetic acid (pH 2.9) presents a relaxation time of 170 ± 10 $\mu$S and does not form significant amount of aggregates (less than 1% of the total birefringence). The same ratio r of 2.2 ± 0.2 is measured on several different preparations, not depending on the applied electric field, up to 500 V/cm and on collagen concentrations up to 200 mg/l. The specific Kerr constant $K_{sp}$ is $1.8 \pm 0.2 \times 10^{-5}$ e.s.u., and according to eq. 1 we get a permanent dipole moment around 10,000 Debyes and a polarizability of $\Delta\alpha_E$ of $10^{-15}$ cm$^3$.

When the pH of the solution is increased to 4, aggregates of relaxation time 2.6 ms appear, corresponding to a length of 2.5 times the monomer length (4). The reverse pulse birefringence signals are modified, and the calculations on decay and reverse responses, com-

Table 2. Electrical parameters obtained on native acid soluble collagen at various pH in acetic acid solutions. Applied field 300 V/cm, collagen concentration 150 mg/1.

| pH | | $\Delta n \times 10^9$ | $\alpha$ | $\tau$(ms) | $K_{sp} \times 10^5$(esu) | r | $\mu$(D) | $\Delta\alpha_E$(cm$^3$) |
|---|---|---|---|---|---|---|---|---|
| 2.9 | monomer | 2.7 | ≃ 1 | . 16 | 1.8 | - 2.2 | 12 000 | $-1.8.10^{-15}$ |
| | aggregates | <.03 | <.01 | 2.5 | non measurable | - | - | - |
| 3.4 | monomer | 4.2 | .97 | .16 | 2.8 | - 3.5 | 13 000 | $-1.15.10^{-15}$ |
| | aggregates | .1 | .03 | 2.7 | | | | |
| 3.7 | monomer | 4.8 | .95 | .17 | 3.2 | - 5.6 | 13 000 | $-.7.10^{-16}$ |
| | aggregates | .25 | .05 | 2.5 | 2.5 | ∿ 0 | ∿ 0 | $2.5.10^{-15}$ |
| 3.9 | monomer | 5.1 | .92 | .18 | 3.4 | - 8 | 13 000 | $-.5.10^{-16}$ |
| | aggregates | .4 | .08 | 2.8 | 3.0 | ∿ 0 | ∿ 0 | $3.\ 10^{-15}$ |
| 4.2 | monomer | 5.25 | .85 | .26 | 3.5 | - 12 | 12 900 | $-.33.10^{-16}$ |
| | aggregates | .8 | .15 | 3. | 3.5 | ∿ 0 | ∿ 0 | $3.6.10^{-15}$ |

Table 3. Electrical parameters of normal, enzyme treated and substituted collagen solubilized in 0.1 M acetic acid. Applied field 300 V/cm, concentration 150 mg/l.

| Collagen | | $\Delta n \times 10^9$ | $\alpha$ | $\tau$(ms) | $K_{sp} \times 10^5$(esu) | r | $\mu$(D) | $\Delta\alpha \times 10^{15}$ ($cm^3$) |
|---|---|---|---|---|---|---|---|---|
| normal | | 2.7 | 1 | .16 | 1.8 | - 2.2 | 12 000 | - 1.8 |
| pepsin treated | | 5.0 | 1 | .14 | 3.3 | <-10 | 12 000 | <-.2 |
| Methylated | monomer | 3.4 | .75 | .28 | 3.0 | - 2.4 | 14 000 | - 2.3 |
| | aggregates | .85 | .25 | 3 | 3.6 | ∿ 0 | ∿ 0 | 3.7 |
| Acetylated | monomer | 4.6 | .8 | .32 | 3.8 | <-10 | 13 000 | <-.2 |
| | aggregates | .9 | .2 | 2 | 4.6 | + 1.2 | 10 500 | + 2.2 |
| Deamidated | monomer | 4.4 | ∿ 1 | .18 | 3.0 | - 6 | 12 500 | - .6 |
| | aggregates | - | <.01 | non | measurable | ...... | ...... | ...... |

bined with static birefringence measurements give the results presented in table 2. In fig. 2 the ratio $r_m$ for the monomer is plotted versus pH and compared to the aspartic and glutamic acid pK range. The electrical parameters are presented at a concentration of 150 mg/l, but have been found to be independent of collagen concentration in the range 20 to 150 mg/l for both monomer and aggregates.

## Enzyme Treated and Chemically Substituted Collagen

Solutions in 0.1 M acetic acid at the same concentrations as for native collagen have been studied; table 3 gives the results of birefringence experiments and the deduced parameters for pepsin-treated, methylated, acetylated and desaminated acid soluble collagen. These results will be discussed later, but it seems useful to look particularly at acetylated collagen values in order to confirm the advantages of the experimental procedure adopted here. Fig. 1b displays transient signals obtained on that type of collagen, from which a straightforward interpretation should have led to a mean relaxation time of 250 $\mu$s and an r value of about + 2.5. These results would have been totally wrong, since the monomer has in fact lost its induced polarizability ($r_m > 10$) and the aggregates ratio r is around + 1.2.

## DISCUSSION

First of all, the negative value of the monomer induced polarizability has to be pointed out: it indicates an excess of polarizability along the transverse axis of the molecule, which was not observed by previous authors (1,2) and has to be explained. As for any polyelectrolyte, the external charges on the molecule make an important contribution to the polarizability; at low pH (such as 0.1 M acetic acid solutions) aspartic and glutamic acid carboxyles are not charged. They become charged above pH 3.5 (see fig. 2). On the other hand, amino groups of lysines are protonized at any acid pH. The important decrease of induced polarizability with acetic acid concentration (table 2) shows the contribution of the negatively charged external carboxyle groups to the polarization. Contrarily, the permanent dipole moment of the molecule does not significantly change in this pH range.

The results obtained on substituted collagen exhibit the complexity of the polarization mechanism: for instance, let us consider deamidation and methylation. Neither changes the charge of the molecule at pH 2.9, since an amide is replaced by a non ionized carboxyle (deamidation) or the same carboxyle is methylated (methylation). Nevertheless, both substitutions give rise to different properties. Deamidation strongly reduces the induced polarization, while methylation does not change it significantly. The aggregation process is very different. Deamidation produces a negligible quantity of aggregates, while methylation is at the origin of a big amount of non polar aggregates.

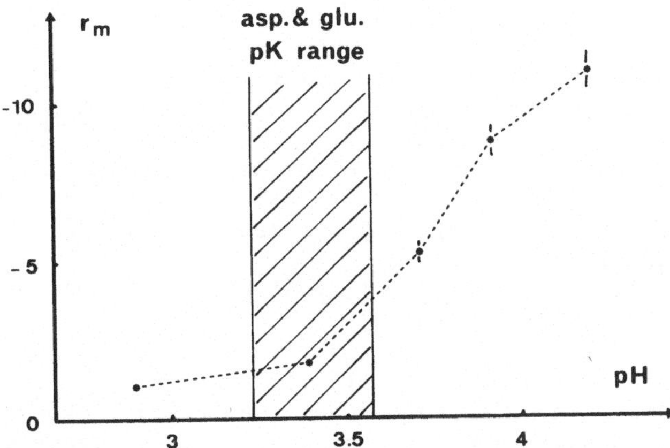


Fig. 2. The ratio $r_m$ for the monomer plotted against pH (results of table 2) compared with aspartic and glutamic acid pK range. (Applied field 300 V/cm, concentration 150 mg/l).

Acetylation is the only substitution which changes the net charge of the molecules, replacing protonated amino groups of lysines by non charged acetyl groups. The monomeric collagen has no more induced moment, and the aggregates which are formed in these conditions have both induced and permanent dipoles. Telopeptide ends of the collagen molecule seem also to contribute to polarization phenomena, since when they are removed by pepsin treatment, the induced polarization disappears.

The most surprising result of this work is the constant value of the monomer permanent dipole moment (around 10,000 D) no matter what happens on the side chains or the non-helicoidal ends of the molecule. It seems to represent an intrinsic property of the triple helix, non affected by the external charges along the rod.

These conclusions have to be confirmed and extended by other measurements on native and substituted collagen, either acid-soluble or neutrosoluble, in various dissolving media, to investigate the origin of the induced and permanent polarization. The experimental procedure using decay, reverse and sine wave experiments, which has been found adequate to measure electrical and hydrodynamical parameters on both monomer and aggregates, will be used for this purpose, in conjunction with dielectric and flow birefringence measurements.


## ACKNOWLEDGEMENTS

This investigation is supported by CNRS (R.C.P. 08-533) and DGRST (contract 78-7-0339).

## REFERENCES


1 Yoshioka K and O'Konski C T, Biopolymers, 4 (1966), 499-507.
2 Khan L D and Witnauer L P, Biochim. Biophys. Acta, 243 (1971), 388-397.
3 Bernengo J C, Roux B and Herbage D, Biopolymers, 13 (1974), 641-647.
4 Bernengo J C, Herbage D, Marion C and Roux B, Biochim. Biophys. Acta, 532 (1978), 305-314.
5 Miller A, in 'Biochemistry of Collagen' (Plenum Press ed.) (1976), p. 85-121.
6 Benoit H, Ann. Phys. (Paris), 6 (1951), 561-609.
7 Benoit H, J. Chim. Phys., 49 (1951), 517-522.
8 Tinoco I and Yamaoka K, J. Phys. Chem., 63 (1959), 423-427.
9 Thurston G B and Bowling D I, J. of Coll. and Interface Sci., 30 (1969), 34-45.
10 Piez K, Lewis M, Martin C and Gross J, Biochem. Biophys. Acta, 53 (1961), 596-598.
11 Drake M P, Davidson P F, Bump S and Schmitt F O, Biochemistry, 5 (1966), 301-312.
12 Green R W, Ang K P and Lam L C, Biochem. J., 54 (1953) 181-187.
13 Bernengo J C, Thesis, Lyon University (1970).
14 Bernengo J C, Roux B and Hanss M, Rev. Sci. Instr., 44 (8) (1973), 1083-1086.
15 Roux B, Thesis, Lyon University, (1976).
16 Coles H J and Jennings B R, Biopolymers, 14 (1975), 2567-2575.

# Polyelectrolytes and Polymers

'He fixed thee mid this dance
Of plastic circumstance.'

ROBERT BROWNING (1812-1889)

Rabbi Ben Ezra

# ELECTRO-OPTICAL CHANGES IN BIOPOLYMERS - CHEMICAL AND ROTATIONAL CONTRIBUTIONS

E. Neumann

Max-Planck-Institut für Biochemie, D-8033 Martinsried/
München, W. Germany.

In a short digression on fundamental principles of electrical-chemical coupling, the analysis of electrically induced optical changes in chemical systems is described. The discussion centers on the optical indication not only of rotational changes in electro-optically anisotropic systems, but also of chemical-conformational transitions in dipolar and ionic equilibria of macromolecules. Some experimentally particular useful criteria are suggested in order to differentiate between chemical and physical contributions to electro-optical signals.

## INTRODUCTION

In experimental physics and physical chemistry electric field methods have been traditionally applied in order to probe electric-ionic as well as electronic and optical properties of atoms and molecules. In particular, for studies of electrical-dynamic properties of chemically interacting molecules, electric field techniques gain increasing importance (1).

In the following account, fundamental principles and the analysis of electrically induced optical changes caused by chemical and orientational processes in anisotropic systems are discussed in a concise form. A more detailed description, with particular emphasis on biological-macromolecular organizations is given elsewhere (1).

## EFFECTS OF ELECTRIC FIELDS

The primary effects of electric fields on molecules involve orientation of dipolar species, deformation of polarizable systems (and subsequent orientation of induced dipoles in electrically anisotropic particles), and movement of ionic species (2,3). These primary events may be coupled to various chemical transformations, such as conformational transitions or dipolar and ionic association-dissociation equilibria or stationary states. In summary, polar structures tend to orient in the direction of the applied electric field; conformations and molecules with large dipole moments increase in concentration at the expense of those configurations with smaller electric moments; electric fields increase the dissociation of weak acids and bases and promote the separation of ion pairs into the corresponding dissociated ions or ionic groups (second Wien effect).

### Biopolymers

Instructive examples for a successful use of electric field techniques to probe ionic structures and electro-optical anisotropies are "linear" polyelectrolytes (3-5). Recently it was found that electric field pulses are capable of producing structural-conformational changes in biopolymers and membranes (5-17).

## CHEMICAL TRANSFORMATIONS IN ELECTRIC FIELDS

The equilibrium constant or the steady-state distribution constant, K(z), of a chemical equilibrium is dependent on intensive variables, z, such as temperature, T, pressure, p, or an external electric field, E. For laboratory conditions the state of zero electric field is taken as a general reference state; thus at E=0, $K=K_o$. When K(E) is the corresponding value in the presence of an electric field, the condition

$$\delta K = K(E) - K_o \ll K_o \tag{1}$$

specifies linear approximation, generally valid for small perturbations caused by the electric field.

### Dipole Equilibria

The isothermal-isobaric displacement of dipolar equilibria in electric fields can be described by

$$\left(\frac{\partial \ln K}{\partial E}\right)_{p,T} = \frac{\Delta M}{RT} \tag{2}$$

where R is the gas constant and $\Delta M$ is the electric reaction moment defined by $\Delta M = \sum \nu_j M_j$ representing the difference in the dipole moments $M_j$ of the interacting species j for one stoichiometric transformation (18,1)

For small equilibrium shifts induced by the application of the electric field, eq. 2 with $\delta \ln K = (\partial \ln K/\partial E)_{p,T} E$ gives:

$$\frac{\delta K}{K_o} = \frac{1}{RT} \cdot \int_o^E \Delta M \, dE \qquad (3)$$

Permanent dipoles. If the reaction partners j carry permanent dipole moments $p_j$, the relative equilibrium shift is given by

$$\frac{\delta K}{K_o} \cong f(\varepsilon, n) \cdot \frac{\sum \nu_j \, p_j^2}{6\,(kT)^2} \cdot E^2 \qquad (4)$$

where k is the Boltzmann constant, $\nu_j$ is the stoichiometric coefficient, and the factor $f(\varepsilon, n)$ depends on the dielectric constant $\varepsilon$ and the refractive index n of the medium (1); see also table 1.

Induced dipoles. For induced dipole moments $m_j$ and polarizabilities $\alpha_j$,

$$\frac{\delta K}{K_o} \cong \frac{1}{kT} \int_o^E \sum \nu_j \cdot \alpha_j \cdot (E_i)_j \, dE = \frac{\Delta\alpha \, E^2}{2\,kT} \qquad (5)$$

where $\Delta\alpha = \sum \nu_j \, \alpha_j$ refers to one stoichiometric transformation; the internal field $E_i$ is approximated by $E_i \cong E$ (ref. 1).

Polyelectrolytes. The induced dipole moment of an individual linear polyelectrolyte depends on the electric field according to a Langevin-like function (10,12). As a consequence of saturation in the counter-ion polarization, the induced moment becomes independent of the external electric field and the molecules behave like permanent dipoles (21,10).

For very long polyelectrolytes the internal field $E_i = E$; we obtain for the interaction between saturated dipole moments $(m_j)_s$ (ref. 1):

$$\frac{\delta K}{K_o} \cong f(\varepsilon, n) \cdot \frac{\sum \nu_j \, (m_j)_s^2}{6\,(kT)^2} \cdot E^2 \qquad (6)$$

Equations 3 - 6 can be applied to estimate relative equilibrium shifts (1). As is seen in table 1, major electrically induced concentration shifts involving dipolar molecules require large reaction moments and high electric fields; such conditions are fulfilled only for macromolecules, in particular by polyelectrolytes.

## Ionic Equilibria (Second Wien Effect)

The field dependence of the distribution constant K of ionic association-dissociation reactions, being non-linear at low fields,

Table 1. Relative displacements $\delta K/K_o$ induced by electric fields, E, in the equilibrium constant $K_o$ (at zero field) of dipolar equilibria at 293 K.

| ΔM', D | f(ε,n) | | E, V cm$^{-1}$ | δK/K$_o$ | | |
|---|---|---|---|---|---|---|
| | (a) | (c) | | (a) | (b) | (c) |
| 5 | 1.8 | 2.4 | $10^4$ | $5x10^{-6}$ | $4x10^{-3}$ | |
| 5 | 1.8 | 2.4 | $10^5$ | $5x10^{-4}$ | $4x10^{-2}$ | |
| 100 | 1.8 | 2.4 | $10^4$ | $2x10^{-3}$ | $8x10^{-2}$ | $3x10^{-3}$ |
| 100 | 1.8 | 2.4 | $10^5$ | 0.21 | 0.83 | 0.27 |

ΔM', reaction moment parameter (defined for convenient comparison),

(a) permanent dipoles: $\Delta M' = \sqrt{\Sigma \nu_j \, p_j^2}$

(b) induced dipoles: $\Delta M' = \Sigma \, \nu_j \, m_j = \frac{1}{2} \, \Delta\alpha . E$,

(c) saturated induced dipoles: $\Delta M' = \sqrt{\Sigma \nu_j \, (m_j)_s^2}$

The factor f(ε,n) is dependent on the dielectric constant ε and the refractive index n; see Eq(4) of the text(1)

is linearly dependent on the absolute value of E at higher fields (22,23). For the reaction $(A^+B^-) = A^+ + B^-$ this linear range is described by

$$\frac{\delta K}{K_o} = \gamma \; |E| \tag{7}$$

where

$$\gamma = \frac{Z_A \, u_A - Z_B \, u_B}{u_A + u_B} \cdot \frac{|Z_A \cdot Z_B| \; e_o^3}{8\pi \cdot \varepsilon_o \cdot \varepsilon (kT)^2} \tag{8}$$

In eq. 8, $e_o$ is the elementary charge, $\varepsilon_o$ is the vacuum permittivity, $Z_j$ is the valency (with sign), and $u_j$ is the ionic mobility.

For polyelectrolytes, we may define an average effective charge number $Z_A^{eff}$ for an ionic chain residue (1); the $\gamma$-factor for $u_A \ll u_B$, B being the counterions, is given by

$$\bar{\gamma}_{res} = \frac{|Z_B^2 \cdot Z_A^{eff}| \; e_o^3}{8\pi . \varepsilon_o \varepsilon (kT)^2} \; ; \tag{9}$$

$\bar{\gamma}_{res}$ is directly accessible from experiment.

Table 2. Relative shifts $\delta K/K_o$ in stationary distribution constants $K_o$ (at zero field) of ionic $Z_A$:$Z_B$ equilibria (second Wien effect)

| $\varepsilon$ | $\gamma, \frac{cm}{V}$ | $E, \frac{V}{cm}$ | $\delta K/K_o$ | | | |
|---|---|---|---|---|---|---|
| | | | 1:1 | 2:2 | $Z_A^{eff}$:1 | $Z_A^{eff}$:2 |
| 80 | $1.4 \cdot 10^{-6}$ | $10^4$ | 0.014 | | | |
| | | $10^5$ | 0.14 | | | |
| 2 | $5.6 \cdot 10^{-5}$ | $10^4$ | 0.56 | | | |
| | | $10^5$ | 5.64 | | | |
| 80 | $1.13 \cdot 10^{-5}$ | $10^4$ | | 0.11 | | |
| | | $10^5$ | | 1.1 | | |
| 80 | $1.4 \cdot 10^{-5}$ | $10^4$ | | | 0.14 | |
| | | $10^5$ | | | 1.4 | |
| 80 | $5.6 \cdot 10^{-5}$ | $10^4$ | | | | 0.56 |
| | | $10^5$ | | | | 5.6 |

$\varepsilon$, dielectric constant; $\gamma$ is defined by the relationship

$\delta K/K_o = \gamma \cdot |E|$, see Eq(8) of the text.

Eqs. 7 - 9 are applied for estimates of field-induced shifts of ionic equilibria. As is seen in table 2, these shifts are appreciably large for polyelectrolyte complexes.

In conclusion, the results of the general analysis of electrically induced chemical overall shifts provide some thermodynamic criteria for the identification of type and extent of electrical-chemical coupling. For instance, equilibria of permanent dipoles show at low fields a dependence of $\delta K/K_o$ on $(E/T)^2$. At larger field intensity the Langevin function $L(r)$ saturates for $r = p.E_d/kT \gg 1$ where then $L(r) = \coth r - \frac{1}{r} = 1$. When the corresponding value of $\Delta M \cong f(\varepsilon, n) \cdot \sum \nu_j p_j$ is inserted into eq. 3 we find that $\delta K/K_o$ is proportional to $E/T$. On the other hand, induced dipolar equilibria at low field intensity show a proportionality between $\delta K/K_o$ and $E^2/T$. At high fields, where the induced dipole moments and thus also the reaction moments are saturated, the field dependence is that of a permanent dipole system. The second Wien effect is characterized by a linear dependence of $\delta K/K_o$ on $E/T^2$ after a non linear part at small field intensities. Thus the dependence of the relative displacements $\delta K/K_o$ on field intensity and temperature may be utilized to differentiate the type of process underlying the measured electric field effect.

## MEASUREMENT OF ELECTRIC FIELD EFFECTS

The analysis of field-induced changes in a chemical system is particularly straightforward, if rectangular pulses are applied. However, if only one process is occurring or if the field induced changes are long-lived and relax with time constants large compared to the field duration, Joule heating temperature jump spectrometers may be used (1,9,10,24-26).

### Chemical Relaxations

The total field-induced concentration shift $\overline{\delta c_j}$ of species j in an interacting system, for linear ranges (i.e. small perturbations) is given by

$$\overline{\delta c_j} = \left(\frac{\partial c_j}{\partial \ln K} \cdot \frac{\partial \ln K}{\partial E}\right)_{p,T} \cdot E = \nu_j \Gamma \cdot \frac{\Delta M}{RT} \cdot E \qquad (10)$$

where $\Gamma = (\sum_{i=1} \nu_j^2/\bar{c}_j)^{-1}$ is the amplitude factor; $\nu_j$ is positive, if j is a reactant; and negative, if j is a reaction product (1).

If an external perturbation is applied faster than the chemical equilibration time, the response to such a step perturbation, is generally a chemical relaxation spectrum containing exponentials of time, t. The time course of the chemical relaxation of a composite chemical system with i (normal mode) processes can be expressed in terms of deviations $\delta c_i\ (t) = c(t) - \bar{c}$ from the new equilibrium $\bar{c}$, $\delta c = \sum \overline{\delta c_i} \cdot \exp(-t/\tau_i)$, where $\tau_i$ is the relaxation time and $\overline{\delta c_i}$ is the amplitude of relaxation mode i.

For an elementary step, the chemical relaxation can be expressed in terms of one reaction partner j as

$$\delta c_j = \overline{\delta c_j} \cdot \exp(-t/\tau) \qquad (11)$$

Eq. 11 is generally applicable for intramolecular elementary steps; for bimolecular steps only if the perturbation is small such that $\delta c_j \ll \bar{c}_j$. In table 3, relaxation parameters for three types of elementary chemical reactions are given.

When systems with several coupled reaction steps have to be considered, the calculation of relaxation times and amplitudes is more elaborate; the main aim is then to calculate the normal modes characterizing the kinetics of the complex system (26-29).

### Indication of Concentration and Orientation Changes

For the measurement of concentration changes and for the recording of orientational changes in solutions of optically anisotropic molecules, light transmission and fluorescence emission appear to cover maximum information on molecular shape, chromophor position

Table 3. Relaxation parameters of elementary chemical reactions (kinetic titration) in terms of rate constants, k, and total (analytical) concentrations (index $^o$); for definition of $\Gamma$ see eq. 10.

| Reaction | Relaxation time | Amplitude factor |
|---|---|---|
| $A + B \underset{k_{-1}}{\overset{k_1}{\rightleftharpoons}} AB$ | $\tau = \dfrac{1}{k_1\sqrt{(c_A^o + c_B^o + K)^2 - 4c_A^o c_B^o}}$ | $\Gamma = \dfrac{K}{2}\left\{\dfrac{1}{\sqrt{1-\dfrac{4\,c_A^o\,c_B^o}{(c_A^o+c_B^o+K)^2}}} - 1\right\}$ |
| $c_B^o$ = const., $c_A^o = 0$ | $\tau_o = \dfrac{1}{k_1(c_B^o + K)}$ | $\Gamma_o = 0$ |
| $c_B^o$ = const., $c_A^o$ | $\tau_m = \dfrac{1}{2k_1\sqrt{Kc_B^o}}$, $c_B^o > K$ | $\Gamma_m = \dfrac{K}{2}\;\dfrac{c_B^o + K}{K}$ |
| | at $c_A^o\,(\tau_m) = c_B^o - K$ | at $c_A^o\,(\Gamma_m) = c_B^o + K$ |
| $2\,B \underset{k_{-1}}{\overset{k_1}{\rightleftharpoons}} BB$ | $\tau = \dfrac{1}{k_1 \quad K(K+8c_B^o)}$ | $\Gamma = \dfrac{K}{8}\left(\dfrac{K + 4\,c_B^o}{\sqrt{K(K + 8c_B^o)}} - 1\right)$ |
| $c_B^o = 0$ | $\tau = k_1 \cdot K = k_{-1}$ | $\Gamma_o = 0$ |
| $B \underset{k_{-1}}{\overset{k_1}{\rightleftharpoons}} B'$ | $\tau = \dfrac{1}{k_1 + k_{-1}} = \dfrac{1}{k_1(1 + K)}$ | $\Gamma = \dfrac{c_B^o \cdot K}{(1 + K)^2}$ |

relative to rotation axis, etc.

Directly correlated to concentration of light absorbing molecules (and to absorption anisotropy) is the absorbance, measured by light transmission and defined by the Lambert-Beer law:

$$A = \sum_j A_j = \ell \sum_j \varepsilon_j c_j \tag{12}$$

where $\varepsilon_j$ is the (decadic) absorption coefficient of component j in a composite system, and $\ell$ is the optical pathway.

## COMPONENT CONTRIBUTIONS TO ABSORBANCE

If $A^E$ is the absorbance in the presence of the field and A is the zero field absorbance, then the electrically induced signal change in a composite system is expressed as

$$\delta A = A^E - A = \sum_j \delta A_j = \ell \; \delta(\sum_j \varepsilon_j c_j), \tag{13}$$

$$\delta A = \delta A^{ch} + \delta A^{rot} + \delta A_{inst} \tag{14}$$

The individual terms in eq. 14 are specified as a chemical, a rotational and an instantaneous contribution, respectively:

$$\delta A^{ch} = \ell \cdot \sum_j \varepsilon_j \cdot \delta c_j, \qquad \delta A^{rot} = \sum_j \delta A_j^{rot} \tag{15}$$

$$\delta A_{inst} = \ell \cdot \sum \varepsilon_j c_j \left( \frac{\sum \varepsilon_j c_j \, \delta \ln \varepsilon_j}{\sum \varepsilon_j c_j} + \delta \ln \rho \right)$$

where $\rho$ is the density of the system; note that $\delta \ln \rho = - \delta V/V$, V being the volume (1).

The intrinsic changes in $\varepsilon$ and $\rho$ are in general rapid compared to chemical and rotational relaxations, and are therefore taken together as an instantaneous change $\delta A_{inst}$ that constitutes a preamplitude preceding slower chemical and rotational relaxations. If only one component j is contributing to absorption,

$$\delta A_{inst} = \ell \, \varepsilon_j c_j \cdot (\delta \varepsilon_j / \varepsilon_j - \delta V/V) = \ell \, c_j (\delta \varepsilon_j - \varepsilon_j \, \delta V/V).$$

## ANISOTROPY EFFECTS (LINEAR DICHROISM)

Dipolar or polarizable molecules aligning in the direction of an external field show linear dichroism at wavelengths corresponding to optical transitions the moments of which are fixed with respect to the main dipole axis. The optical effects of rigid macromolecules of cylindrical symmetry such as linear (rodlike) polyelectrolytes in dilute solutions of low ionic strength are quantitatively fairly well analyzable (5,6).

Following eq. 13 the electrically induced absorbance change $\delta A_\gamma$ for any angle of polarization $\gamma$ relative to E is given by

$$\delta A_\gamma = A^E_\gamma - A \tag{16}$$

Three polarization angles are of practical interest for the analysis of electric field effects. At $\gamma = 0$, the absorbance change of the parallel mode $\delta A_{\parallel}$ is:

$$\delta A_{\parallel} = A^E_{\parallel} - A = \frac{2}{3}\Delta A, \tag{17}$$

where $\Delta A = A_{\parallel} - A_{\perp}$ is defined as dichroism. The perpendicular mode specified by $\gamma = 90^o$ is represented by:

$$\delta A_{\perp} = A^E_{\perp} - A = -\frac{1}{3}\Delta A \tag{18}$$

Comparison of eqs. 17 and 18 demonstrates that the rotational contributions are characterized by:

$$\delta A^{rot}_{\parallel} = -2\,\delta A^{rot}_{\perp} \tag{19}$$

Corresponding to eq. 16 and with $\delta I = I\,\{\exp(-2.3\,\delta A) - 1\}$, where the factor 2.3 accounts for the decadic absorbance scale of conventional spectrophotometers, the electrically induced change of the transmitted light $\delta I_\gamma = I^E - I$ can be expressed by:

$$\left(\frac{\delta I}{I}\right)_\gamma = \cos^2\gamma\left(e^{-\frac{2}{3}2.3\Delta A}\right) + \sin^2\gamma\left(e^{\frac{1}{3}2.3\Delta A}\right) - 1 \tag{20}$$

For the parallel polarization mode, specified by $\gamma = 0$, hence $\cos^2\gamma = 1$ and $\sin^2\gamma = 0$,

$$\left(\frac{\delta I}{I}\right)_{\parallel} = e^{-2.3\,\delta A_{\parallel}} - 1 \cong -2.3\,\delta A_{\parallel} \tag{21}$$

and, correspondingly, for the perpendicular mode where $\gamma = 90^o$,

$$\left(\frac{\delta I}{I}\right)_{\perp} = e^{-2.3\,\delta A_{\perp}} - 1 \cong 2.3\,\delta A_{\perp} \tag{22}$$

the approximation sign being valid for $\delta I \ll I$, i.e. for small perturbations.

Series expansion of eq. 20 for $\bar{\gamma} = 54.75^o$ where $\cos^2\bar{\gamma} = 1/3$ and $\sin^2\bar{\gamma} = 2/3$ leads to:

$$\left(\frac{\delta I}{I}\right)_{\bar{\gamma}} = \frac{1}{9}(2.3\cdot\Delta A)^2 - \frac{1}{81}(2.3\cdot\Delta A)^3 + \ldots\ldots \tag{23}$$

Table 4. **Rotational relaxation (dichroism) at small perturbations caused by a step field pulse starting at $t_o$ and ending at $t_e$** (E = constant):

| Signal built-up | $\frac{\delta A(t)}{\overline{\delta A}}$, $t_o \leq t \leq t_e$ | $(\frac{d\delta A}{dt})_o$ | Decay, $t \geq t_e$ | $\frac{\int_o^{t_e} \delta A(t)dt}{\int_{t_e}^{\infty} \delta A(t)dt}$ |
|---|---|---|---|---|
| (a) general, $r \neq 0$ | $\frac{3r\, e^{-2D_r \cdot t} - (r-2)e^{-6D_r \cdot t}}{2(r+1)}$ | 0 | $e^{-6D_r \cdot t}$ | $\frac{4r+1}{r+1}$ |
| (b) permanent and saturated induced dipoles, $r = \infty$ | $e^{-2D_r \cdot t} + e^{-6D_r \cdot t}$ | 0 | $e^{-6D_r \cdot t}$ | 4 |
| (c) induced dipoles, $r = 0$ | $e^{-6D_r \cdot t}$ | $\frac{1}{6D_r}$ | $e^{-6D_r \cdot t}$ | 1 |

$\delta A(t)/\overline{\delta A}$, relative deviation from steady-state or equilibrium value, $\overline{\delta A}$ being the maximum deviation at $t_o$ for the built-up of $\delta A$, and at $t_e$ for the field-free relaxation, respectively; $(d\, \delta A/dt)_o$ represents the initial slope at $t_o$ of the signal built-up; $D_r$ is the rotational diffusion coefficient ($\alpha L^3$, L being the length of the (dipolar) molecule; r is the ratio between permanent and induced dipole terms; $r = \beta^2/2\gamma$ where $\beta = B\, p\, E/kT$ ($B \approx 1$ for elongated particles) and $\gamma = 0.5 \cdot \Delta\alpha \cdot E^2/kT$ ($\Delta\alpha$ being the excess polarizability); the integral ratio of the last row represents the ratio between the area above the rise curve ($t_o \leq t \leq t_e$) and that below the zero-field relaxation curve ($t_e \leq t < t_\infty$). (See, e.g. ref.(**31**) and (**30**)).

For $\delta I \ll I$, i.e. $2.3\ \Delta A \ll 1$, we obtain

$$(\frac{\delta I}{I})_{\bar{\gamma}} = - \delta A^{rot}_{\bar{\gamma}} = 0, \qquad (24)$$

corresponding to zero dichroism at $\bar{\gamma} = 54.75$.

In summary, eqs. 19 and 24 express characteristic features of the rotational contributions, $\delta A^{rot}$; and are useful to differentiate between chemical and orientational contributions to electrically induced absorbance changes in solutions of electro-optically anisotropic molecules. Provided that field induced intensity changes are small, chemical relaxations are directly measurable at $\bar{\gamma} = 54.75^o$ without interference by orientational changes; see, however, references 25 and 1.

Table 4 summarizes a number of specific relationships to differentiate between permanent and induced dipole mechanisms.

## DIFFERENTIATION OF COMPONENT CONTRIBUTIONS

If chemical transformations are associated with isosbestic or isochromic wavelengths, $\lambda_{iso}$ (where the absorption coefficients of all absorbing reaction partners are equal, i.e. $\sum \varepsilon_j = 0$), chemical contributions to absorbance are zero at $\lambda_{iso}$. Rotational contributions cancel when plane polarized light is used under the angle $\bar{\gamma}$ (= 54.75°, or $\bar{\gamma}$ equals the corresponding apparatus-specified angle) to the field direction. The parallel and perpendicular modes of polarization serve to further identify rotational contributions (14).

From a practical point of view, the analysis of the time course of the measured signal is indispensable to determine the various components and their amplitudes (14). As is demonstrated in fig. 1, the component contributions may have opposite signs and instantaneous signal changes must usually be considered.

If the time constant of the chemical process is small compared to the rotation times of the interacting molecules ($\tau^{ch} \ll \tau^{rot}$), the chemical relaxation is rapidly established before the orientational changes occur; thus $\delta A^{ch}_{\parallel} = \delta A^{ch}_{\perp} = \delta A_{\bar{\gamma}}$, see fig. 1(a). The rotational contributions always obey eq. 19 and must show the same time constants in all three polarization modes. The same is true for the chemical contribution at the various polarization modes. For $\tau^{ch} \gg \tau^{rot}$, the rotational changes are established before the chemical change. Therefore the more rapid signal change follows eq. 19. The chemical contribution of the parallel mode $\delta A^{ch}_{\parallel}$ will in general be different from $\delta A^{ch}_{\perp}$, but the time constant of both modes are the same; see fig. 1(b).

In conclusion of the main results of this account, we may summarize that there are a number of possibilities to differentiate

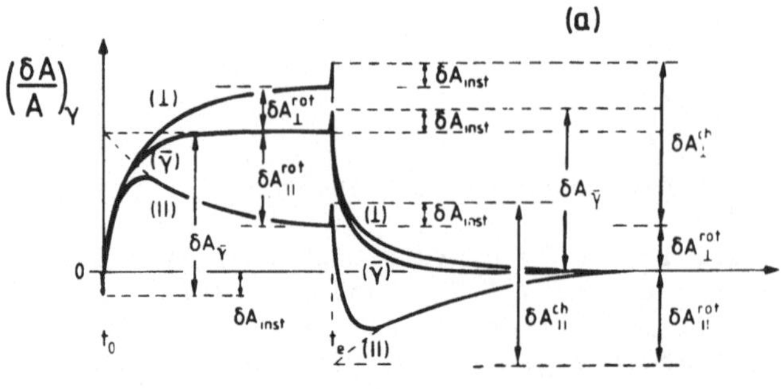

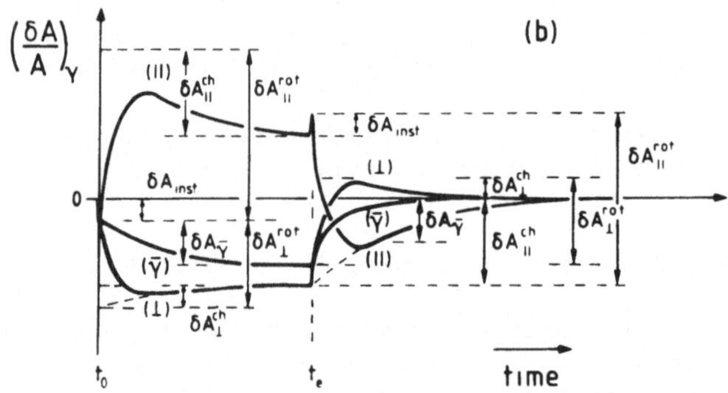

Fig. 1. Schematic representation of the relative absorbance change $(\delta A/A)_\gamma$ as a function of time at three different polarization modes; $\parallel$, $\gamma = 0^o$; $\perp$, $\gamma = 90^o$; $\bar{\gamma}$ (=54.75°) for the two limiting cases: (a) $\tau^{ch} \ll \tau^{rot}$, with $\delta A^{ch}_{\parallel} = \delta A^{ch}_{\perp} = \delta A_{\bar{\gamma}} < \delta A^{rot}_{\parallel}, \delta A^{rot}_{\perp}$; $\delta A^{rot}_{\parallel} = -2\,\delta A^{rot}_{\perp}$.
(b) $\tau^{ch} \gg \tau^{rot}$ with $\delta A^{ch}_{\parallel} > \delta A^{ch}_{\perp}$; $\delta A^{rot}_{\parallel} = -2\,\delta A^{rot}_{\perp}$.
Note that an initial rapid absorbance change $\delta A_{inst}$, due to a density change and/or a change in the intrinsic optical properties of the system is taken into account (1); see ref.14.

between chemical and orientational contributions to absorbance changes produced by electric fields in solutions of anisotropic molecules and particles. Chemical contributions are usually negligibly small if simple dipolar equilibria are concerned; appreciable field effects are encountered only in macromolecular dipolar systems at high field intensities ($>10$ kV cm$^{-1}$). On the other hand, the second Wien effect and structural changes coupled to ionic dissociation-association processes may occur at already low field intensities.

## REFERENCES


1 Neumann E, in "Topics of Bioelectrochemistry and Bioenergetics" (ed. Milazzo G), Vol. 4, John Wiley, New York (1979), in press.

2 DeMaeyer L C M and Persoons A, in "Techniques of Chemistry", 6(2) (1973) 211-235.

3 DeMaeyer L C M, Methods Enzymol. 16 , (1969) 80-118.
4 O'Konski C T and Haltner A J, J. Am. Chem. Soc. 79, (1957) 5634-5649.
5 Fredericq E and Houssier C, "Electric Dichroism and Electric Birefringence", Oxford University Press, London (1973).
6 Tricot M and Houssier C, in "Polyelectrolytes" (eds. K C Frisch, D. Klempner and A V Patsis) Technomic, Westport (1976) 43-90.
7 O'Konski C T and Stellwagen N C, Biophys. J., 5, (1965) 607-613.
8 Schwarz G and Seelig J, Biopolymers, 6, (1968) 1263-1277.
9 Neumann E and Katchalsky A, Proc. 1st Eur. Biophys. Congr. (Austria) 6, (1971) 91-95.
10 Neumann E and Katchalsky A, Proc. Natl. Acad. Sci. USA, 69, (1972) 993-997.
11 Kikuchi K and Yoshioka K, Biopolymers, 12 (1973) 2667-2679.
12 Kikuchi K and Yoshioka K, Biopolymers, 15 (1976) 583-587.
13 Yasunaga T, Sano T, Takahashi K, Takenaka H and Ito S, Chem. Lett. (Jap.) (1973) 405-408.
14 Revzin A and Neumann E, Biophys. Chem. 2, (1974) 144-150.
15 Pörschke D, Nucleic Acid Res., 1, (1974) 1601-1618.
16 Pörschke D, Biopolymers, 15, (1976) 1917-1928.
17 Pörschke D, Biophys. Chem. 4, (1976) 383-394.
18 Schwarz G, J. Phys. Chem., 71, (1967) 4021-4030.
19 Böttcher C J F, Van Belle O C, Bordewijk P and Rip A, "Theory of electric polarization", Vol. 1, Elsevier, Amsterdam, (1973).
20 Eigen U and Schwarz G, in "Electrolytes" (ed. B. Pesce), Pergamon Press, Oxford (1962) 309-335.
21 Ding D-W, Rill R and Van Holde K E, Biopolymers 11, (1972) 2109-2124.
22 Onsager L, J. Chem. Phys. 2, (1934) 599-615.
23 Manning G S, Biophys. Chem., 9 (1977) 189-192.
24 Neumann E and Rosenheck K, J. Membrane Biol., 10, (1972) 279-290.
25 Dourlent M, Hogrel J F and Hélène C, J. Amer. Chem. Soc., 96, (1974) 3398-3406.
26 Yapel A F and Lumry R, Methods of Biochem. Analysis, 20, (1971) 169-350.
27 Eigen M and DeMaeyer L C M, in "Techniques of Organic Chemistry", Vol. 8(2), Wiley, New York (1963) 895-1054.
28 Eigen M and DeMaeyer L, in "Techniques of Chemistry", Vol. 6(2), Wiley, New York (1973) 63-146.
29 Jovin T M, in "Biochemical Fluorescence: Concepts", Marcel Dekker, New York, (1976) 305-374.
30 Benoit H, Ann. Phys., 6, (1951) 561-609.
31 Nishinari K and Yoshioka K, Kolloid Z. u. Z. Polymere, 240, (1970) 831-836.

# REVERSAL OF BIREFRINGENCE SIGN OF NATURAL AND SYNTHETIC POLYELECTROLYTES IN THE PRESENCE OF METAL CATIONS AND COORDINATION COMPLEXES

C. Houssier and M. Tricot

Laboratoire de Chimie Physique, Université de Liège
Sart-Tilman, B 4000 Liège, Belgium

The interaction of $Mg^{2+}$, $Mn^{2+}$, $K_2PtCl_4$ and cis-$Pt(NH_3)_2\ Cl_2$ with DNA, chromatin and polyvinylimidazole has been investigated using electric birefringence measurements. Reversals of sign of the birefringence have been observed in a number of experimental conditions. Their interpretation is discussed in terms of changes in the intrinsic optical anisotropy due to the formation of compact structures, form anisotropy and in the appearance of new absorption bands.

## INTRODUCTION

Interactions of metal cations and coordination complexes with macromolecules are of primary importance in a number of chemical and biological processes. They are frequently responsible for drastic conformational changes encountered by these macromolecules.

The electric birefringence method is very appropriate for such studies because of its high sensitivity for the determination of the optical, electrical and hydrodynamical parameters of the oriented entities. When the ligand attached to the macromolecule has absorption bands in accessible wavelength regions, the measurement of the dichroism is particularly adequate to determine the orientation of its chromophoric groups with respect to the macromolecular axis, as for example in the case of the interactions with dyes. This was not the case in the present study which deals with the interaction of $Mg^{2+}$, $Mn^{2+}$, $K_2PtCl_4$ and cis-$Pt(NH_3)_2Cl_2$ with DNA, chromatin and polyvinylimidazole. Apart from $Mg^{2+}$, the other interactions have in common a coordinative binding to heterocyclic bases and this is why a comparison of the resulting observed behaviour appeared opportune.

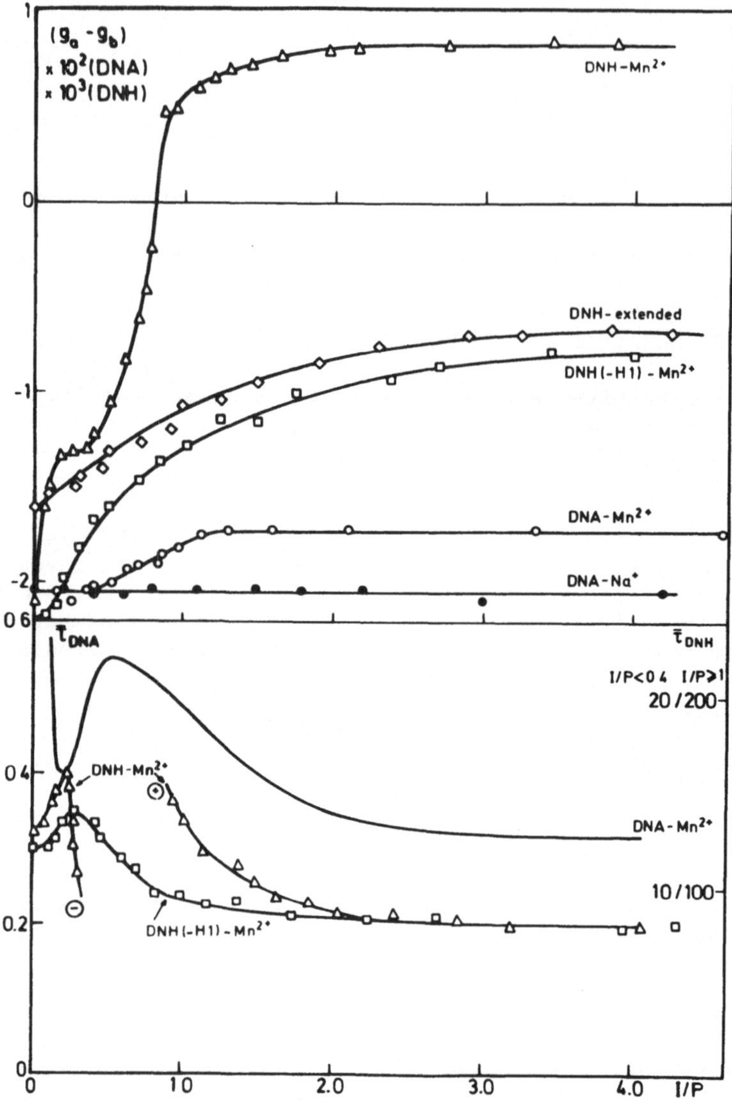


Fig. 1. Dependence of the optical anisotropy $(g_a - g_b)$ and mean $\bar{\tau}$ of DNA and chromatin (DNH) on the ratio I/P of $Mn^{2+}$ to DNA-phosphate molar concentrations.
DNH-extended: fraction obtained by ECTHAM chromatography of sonicated DNH.
DNH(-H1): H1 depleted chromatin
$\bar{\tau}_{DNA}$: mean relaxation time of DNA in ms.
$\bar{\tau}_{DNH}$: of chromatin (sonicated samples) in $\mu$s (scales 10, 20 $\mu$s for I/P$<$0.4 (-) and 100, 200 $\mu$s for I/P$>$1 (+) for whole chromatin, DNH).

We shall focus our attention on the interpretation of the changes in the optical anisotropy of the sample solutions.

## RESULTS

### Interactions of DNA and Chromatin with $Mg^{2+}$, $Mn^{2+}$ (ref. 1,2).

The results of these electric birefringence observations may be summarized as follows.
(a) The intrinsic optical anisotropy of chromatin is appreciably smaller than that of DNA. The removal of the proteins from chromatin progressively restores the DNA anisotropy, but the difference between H1-depleted chromatin and whole chromatin anisotropy is not significant.
(b) The addition of $Mn^{2+}$ cations (fig. 1) produces a decrease of the optical anisotropy of DNA and of the extended chromatin subunit, and a reversal of the birefringence sign for whole chromatin and for its compact subunit. This change of birefringence sign is prevented by the removal of histone H1.
(c) $Mg^{2+}$ ions do not appreciably affect the optical anisotropy of DNA and of the extended chromatin subunit but produce for whole chromatin the appearance of complex signals characterized by the presence of positive and negative optical anisotropy contributions.

The shape of the decrease of mean relaxation time with the ratio I/P (ion over mononucleotide molar concentration) is characterized by the appearance of an extension phase for some of the samples, in the range of I/P of 0.2 to 0.5 (fig. 1.)

Small and irregular variations of the electric polarizability with I/P were observed. In the case of chromatin, the positive birefringence component at high I/P was characterized by a significantly higher polarizability and about 10 times larger relaxation times than the negative birefringence at low I/P or in the presence of $Na^+$.

### Interaction of DNA and Chromatin with Platinum Compounds (ref. 3,4).

One of the simplest square-planar platinum coordination complex, cis-Pt $(NH_3)_2Cl_2$ (fig. 2) is one of the most active antitumor drugs known today and discovered by the Rosenberg's group in 1969. Its activity is believed to arise from its interaction with the nuclear DNA with which it would form a bidentate coordination complex involving the N-7 and O-6 atoms of guanine. The cis-Pt $(NH_3)_2Cl_2$ also shows a marked antimitotic activity characterized by an accumulation of cells in the G2 or premitotic phase, which would indicate an inhibition of the entrance in mitosis rather than a blocking of the DNA synthesis. Some condensation of the chromatin in the nuclei of cells treated with this compound has also been detected. It thus appeared interesting to investigate the kind of structural alteration produced on DNA and chromatin upon their interaction with platinum compounds.

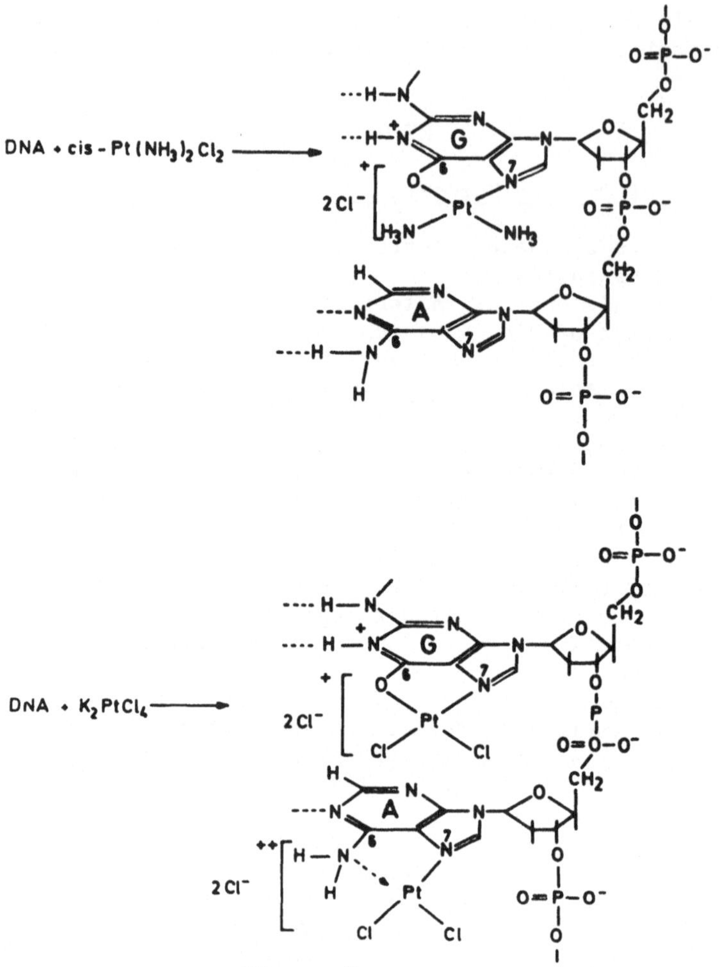

Figure 2

Fig. 3 shows that a drastic decrease of the DNA birefringence as the amount of bound Pt increases, is present. For chromatin, the birefringence becomes positive when the binding ratio r exceeds 0.12, similarly to DNA upon reaction with $K_2PtCl_4$ which attacks the same binding site as cis-$Pt(NH_3)_2Cl_2$, but leaves two labile Cl groups on the Pt atom (fig. 2.)

We also detected significant decreases of the relaxation times with increasing binding ratios. This is in contrast with the observation made with chromatin + $Mn^{2+}$ where the samples showed a noticeable turbidity and much larger relaxation times for the positive birefringence signals.

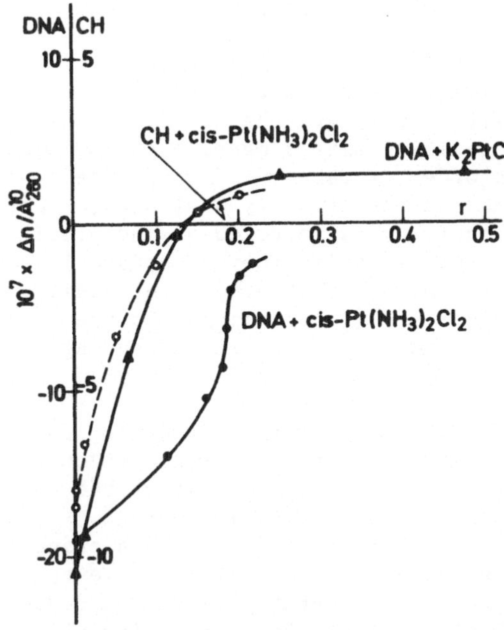


Fig. 3. Dependence of the optical anisotropy $\Delta n/A_{260}^{10}$ of DNA and chromatin (CH) on the binding ratio $r$ , after reaction with cis-$Pt(NH_3)_2Cl_2$ and $K_2PtCl_4$.

Interaction of Partially Quaternized Polyvinylimidazole with cis-$Pt(NH_3)_2Cl_2$ and $K_2PtCl_4$ (ref. 5)

In order to gain a better insight into the conformational changes induced by the platinum complexes in DNA and chromatin, we also examined the solutions properties of complexes of poly-1-vinylimidazole (partially quaternized with ethyl bromide) with the above mentioned platinum compounds. Despite the fact that this synthetic polyelectrolyte is a polycation whereas DNA and chromatin are polyanions, we observed some similitude of behaviour between the binding of cis-

Figure 4

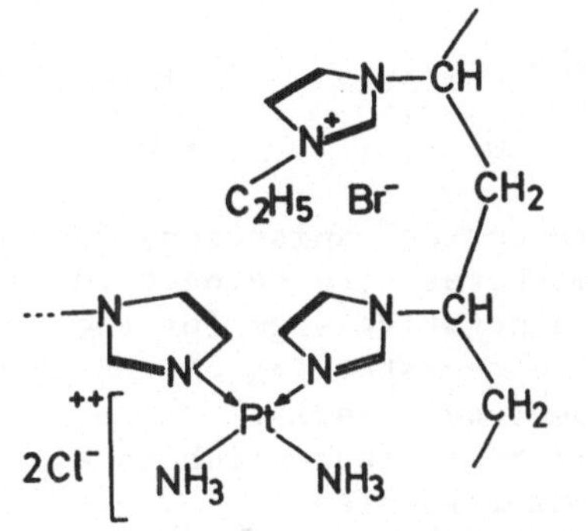

PVI.EtBr / cis-$Pt(NH_3)_2Cl_2$

$Pt(NH_3)_2Cl_2$ to the free imidazole rings of the polyelectrolyte and to the guanine bases of DNA (fig. 4).

A decrease of the negative birefringence and a reversal of its sign were observed (fig. 5A,B) for ratios r larger than 0.18. The limiting values of the negative and positive birefringences at low and high r respectively, are of the same order of magnitude, while the shape of the birefringence curves remained practically unaltered; only a slight increase of electric polarizability is detected as the platinum content increases. The measured optical anisotropy at saturation $\Delta n_s$ which becomes equal to zero at r = 0.18, reaches constant positive values for r larger than 0.4; a ratio r of 0.5 corresponds to one Pt group for two free imidazole bases. A small decrease of the average relaxation times is also noticed.

As for the binding of cis-$Pt(NH_3)_2Cl_2$ to polyvinylimidazole, the interaction with $PtCl_4^{2-}$ ions induces a change of sign of the electric birefringence above r = 0.18, but here a significant change of the shape of the orientation function is observed, saturation being reached at low fields for r larger than 0.2 (fig. 6A,B). The average electric polarizability sharply increases by a factor of about seven when r increases from 0.15 to 0.25, while the average relaxation times increase from 12 - 13 $\mu$s for the polyelectrolyte alone or at low platinum contents to values of the order of 200 - 300 $\mu$s for r higher than 0.2. The solutions of PVI/$PtCl_4^{2-}$ became opalescent when r exceeded 0.3 and precipitation occurred for r $\sim$ 0.35.

## GENERAL INTERPRETATION

If we consider that the changes observed mainly arise from the alteration of the optical anisotropy term with no important modification of the degree of orientation of the macromolecule in the electric field, then the decrease and reversal of sign of the birefringence must be originated in the change of the relative importance of the three following contributions to the total measured anisotropy:

$$\Delta n_s = \underset{<0}{\Delta n_{\text{intrinsic}}} + \underset{>0}{\Delta n_{\text{form}}} + \underset{\text{unknown sign}}{\Delta n_{\text{new transitions}}} \qquad (1)$$

- the intrinsic optical anisotropy due to the orientation of the heterocyclic chromophores with respect to the macromolecular axis, a term which shows a negative sign for the three polyelectrolytes investigated owing to the stacking of the chromophoric groups perpendicularly to the macromolecular axis.
- the form anisotropy term which is expected to be positive for highly elongated macromolecules.
- and a contribution arising from new absorption bands of unknown polarization of their transition moments, and whose sign is thus also unknown.

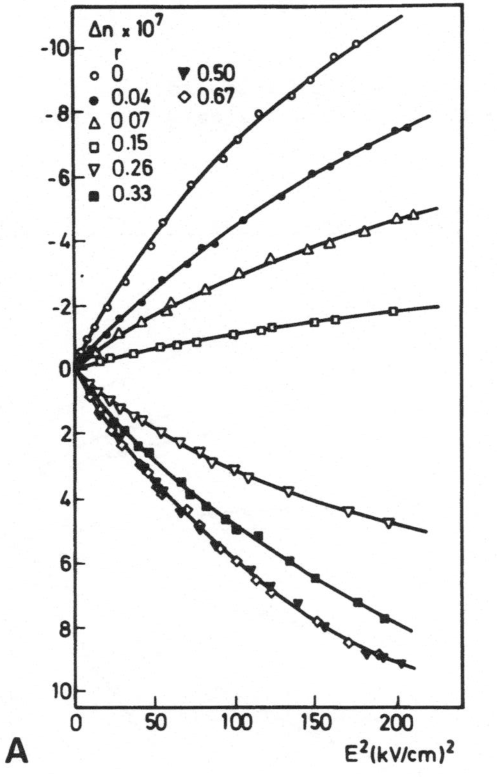


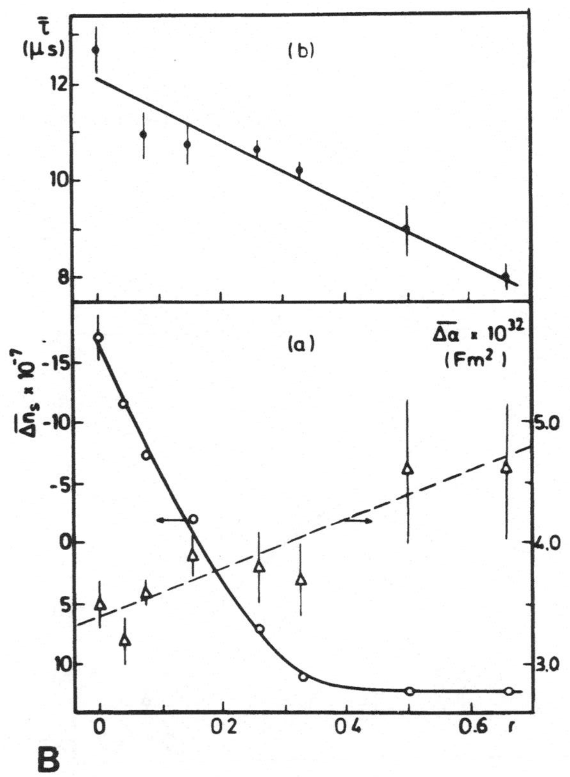


Fig. 5. Electric birefringence of partially quaternized polyvinylimidazole in the presence of cis-$Pt(NH_3)_2Cl_2$. Part A: field strength dependence of the electric birefringence $\Delta n$ at various binding ratios r. Part B: dependence of the birefringence at saturation $\Delta n_s$, mean relaxation time $\bar{\tau}$ and polarizability $\Delta\alpha$ on the binding ratio r.

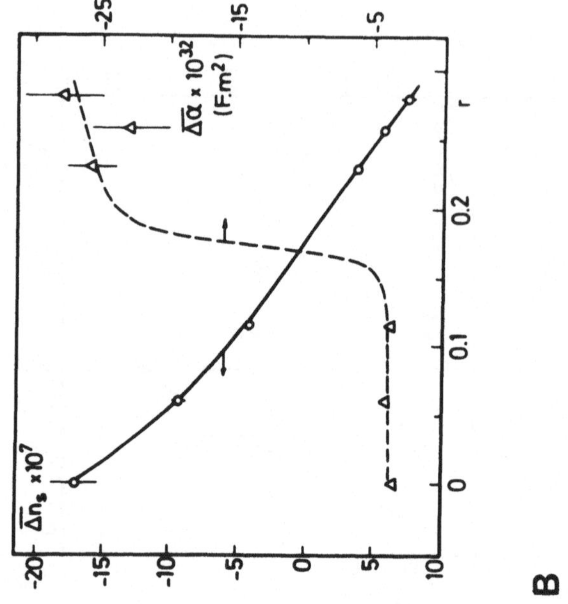


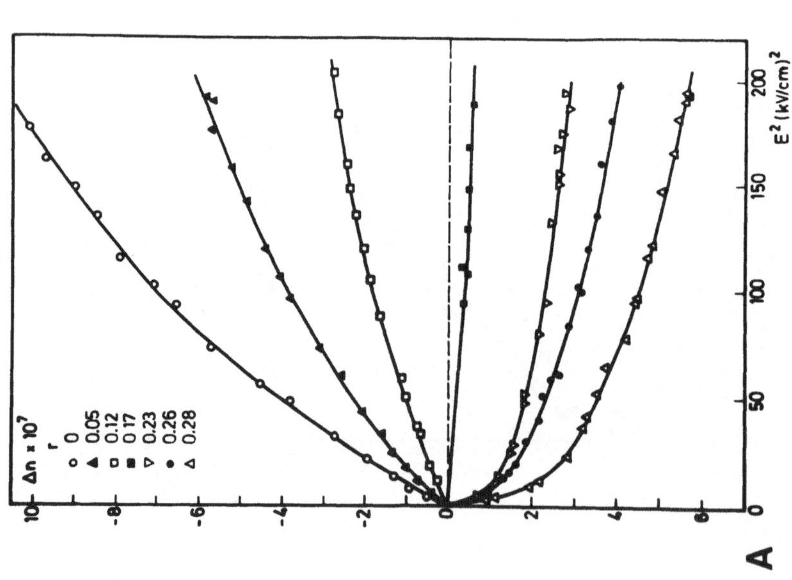


Fig. 6. Electric birefringence of partially quaternized polyvinylimidazole in the presence of $K_2PtCl_4$. Same conditions as in fig. 5 for Part A and B.

There is no evidence for the appearance of new absorption bands upon interaction of nucleic acids constituents with $Mg^{2+}$ and $Mn^{2+}$, or with Pt compounds of the type used here. Of course, such new transitions could be present in the far ultraviolet and have nevertheless an influence on the amplitude of the birefringence measured in the visible region. If we had to consider this possibility, we would however be led to pure speculations. In addition, it does not appear reasonable to put all the responsibility of the changes observed on this contribution and to neglect conformational changes which are also evidenced by other methods. Thus, we will discard the occurrence of the third contribution of eq. 1 for the time being. This is in opposition with the interpretation presented by Ding (6) for his electric dichroism observations on DNA-silver and DNA-mercuric ions complexes, explained by the presence of charge-transfer bands.

We will thus consider two possible interpretations whose appropriateness will be discussed for the various experimental findings above described.

A. If we consider that the form anisotropy contribution cannot be estimated with sufficient confidence, or measured, or eliminated in some particular experimental conditions, we simply omit it and attribute all the changes observed to modifications of the intrinsic anisotropy originated in alterations of the heterocyclic chromophore arrangement in the macromolecule.

(a) This is the interpretation that we favour in the case of the interaction of DNA and chromatin with $Mn^{2+}$ since there is no indication of the disorganization of the double-stranded hydrogen-bonded structure in these cases. Coordination of $Mn^{2+}$ to the DNA bases would bring about a specific organization of this molecule into a more compact and more rigid structure. This condensation of the DNA chain would occur through the introduction of bends by means of a structural transition of some guanine from the anti- to the syn-conformation, or through the formation of kinks in the backbone as in the models of Crick and Klug (7) and of Sobell et al. (8).
In the case of chromatin, the action of divalent cations would produce a folding of the linear arrangement of nucleosome strings into a superhelicoidal arrangement (2), in the presence of histone-H1 fraction. On the basis of the structural parameters presently available for the nucleosome particle and for the condensed superhelicoidal array of nucleosomes, we estimated tentatively that the superhelicoidal axis of the nucleosome DNA should be perpendicular to the backbone axis of the array in order for the resulting optical anisotropy to be positive.

(b) This interpretation is also acceptable for the observations made on the complexes of DNA and chromatin with cis-$Pt(NH_3)_2Cl_2$ and $K_2PtCl_4$ although in these cases there are clear indications of a denaturation of the double-stranded structure from the absorption spectra and thermal denaturation profiles. This will give some support to the other interpretation presented below, for these complexes.

(c) For the interaction with polyvinylimidazole, the occurrence of quaternary or superhelicoidal structures is difficult to imagine.

B. We can also consider that, as a consequence of the disorganization or coiling of the stacked structure of the macromolecule (produced by the denaturation in the case of DNA and chromatin interacting with Pt) the intrinsic optical anisotropy term falls to zero, and that the positive anisotropy observed is the form anisotropy term only.

We found this interpretation not very convincing for the DNA and chromatin interactions with the divalent cations and with the platinum compounds since the relaxation times and electric polarizability results seem to indicate the formation of a more compact and more rigid structure.

It is probably the best explanation however for the observations on the polyvinylimidazole-Pt interactions. The change of sign of the measured optical anisotropy would be the consequence of a coiling of the polyelectrolyte chain caused by the bridging of non-adjacent segments through the platinum compound (fig. 4). Formation of large aggregates also occurs at high r values for the polyvinylimidazole - $K_2PtCl_4$ interaction, and perhaps also in the case of the chromatin-$Mn^{2+}$ interaction.

We have been trying to find more direct evidence for the formation of compact arrangements in our experimental conditions. Recent electron microscopic observations reveal the presence of such arrangements in the case of our chromatin-Pt complexes.

We have not been able to detect any reversal of the sign of the electric dichroism in the ultraviolet absorption band of the purine and pyrimidine bases but the sensitivity of our instrumentation in this wavelength region is not sufficient to allow the detection of very small dichroism signals (below 0.05). We prepared dye complexes of the chromatin-platinum complexes and looked at their dichroism in the visible absorption band of the dye where the sensitivity is much better. Until now, such mixed complexes showed either no detectable dichroism or, at lower Pt contents, a small negative dichroism. It is however probable that the dye molecules can only bind to those sites where no platinum is present and which may have retained their native base-stacking arrangement.

Further experimental investigation is required to clear up all these problems.

## REFERENCES

1 Emonds-Alt X, Thesis, University of Liège, (1976).

2 Houssier C, Bontemps J, Emonds-Alt X and Fredericq E, Ann. N.Y. Acad. Sci., 303, (1977) 170.

3 Depauw-Gillet M C, Houssier C and Fredericq E (1978), submitted for publication.

4 Dupont A and Houssier C (1978), unpublished observations.

5 Dellicourt C and Tricot M (1978), unpublished observations.

6 Ding D, Thesis, University of New Mexico, (1977).

7 Crick F H C and Klug A, Nature, 255, (1975) 530.

8 Sobell H M, Tsai C C, Jain S C and Gilbert S G, J. Mol. Biol., 114, (1977) 333.

# ELECTRO-OPTIC STUDY OF THE CONFORMATIONAL CHANGES INDUCED IN PARTIALLY CHARGED POLY-4-VINYLPYRIDINE BY IONS OF HEAVY METALS

M. Tricot

Laboratoire de Chimie-Physique, Université de Liège (Sart-Tilman), B 4000 LIEGE, Belgium

We first studied, in aqueous solution, the effect of the degree of charge on the electro-optical parameters and on the conformation of poly-4-vinylpyridine partially quaternized by ethylbromide. We then investigated the influence of $CaBr_2$ and $CuBr_2$, showing that the former salt only acts through the increased ionic strength. On the contrary, the coordination of $Cu^{++}$ ions on the uncharged pyridine units produces, with increasing ion concentration, an increase of the electric polarizability and a much more pronounced decrease of the optical anisotropy. Birefringence relaxation time and reduced viscosity data display for $Ca^{++}$ and $Cu^{++}$ ions a sharp decrease of the extension of the polyelectrolyte. The influence of the degree of charge and of the $Cu^{++}$ ions attached to the uncharged pyridine units on the electric and optical parameters is tentatively explained on the basis of a segmental chain model and of the polarization of the counterionic atmosphere along rigid subunits.

## INTRODUCTION

In recent years, attention has been paid to the physico-chemical properties of complexes of transition metal ions with partially quaternized poly-4-vinylpyridine (4-PVP)(1-3). The interest of several workers in such systems is justified by the high catalytic activity in redox reactions of the polymer-metal complexes (3). Furthermore, the ability of 4-PVP partially crosslinked with 1,4 dibromobutane to act as chelate-forming resins has been recently evidenced (4). A large efficiency of the binding capacity for a specified ion is observed when the crosslinking by 1,4 dibromobutane is performed in the presence of this metal ion as template. The polymeric ligand is thus maintained at the optimal conformation of the coordination sphere of the metal ion (5).

The general features of the complexation of metal ions with polymeric ligands are a drastic conformational change of the polymer which adopts a compact form because of the intra-polymer chelation on one hand and an enhancement of the complex formation constants (as compared to those of monomeric models) due to the polymer effect on the other hand (4). Soluble complexes of copper (II) with 4-PVP partially quaternized with methyl and ethylbromide have been more extensively studied in aqueous solution. In the case of 4-PVP quaternized up to 35 % by ethylbromide, Nishikawa and Tsuchida displayed the existence of complexes involving four free pyridine groups for one cupric ion (3). However it has been reported that, because of electro-static and steric hindrance effects, not all the free pyridine groups are able to be chelated to the metal ion (1).

The purpose of the present paper is to present some results about the influence of the copper (II) ions on the conformational and electro-optical behaviour of partially charged 4-PVP samples. A comparison will be made with the simple ionic strength effect due to calcium ions. This investigation requires a previous knowledge of the strong polyelectrolytic behaviour of the 4-PVP quaternized to different extents by ethylbromide.

## RESULTS AND DISCUSSION

The poly-N-ethyl-4-vinylpyridinium bromides (4-PVP-EtBr) were prepared as mentioned in reference 6 from a fractionated 4-PVP sample whose average viscometric molecular weight is of the order of $2.4 \times 10^5$. The degrees of quaternization, determined by potentiometric titration of the bromide counterions, are collected in table 1. The electric birefringence experiments were performed with the previously described instrumentation (7,8) and the electro-optical parameters, i.e. the mean values of the birefringence at saturation $\overline{\Delta n_s}$, of the electric polarizability $\overline{\Delta \alpha}$ and of the relaxation time $\bar{\tau}$, were calculated as recently reported (9).

### Influence of the Degree of Charge

For the various 4-PVP.EtBr samples in water, in the absence of added salt, the reversing-pulse method did not show at low electric field any transient at the field reversal, so that the analysis of our data will be made in terms of a pure induced dipole moment mechanism.

As always observed with such polyelectrolytes at high dilution, the field strength dependence of the electric birefringence approaches saturation at high fields; the saturation effect is more pronounced at low concentration and the birefringence, which always remains

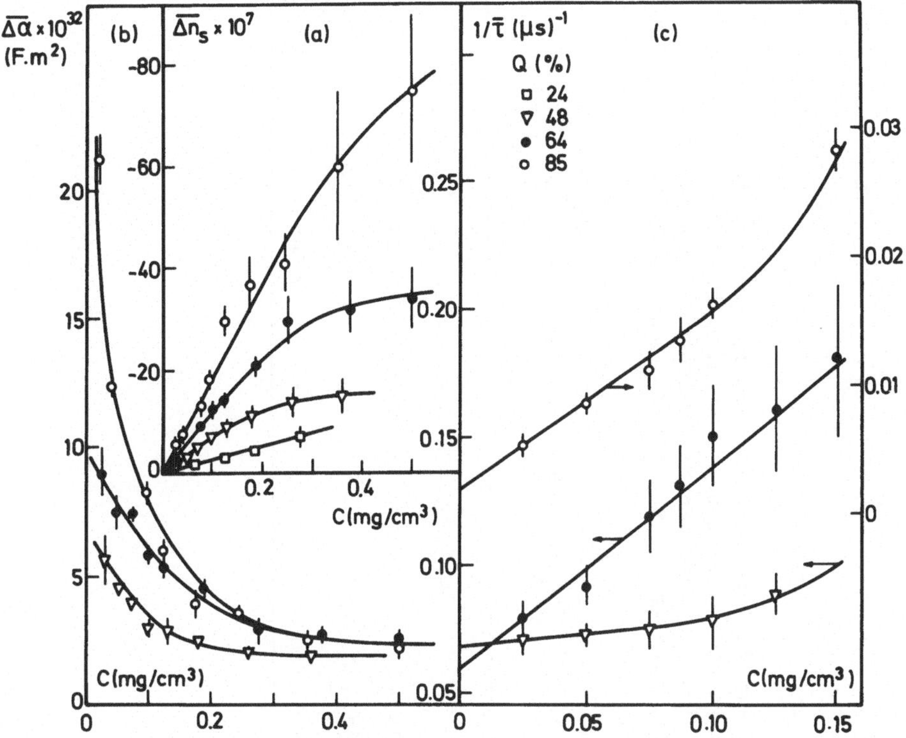


Fig. 1. Influence of concentration on (a) birefringence at saturation and (b) the electric polarizability for various 4-PVP.EtBr samples; (c) extrapolation of the inverse of the average relaxation time to zero concentration.

negative, increases with the degree of charge. In this paper, only the measured negative optical anisotropy factor $(g_a - g_b)$ or $n\Delta n_s/2\pi c\bar{v}$ is considered, but the existence of the positive form anisotropy term, arising from the difference of refractive indices between the polymer and the solvent, must be kept in mind; the large influence of this term has been clearly evidenced in an earlier report (10). From the linear variation of $\Delta n_s$ vs.c, obeyed for concentrations lower than $\simeq 0.3$ mg/cm³ (fig. 1a) and from the partial specific volume $\bar{v}$ values collected in table 1, it is shown that the optical anisotropy factor $(g_a - g_b)$ increases with the degree of charge (fig. 2a), ranging from $-0.7 \times 10^{-3}$ to $-6.3 \times 10^{-3}$ when Q changes from 24 to 85%. Since the optical polarizability values of the pyridine and pyridinium ring only differs by a factor of about 1.4 ($10 \times 10^{-40}$ $Fm^2$ to $14 \times 10^{-40}$ $Fm^2$ respectively), we must conclude that the increase of the optical anisotropy factor results more directly from drastic conformational change of the polymer rather than from a modification of the optical polarizability of the heterocyclic chromophore.

Table 1. Values of the degree of quaternization, partial specific volume, intrinsic viscosity, axial ratio, chain length at infinite dilution, optical anisotropy factor, linear charge density, electric polarizability at infinite dilution and length of the rigid subunit for the various 4-PVP.EtBr samples.

| Q ( %) | $\bar{v}$ $(cm^3 g^{-1})$ | ($\eta$) $(dl.g^{-1})$ | p | $L_o$(A) | $(g_a-g_b).10^3$ | $\lambda_{effective}$ | $\phi$ | $\overline{\Delta\alpha}_\delta x10^{32}$ $(F.m^2)$ | b(A) |
|---|---|---|---|---|---|---|---|---|---|
| 0.24 | 0.76 | | | | - 0.75 | | | | |
| 0.48 | 0.72 | 3.9 | ~100 | 1250±40 | - 2.00 | 1.7 | 0.42 | 7.1 | 550 |
| 0.64 | 0.63 | 3.5 | ~100 | 1200±30 | - 4.10 | 2.3 | 0.56 | 10.5 | 520 |
| 0.85 | 0.63 | 12.8 | ~200 | 4100±300 | - 6.30 | 3.0 | 0.67 | 35.0 | 670 |

The so-called polyelectrolytic effect, or the extension of the polyelectrolyte with the dilution, is reflected in the increase of the reduced viscosity, relaxation time and electric polarizability (fig. 1b) when the concentration decreases; this effect appears even for the low degree of charge. The average relaxation time at infinite dilution, obtained by a linear extrapolation of $(\bar{\tau})^{-1}$ vs.c (fig. 1c) allows us to estimate the chain length $L_o$ on the basis of the rigid rod model, making use of the axial ratio values p roughly evaluated from the intrinsic viscosity (table 1). The $L_o$ values only reflect the degree of extension of the polyion. We have indeed proved earlier that the conformation of these polyelectrolytes at very high dilution can be better identified to semi-rigid chain models such as the worm-like chain or the weakly bending rod (6,9,10). It appears nevertheless that $(\eta)$ and $L_o$ present a similar variation as a function of the degree of charge (fig. 2). $L_o$ remains practically constant, of the order of 1200 A, up to Q $\sim$ 60% and sharply increases for higher values of Q, whereas we have shown that the optical anisotropy increased almost linearly with the degree of charge. The fact that the extension of the polyion at infinite dilution does not change appreciably when Q varies from 40 to 60% could have two possible origins: hydrophobic interactions between uncharged pyridine groups on one hand, electrostatic interactions between positively charged pyridinium groups and uncharged pyridine groups (carrying free p electron pairs) located on non-adjacent segments on the other hand. Both interaction effects would restrict the extension of the polyion and the electrostatic repulsion between neighbour charged sites would start to stretch markedly the polyion when the degree of charge exceeds 60%

The average values of the electric polarizability have been extrapolated to zero concentration with linear $(\overline{\Delta\alpha})^{-1}$ vs.c plots. The $\Delta\alpha_o$ at infinite dilution has similar dependence on Q as those of $(\eta)$ and $L_o$ and starts only to increase when Q overcomes 50 percent (fig. 2b). We previously reported (6,10) that the average electric polarizability of such polyelectrolytes appears to be nearly independent of the molecular weight, like the specific dielectric increment measured in the high frequency range of the dispersion curve (11). It is thus assumed that $\Delta\alpha$ and $\Delta\varepsilon_2/c$ must arise from the same molecular phenomenon, i.e. the delocalization of the bound counterion fraction $\phi$ along rigid subunits b whose length would not change with the molecular weight (10). The polyion would thus be represented as a broken chain model with rigid subunits, somewhat similar to the weakly bending rod or wormlike chain model. By analogy with Mandel's theory of polarization (12), where the electric polarizability is expressed by the relation:

$$\Delta\alpha = \frac{Nz^2 e^2 \phi BL^2}{12 kT} \tag{1}$$

We expressed the electric polarizability derived from electro-optical measurements in the form (for monovalent counterions, z = 1):

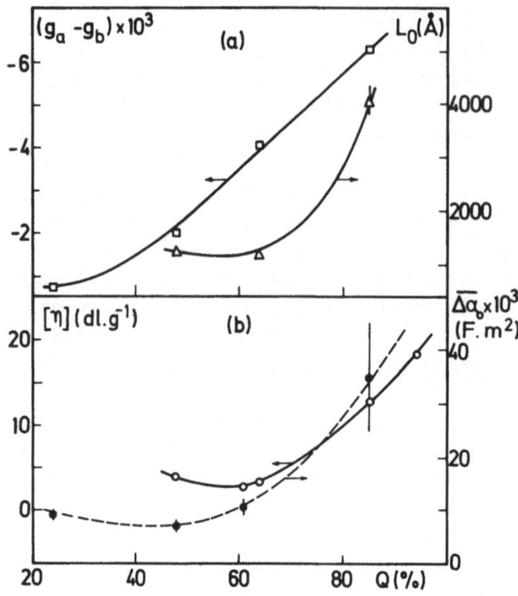


Fig. 2. Influence of the degree of quaternization (a) on the optical anisotropy factor and the chain length, (b) on the intrinsic viscosity and the electric polarizability, both at infinite dilution.

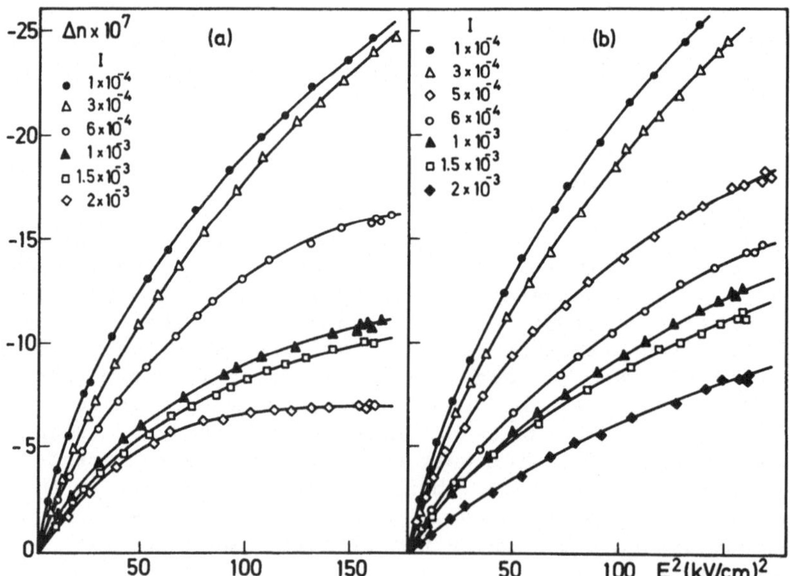


Fig. 3. Field strength dependences of the electric birefringence of the 4-PVP.EtBr 85% (at C = 0.25 mg/cm$^3$) at various ionic strengths in the presence of (a) $Cu^{++}$ and (b) $Ca^{++}$ cations.

$$\Delta\alpha = \frac{n\ e^2\ \phi\ b^2}{12\ kT} \qquad (2)$$

where n is the number of charged sites on rigid subunits of length b, assuming that the ratio B between the effective and applied electric field is equal to unity. The parameter n can be replaced by the ration bQ/h (where h is the length of the monomer unit), so that the experimental values of $\Delta\alpha_o$ yield b parameters of the order of 520 A for Q = 64% and of 670 A for Q = 85%. It should be noted that the fraction of bound counterions $\phi$ was not obtained experimentally but from an approximative calculation procedure, based on the correlation of $\phi$ with the effective linear charge density:

$$\phi = 1 - (\lambda_{effective})^{-1} \ (\text{for} \quad \lambda_{effective} > 1) \qquad (3)$$

where

$$\lambda_{effective} = \lambda Q = \frac{Q\ e^2}{4\pi\ \varepsilon_o'\ \varepsilon_s\ hkT} \qquad (4)$$

$\varepsilon_s$ designates the dielectric permittivity of the solvent and $\varepsilon_o'$, that of a vacuum, or $8.85 \times 10^{-12}$ $F.m^{-1}$.

This treatment, which is essentially valid for highly charged polyions, would explain the increase of electric polarizability with the degree of charge by an increase of the number of charged sites, of the fraction of bound counterions and of the length of the subunit along which these counterions can be delocalized under the action of the external electric field. A satisfactory agreement is noticed between the b values found for the 4-PVP.EtBr and those determined from dielectric experiments for the sodium salts of polystyrenesulfonic and polymethacrylic acids (13), for which b remains of the order of 400-600 A. Further experiments are required to attempt to define the physical meaning of this subunit b and its correlation with the parameters issued from other semi-rigid chain models.

## Effect of Divalent Cations

We studied the effect of $Cu^{++}$ and $Ca^{++}$ on the electro-optical behaviour of partially charged 4-PVP by adding $CuBr_2$ and $CaBr_2$ salts in order to avoid specific effects of counterions. Two 4-PVP.EtBr samples (Q = 48 and 85%) were examined at a constant polyelectrolyte concentration: c = 0.25 $mg/cm^3$. The concentrations of salts can be expressed in two ways: the total added ionic strength and the ratio $(Me^{++})/(Py)$ of the molar concentrations of metal ion and free pyridine base; the latter parameter is evidently dependent on the degree of charge Q and on the polyelectrolyte concentration.

It is seen that the field strength dependences of the birefringence differs strongly with the nature of the added salts (fig. 3).

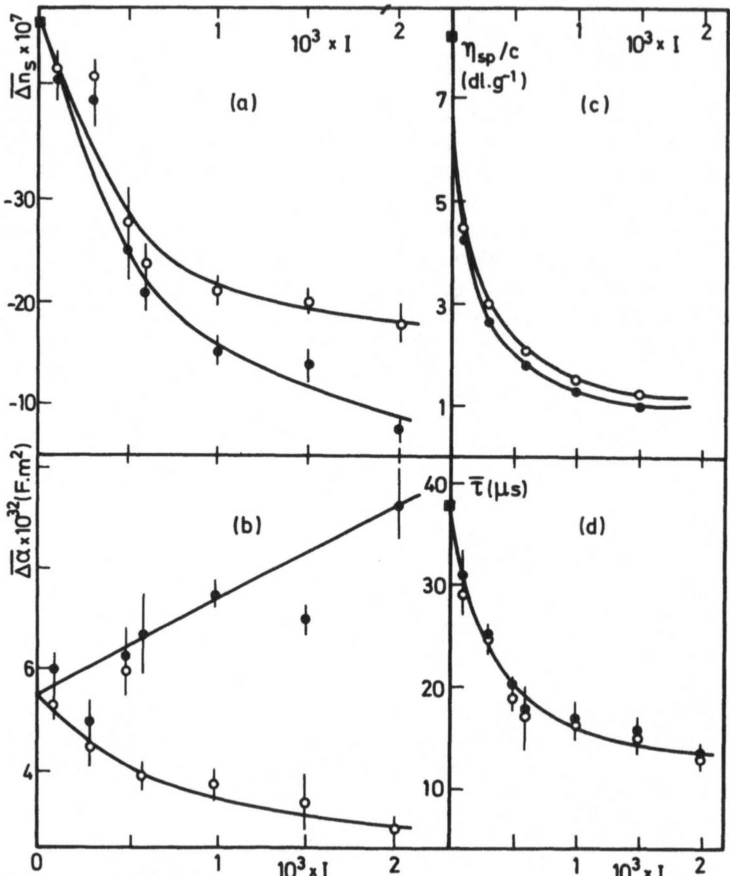


Fig. 4. Influence of the ionic strength ($Cu^{++}$ cation) on (a) the birefringence at saturation, (b) the electric polarizability, (c) the reduced viscosity and (d) the average relaxation time of the 4-PVP.EtBr 85%.

If the birefringence is always decreasing as I increases, the tendency to saturation at high fields is more pronounced with $Cu^{++}$. This results in lower values of the optical anisotropy $\overline{\Delta n_s}$ in the presence of $Cu^{++}$. It clearly appears that for 4-PVP.EtBr 85%, $\overline{\Delta n_s}$ at I = 2 x $10^{-3}$ is about two times lower with $Cu^{++}$ than with $Ca^{++}$ (fig. 4a). In a similar way, the changes in the shape of the orientation function reflect large differences in the electric polarizability values: $\overline{\Delta\alpha}$ increases with $Cu^{++}$ and decreases with $Ca^{++}$, as the concentration of added salt increases (fig. 4b). Similar effects were noticed but to a lower extent with the 4-PVP.EtBr 48%. On the contrary, the dependences of the average relaxation time and of the reduced viscosity do not allow us to distinguish significantly specific effects of

$Cu^{++}$ and $Ca^{++}$ (fig. 4c and d): both ions cause a pronounced decrease of $\bar{\tau}$ and of $\eta_{sp}/c$. This means that the polyion adopts a more coiled conformation in the presence of added salt. The decrease of length is of the order of 30 to 40% when I changes from 0 to 2 x $10^{-3}$: at C = 0.25 mg/$cm^3$, L changes from 1600 to 1200 A and from 900 to 560 A for the samples quaternized at 85% and 48% respectively. We do not observe with the relaxation experiments, the presence of associates previously detected by Kirsh et al. (1) on the basis of the variation of the sedimentation coefficient which increases by a factor of about six when the concentration of added $Cu(NO_3)_2$ changes from 0 to 3 x $10^{-2}$ M. We must however add that we always worked at concentrations of added salt which did not exceed 2 x $10^{-3}$, owing to the restriction imposed by the conductivity of the solutions in electro-optical experiments.

We must now try to explain the difference in the electric parameters of the polyion, the degrees of extension being almost identical in the presence of $Ca^{++}$ and $Cu^{++}$. The effect of $Ca^{++}$ can be identified to a simple ionic strength effect analogous to that observed with potassium bromide, involving a decrease of the extension, of the degree of rigidity and of the electric polarizability (6). On the contrary, the copper ions are strongly chelated by the free pyridine bases located on neighbour monomer units or on non-adjacent segments of the polyion, some intermolecular bonding being not excluded. This chelation of $Cu^{++}$contributes to charge the free pyridine units, so that the effective number of charged sites is larger. The linear charge density is also increased, so that the fraction of bound counterions, responsible for the induced dipole moment, becomes more important. However, it does not seem possible to estimate to what extent the length b, along which the counterions are delocalized, is influenced by the chelation of the $Cu^{++}$ ions and the conformational change of the polyion. We must indeed keep in mind that this length b is the preponderant term determining the magnitude of the induced dipole moment, and hence of the electric polarizability value.

We did not investigate the local configuration of the pyridine-copper complexes in the chain, but we do not believe however that the main part of the chelated copper ions do exist as the tetrapyridinate -$Cu^{++}$ species (1,3) when the degree of quaternization overcomes 60%. Complementary dialysis experiments would be necessary to determine the fraction of metal ions bound to the polyelectrolyte. Furthermore, only careful investigations with the UV and ESR spectroscopy would allow one to define the nature of the existing complexes.

## REFERENCES


1 Kirsh Yu E, Kovner V Ya, Kokorin A I, Zamaraev K I, Chernyak V Ya and Kabanov V A, Eur. Polymer J., 10 (1974) 671.

2 Kokorin A I, Zamaraev K I, Kovner V Ya, Kirsh Yu E and Kabanov V A, Eur. Polymer J. 11 (1975) 719.

3 Nishikawa H and Tsuchida E, J. Phys. Chem. 79 (1975) 2072.
4 Nishide H and Tsuchida E, Makrom. Chem. 177 (1976) 2295.
5 Nishide H, Deguchi J and Tsuchida E, J. Polymer Sci., Polymer Chem. 15 (1977) 3023.
6 Tricot M, Houssier C and Desreux V, Eur. Polymer J. 12 (1976) 575.
7 Fredericq E and Houssier C, "Electric Dichroism and Electric Birefringence", Oxford University Press, London (1973).
8 Tricot M and Houssier C, in "Polyelectrolytes", 43-90, Technomic Publ., Westport (1976).
9 Tricot M, Houssier C, Desreux V and Van der Touw F, Biophys. Chem. 8 (1978) in press.
10 Tricot M, Houssier C and Desreux V, Eur. Polymer J. 14 (1978) 307.
11 Van der Touw F and Mandel M, Biophys. Chem. 2 (1974) 218.
12 Mandel M, Mol. Phys. 4 (1961) 489.
13 Van der Touw F and Mandel M, Biophys. Chem. 2 (1974) 231.

# KERR CONSTANTS OF NATURALLY-OCCURRING $\alpha$-AMINO ACIDS IN AQUEOUS SOLUTION

M. S. Beevers, G. Khanarian and W. J. Moore

School of Chemistry, University of Sydney, N.S.W.
Australia, 2006

The electro-optical Kerr constants of equeous solutions of some naturally-occuring L-$\alpha$-amino acids have been measured. The Kerr effect technique is well able to discriminate between solutions of different amino acids despite the relatively small differences between their dielectric constant increments. L-valine and L-isoleucine, whose structures are very similar, have Kerr constant increments of 3.6 and 2.28, respectively.

## INTRODUCTION

Although the electro-optical Kerr effect is well established as a sensitive technique for studying the structure and conformation of small to medium-sized molecules in solution the majority of studies have been confined to non-aqueous systems (1-3). This is partly due to experimental problems encountered in the measurement of Kerr constants of aqueous solutions; which generally possess appreciable electrical conductivity. However, perhaps the most difficult aspect concerning electro-optical studies of molecules in aqueous solution, lies in the interpretation of experimental data in terms of structural parameters, permanent electric moments and optical polarisabilities. Success in this particular sphere will depend on how much is known about the dielectric properties of the solvent-solute system chosen for study. The Le Fèvres and their many associates, over a period of two decades or more, successfully applied the electro-optical Kerr effect in their conformational studies of a large number of molecules in non-aqueous solution, where a simple dielectric theory sufficed (1,2). However, the study of aqueous solutions of molecules, using electro-optical techniques, places much greater demands on experiment and theory.

Some early investigations, which highlight the above mentioned problems, were carried out by Wyman in the 1930's, who measured the dielectric constants of aqueous solutions of a number of $\alpha$, $\omega$-amino acids, and concluded that the alkyl chains of the amino acids possessed a considerable degree of internal rotational freedom (4-6). A similar study was undertaken by Edward and co-workers on aqueous solutions of an homologous series of $\alpha,\omega$-amino acids (7,8). Their conclusions were disputed by Walder, who re-examined and interpreted their data to indicate that internal rotational motion occurred in the long chain amino acids ($>C_6$), but not in the lower members of the series which existed in extended conformations (9). These findings have been largely substantiated by carbon - 13 NMR relaxation studies (10-13). Electro-optical studies of small molecules in aqueous solution, carried out with the intention of extracting information about their conformations and structure, are very scarce indeed. Orttung and Meyers have measured the Kerr constants of aqueous solutions of glycine, DL-alanine and $\alpha$-amino isobutyric acid (14). Applying Scholte's ellipsoidal cavity model (15) they obtained a reasonably good correspondence between optical polarisability components derived from crystal refractive indices and components calculated from electro-optical data of solutions of the amino acids. It has been recently suggested that the optical polarisabilities of carbon-carbon and carbon-nitrogen single bonds and carbon-hydrogen bonds in alkyl amides in solution in 1,4-dioxane are essentially unchanged in aqueous solution (16).

In this paper we report on the Kerr constant increments, $\delta'$, of 19 naturally-occurring L-$\alpha$-amino acids measured in solution in water at 298 K and 633 nm.

## APPARATUS AND METHOD

The apparatus used to measure the electro-optical Kerr effect is to be fully described elsewhere (17). However, for the sake of completeness, a few general comments will be made here. The optical train consisted of a 5 mW helium-neon laser, polarising prism, Kerr cell, $\lambda/4$ retarder plate, analysing prism and a photo-electron multiplier tube. High voltage pulses of very short duration ($\sim 1.5\ \mu s$) were applied to the electrodes of the Kerr cell using an electronic switch built according to the circuit details published by Krause and O'Konski (18). The optical signals were retained in the solid-state memory of a transient recorder and continuously displayed on an oscilloscope for measurement. The Kerr cell was constructed from perspex and fitted with stainless steel electrodes. The temperature of the cell could be accurately controlled. It was possible to obtain useful Kerr effect data for solutions with resistivities as low as 1.5 ohm m; equivalent to an inter-electrode resistance of about 30 ohms.

In order to obtain reliable Kerr constants it is crucial to employ a detection system which is independent of 'random' fluctuations in the level of light emitted from the source and which is unaffected by attenuation of the intensity of light by the optical components. Serious errors can also occur in the measurement of the amplitude of the electrically-induced birefringence due to the non-uniform nature of the surface coating of the photo-cathode and to fatigue and saturation effects in the photomultiplier tube (19-21). In an attempt to try and overcome these problems several different detection systems were carefully examined and tested using organic liquids whose Kerr constants had been accurately determined using the conventional dc null method.

As a result of these studies the pulsed signal-nulling method of detection, similar to that used by Kaye and Devaney, was found to be the most satisfactory (22). This method yields accurate and reproducible results and is relatively simple; it requires only one Kerr cell and no sophisticated electronics other than an oscilloscope and a transient recorder. The measurement procedure involves arranging the optical components as for linear detection (23) and then applying a sequence of high voltage pulses, of constant magnitude, to the Kerr cell. The analysing prism is rotated to a position which reduces the amplitude of the optical pulse to zero magnitude. The optical retardation, $\delta$ , is equal to $-4\alpha$ (not $-2\alpha$ which is the case for dc null method), where $\alpha$ is the rotation of the analysing prism from the position of extinction in the absence of the applied electric field.

The pulsed signal-nulling method of detection is, for all practical purposes, completely independent of fluctuations in the level of light and is not subject to vagaries associated with the photomultiplier tube and amplifying circuitry. The main source of error in the determination of the solution Kerr constants, estimated to be $\pm$ 5%, occurs in the measurement of the amplitude of the high voltage pulses. These were measured using a 1000 : 1 high-voltage probe, in conjunction with a transient recorder and oscilloscope.

## MATERIALS

The L-$\alpha$-amino acids were high grade materials of analytical quality obtained from Merck Chemicals Ltd. Solutions of the amino acids were freshly prepared immediately before they were required using water which had been distilled and deionised.

## RESULTS

Kerr effect measurements were made on several aqueous solutions of each amino acid with concentrations not exceeding 3% by weight.

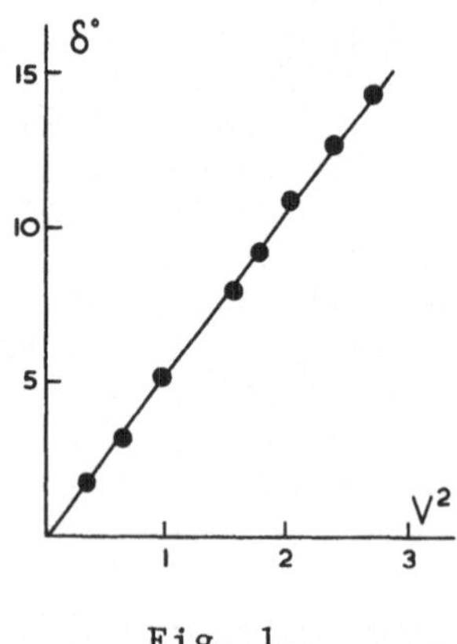


Fig. 1.

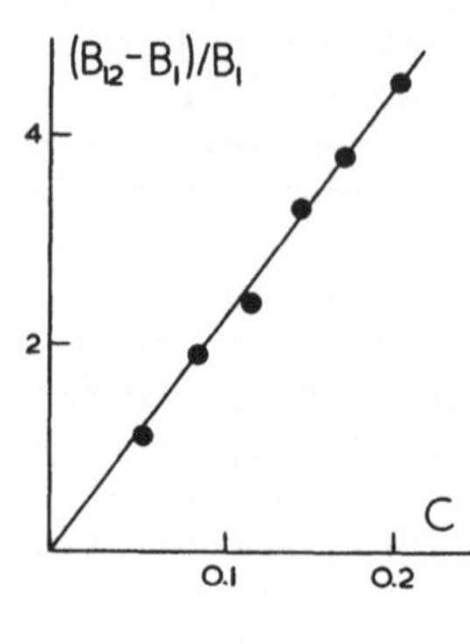


Fig. 2.

Fig. 1. Optical phase retardation $\delta$ versus square of the applied voltage for a 0.116 mol $kg^{-1}$ aqueous solution of L - arginine at 298 K. Magnitude of $V^2$ is relative.

Fig. 2. Normalised Kerr constant, $(B_{12} - B_1)/B_1$ for aqueous solutions of L - arginine at 298 K and 633 nm.

Fig. 1 shows a graph of the optical retardation plotted as a function of the square of the applied voltage for an aqueous solution of L - arginine. Adherence to Kerr's law was observed by all the amino acids taken for study. It is convenient to present the Kerr effect data in the form of incremental solution Kerr constants, $\delta'$, defined by (1)

$$B_{12} = B_1 (1 + \delta' C) \qquad (1)$$

where $B_1$ is the Kerr constant of the pure solvent; which for water has been recently determined to be $3.0 \times 10^{-14}$ $V^{-2}m$ at 298 K and 633 nm (17). C is the concentration of the solute expressed as a molality. For the majority of amino acids the graph of $(B_{12} - B_1)/B_1$ versus concentration varied linearly as shown in fig. 2 for L - arginine. A summary of the Kerr effect data obtained for aqueous solutions of the $\alpha$ - amino acids is presented in table 1 along with some dielectric constant increments taken from reference 5.

## DISCUSSION

The broad range of Kerr constant increments listed in table 1 serves to emphasise the discriminating nature of the Kerr effect. Amino acids which differ little in structure generally possess substantially different values of $\delta'$. Thus, glycine, L - alanine and L - valine have Kerr constant increments of 4.8, 4.1 and 3.6, respectively. The gradual decrease in $\delta'$ with increase in the size of the side-chain

Table 1. Kerr constant increments, $\delta'$, and dielectric constant increments $d\epsilon/dc$ of L-$\alpha$-amino acids in aqueous solution

| Substance | $\delta'$ $mol^{-1}$ kg | $d\epsilon/dc$ $mol^{-1}$ |
|---|---|---|
| Glycine | 4.8 | 23 |
| Alanine | 4.1 | 24 |
| Valine | 3.6 | 25 |
| Leucine | 1.85 | 25 |
| Isoleucine | 2.28 | |
| Phenylalanine | 9.0 | |
| Serine | 4.0 | |
| Threonine | 3.56 | |
| Proline | 5.0 | 21 |
| Hydroxyproline | 6.6 | |
| Methionine | 1.36 | |
| Tryptophane | 15.1 | |
| Histidine | 12.0 | |
| Arginine | 24.0 | 62 |
| Aspartic acid | 0.0 | 28 |
| Asparagine | -0.5 | 20 |
| Glutamic acid | 1.5 | 26 |
| Glutamine | 3.2 | 21 |

of the amino acid probably reflects a preferential increase in the optical polarisability along a direction approximately normal to the zwitterionic electric moment. L - leucine and L - isoleucine have Kerr constant increments of 1.85 and 2.28, respectively. The structure of L - valine closely resembles that of L - isoleucine, differing only in a terminal methyl group in the side-chain, but their Kerr constant increments are quite different. Since the replacement of a hydrogen atom by a methyl group does not appreciably change the electric moment of the amino acid the difference in the Kerr constant increments of L - valine and L - isoleucine is probably due to a difference in their optical polarisabilities. As expected, the Kerr constant increment of L - alanine is well removed from that of L - phenylalanine due to the highly anisotropic optical polarisability of the phenyl group.

For reasons which will not be elaborated here it is not expected that there would be a high degree of correlation between Kerr constant increments and dielectric constant increments. Thus, L - aspartic acid, which possesses the second largest dielectric constant increment in table 1, has a Kerr constant increment close to zero. However, for

L - glutamic acid, whose structure differs from that of L - aspartic acid by only an additional methylene unit, the Kerr constant increment is equal to 1.5. The Kerr constant increments of the amides, L - asparagine and L - glutamine, also differ quite markedly, but to a much greater extent than those of the parent acids.

It has thus been established that the electro-optical Kerr effect is readily able to distinguish between solutions of different L - $\alpha$ - amino acids. Since the dielectric constant increments generally vary little from one amino acid to another it is concluded that the broad range of Kerr constant increments is mainly due to differences in the optical polarisability of the amino acids in aqueous solution.

Calculation of the molecular Kerr constants of some naturally-occurring L - $\alpha$ - amino acids, using tabulated bond optical polarisabilities and atom partial charges, is currently being undertaken in this laboratory.

## ACKNOWLEDGEMENT

This work forms part of a programme directed by Professor W. J. Moore and supported by the Australian Research Grants Committee on the properties of biological molecules in high electric fields. Our thanks are extended to Professor M. J. Aroney and Dr. R. K. Pierens for advice and encouragement. One of us (G.K.) is the recipient of a Commonwealth Postgraduate Scholarship.

## REFERENCES

1 Le Fèvre C G and Le Fèvre R J W, Rev. Pure and Appl. Chem., 5, (1955) 261.
2 Le Fèvre R J W, Advan. Phys. Org. Chem., 3, (1965) 1.
3 Aroney M J, Angew. Chem. Int. Ed. Engl., 16, (1977) 663.
4 Wyman J and McMeekin T L, J. Amer. Chem. Soc., 55, (1933) 908.
5 Wyman J, Chem. Rev., 19, (1936) 213.
6 Wyman J, J. Phys. Chem., 43, (1939) 143.
7 Edward J T, Farrell P G and Job J L, J. Phys. Chem., 77, (1973) 2191.
8 Edward J T, Farrell P G and Job J L, J. Amer. Chem. Soc., 96, (1974) 902.
9 Walder J A, J. Phys. Chem., 80, (1976) 2777.
10 Cutnell J D, Glasel J A and Hruby V J, Org. Magn. Resonance, 7, (1975) 256.
11 Deslauriers R and Somorjai R L, J. Amer. Chem. Soc., 98, (1976) 1931.
12 Yasukawa T, Ghesqiere D and Chachaty C, Chem. Phys. Letters, 45, (1977) 279.
13 Beierbeck H and Saunders J K, Can. J. Chem., 55, (1977) 771.

14 Orttung W H and Meyers J A, J. Phys. Chem., 67, (1963) 1911.
15 Scholte T G, Physica, 15, (1949) 436.
16 Beevers M S, submitted for publication.
17 Beevers M S and Khanarian G, submitted for publication.
18 Krause S and O'Konski C T, J. Amer. Chem. Soc., 81, (1959) 5082.
19 Pietri G, Acta Electronica, 1, (1956) 35.
20 Rodman J P and Smith H J, Appl. Opt., 2, (1963) 181.
21 Keene J P, Rev. Sci. Instr., 34, (1963) 1220.
22 Kaye W and Devaney R, J. Appl. Phys., 18, (1947) 912.
23 Fredericq E and Houssier C, 'Electric Dichroism and Electric Birefringence', Clarendon Press, Oxford (1973).

# A BRIDGE METHOD FOR MEASURING THE DIELECTRIC RELAXATION OF CONDUCTING SOLUTIONS

M. M. Springer and J. Blom

Laboratory for Physical and Colloid Chemistry,
Agricultural University, Wageningen, The Netherlands

A method is described for measuring the dielectric relaxation of conducting solutions in the kHz region with commercially available bridges and a cell with large variable electrode spacings. By calculating the difference of the measured impedances at successive frequencies and different electrode distances, impedances due to electrode polarization, lead inductance and stray fields could be eliminated from the measurements. The literature value of $\varepsilon'$ of 1 mM KCl has been determined within two units down to 4.00 kHz. The complex dielectric and conductivity data of a dilute solution of polystyrene particles with high ionic conductivity have been measured.

## INTRODUCTION

Up to now there are only accurate dielectric relaxation measurements in the kHz region on concentrated solutions of spherical charged polystyrene particles (1) and on dilute solutions with low ionic conductivity (2).

The interpretation of the relaxation spectrum in terms of parameters of the diffuse electric double layer is mathematically only convenient if there exists no overlap of the diffuse double layers of the particles and if mutual polarization of the particles is absent (3). This requires measurements on dilute polystyrene solutions with high ionic conductivity in the kHz region. Then the relaxation effects are small, however. They amount to less than 1% of the ionic conducting current. Moreover, the polarization of the solution is screened by polarization effects near the electrodes and systematic errors could arise from lead inductances.

Different techniques have been developed to overcome these difficulties. We used a method similar to the one of Schwan et al. (1); however, with some modifications which make the elimination of the disturbing factors easier. The method could, in general, be used for the determination of the relaxation of any conducting liquid in this frequency region.

The used bridges were the transformator ratio arm bridges General Radio 1621 for frequencies below 100 kHz and the Wayne Kerr B201 for frequencies beyond 100 kHz up to 1 MHz. The advantage of transformator ratio arm bridges relative to those of the Schering type is that impedances from the terminals to earth are not seen by the bridge (4). The General Radio bridge was extended by a conductance box in order to be able to balance at high conductivities.

A schematic diagram of the three terminal cell is shown in fig. 1. The electrodes are sandblasted platinum discs, 2.5 cm in diameter. In order to diminish the effect of the electrode polarization on the capacitance readings of the bridge, large electrode separations were used. The upper electrode could be moved to a distance of 100.0 mm from the lower one. The solution was kept between the electrodes in a cylindrical teflon holder. In order to control stray fields, a shield was put around the holder and short-circuited with the upper electrode. This point of the cell was connected to the low terminal of the bridge.

Because of the high value of the ratio of the measured conductance, $G_m$, to the measured susceptance, $\omega C_m$, the potential in the solution will be practically determined by the conductivity of the solution and the distance to the lower electrode. So the current from the lower electrode to the shield will be capacitive and frequency-independent. This is confirmed by the following measurements. The upper electrode was connected to earth, the shield to the low terminal and the lower electrode to the high terminal of the bridge. Then, the potentials of the electrodes and the shield are the same as in the actual measurements. The effective stray capacitance could be determined by balancing the bridge. In fig. 2 it is shown that this stray capacitance is frequency-independent.

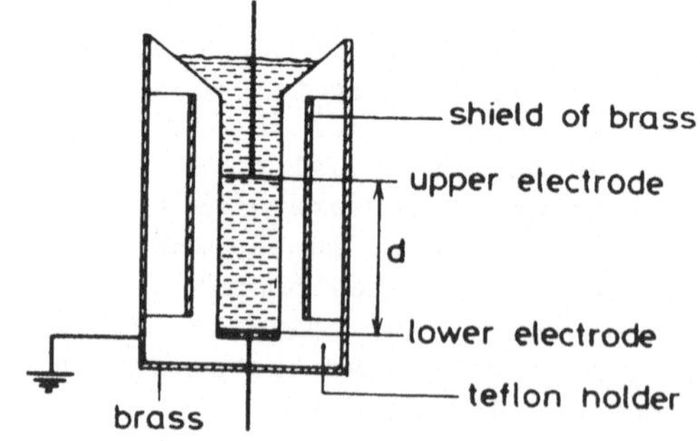


Fig. 1. Schematic diagram of the cell.

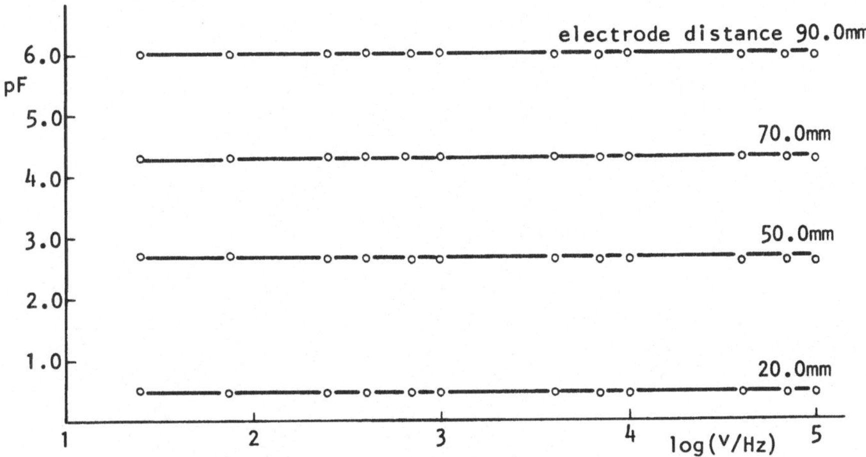


Fig. 2. Capacitance from lower electrode to shield as a function of the frequency. Upper electrode on earth, lower electrode on the high terminal and shield on the low terminal of the bridge; cell filled with 1 mM KCl in $H_2O$.

Because it is generally accepted that the electrode impedance is independent of the electrode separation and, consequently, of the admittance of the solution, the equivalent circuit of the cell may be represented as is shown in fig. 3. First, $\varepsilon_\nu$ and $k_\nu$ were determined at 400 kHz using eight electrode distances between 4.00 and 7.50 mm. At these small distances, stray effects vanish, so the following complex equation may be written:

$$Z_m = \frac{1}{k_\nu + j\omega\varepsilon_\nu\varepsilon_0} \cdot \frac{d}{A} + Z_{e\ell} + Z_{eu} + j\omega L_c$$

Because $Z_{e\ell}$, $Z_{eu}$ and $j\omega L_c$ are independent of d, the real and imaginary part of $Z_m$ versus d are straight lines. From the slopes of these lines, $k_\nu$ and $\varepsilon_\nu$ could be calculated if the area of the electrodes is known. It appears that in using for A the area of a cross-section of the space filled with the liquid, better agreement with the literature value of $\varepsilon_\nu$ of 1 mM KCl was obtained than in using the somewhat smaller value of the electrode area.

At lower frequencies where the electrode impedance becomes high, larger electrode separations are needed. Moreover, in the balance equation of the bridge one has now to take account of a lead inductance, $L_s$, in series with the low resistances on the standard side of the bridge. It heightens the capacitance readings on the bridge with an amount of $L_sG_m^2$. This inductance changes with $G_m$ and consequently with

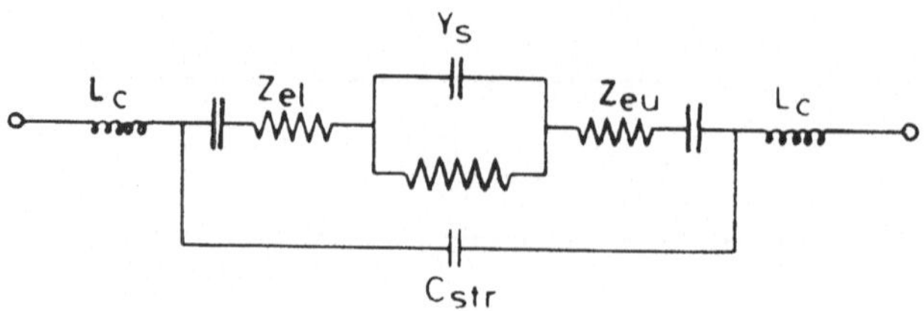


Fig. 3. Equivalent circuit of the cell.

the distance of the electrodes. Because $G_m \gg \omega C_m$, the stray capacitance and the lead inductance on the standard side of the bridge enters into the equation as follows:

$$Z_m = \frac{1}{k_\nu + j\omega\, \varepsilon'_\nu \varepsilon_0} \cdot \frac{d}{A} + Z_{el} + Z_{eu} + j\left\{\omega(L_c - L_s) - \frac{\omega\, C_{str}}{G_m^2}\right\}$$

The variation of $G_m$ with the frequency is less than 1%, so $C_{str}/G_m^2$ is in a good approximation frequency-independent. Dividing the r.h.s. and l.h.s. by $\nu$, and taking the difference between $Z_m(\nu_1)/\nu_1$ and $Z_m(\nu_2)/\nu_2$, the frequency-independent terms of the inductance and the stray capacitance vanish. We arrive at the equation:

$$\frac{Z_m(\nu_1)}{\nu_1} - \frac{Z_m(\nu_2)}{\nu_2} = \left\{\frac{1/\nu_1}{k_{\nu_1} + j\omega\varepsilon'_{\nu_1}\varepsilon_0} - \frac{1/\nu_2}{k_{\nu_2} + j\omega\varepsilon'_{\nu_2}\varepsilon_0}\right\} \cdot \frac{d}{A} + \frac{Z_{el}(\nu_1) + Z_{eu}(\nu_1)}{\nu_1} - \frac{Z_{el}(\nu_2) + Z_{eu}(\nu_2)}{\nu_2}$$

Plots of the real and imaginary part of the l.h.s. versus d give again straight lines. From the slopes, the difference in the specific impedance divided by $\nu$ at the two frequencies may be calculated. Once the specific impedance at one frequency (400 kHz) is known, the specific impedance at other frequencies may be calculated with the found differences and converted to values of $k_\nu$ and $\varepsilon'_\nu$. In this way we determined the dielectric constant of 1 mM KCl at different frequencies, as shown in the table.

In this method, one avoids the cumbersome task of determining the reactive behaviour of the conductance box on the standard side of the bridge (5).

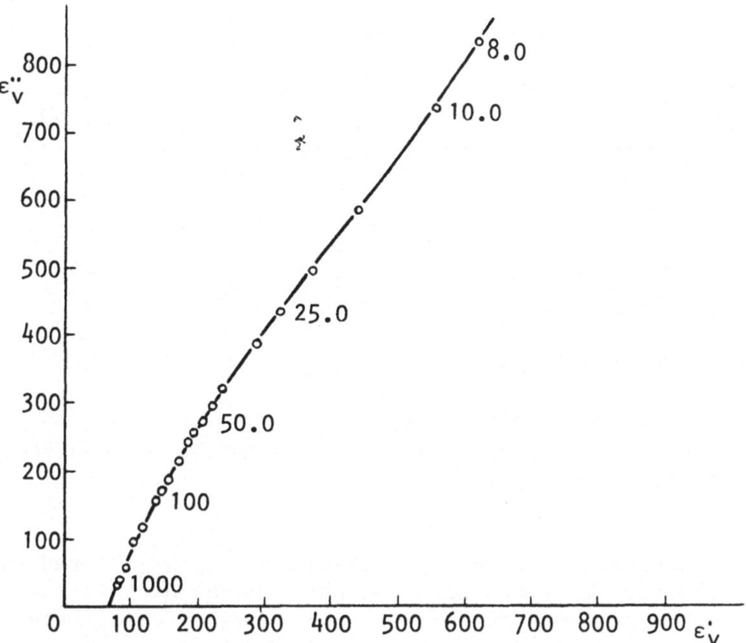


Fig. 4. $\varepsilon_\nu''$ versus $\varepsilon_\nu'$ plot of a polystyrene latex. The numbers refer to frequencies in kHz. For further specifications see text.

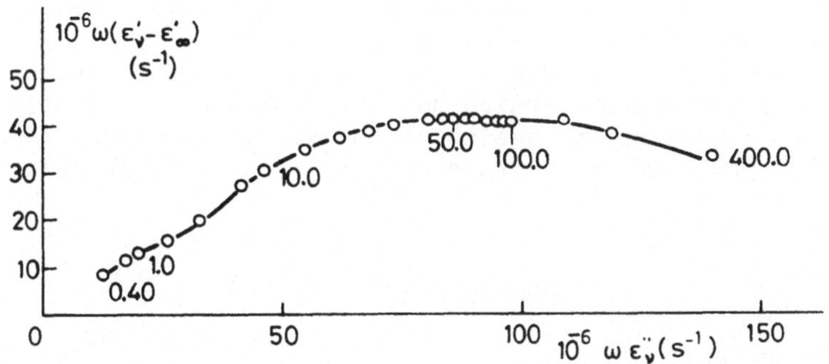


Fig. 5. Complex conductivity plot of a polystyrene latex. The numbers refer to frequencies in kHz. For further specifications see text.

Table. Range of d values, dielectric constants and tan $\delta$ values of 1 mM KCl.

| $\nu$ kHz | range of d (mm) | $\varepsilon'_\nu$ | $\pm\Delta\varepsilon'_\nu$ | tan $\delta$ $(=\frac{K_\nu}{\omega\varepsilon'_\nu\varepsilon_0})$ |
|---|---|---|---|---|
| 800 | 4.00 - 7.50 | 79.0 | 1.4 | 4.3 |
| 400 | " | 78.4 | 1.1 | 8.5 |
| 200 | " | 78.7 | 1.1 | 16.9 |
| 100 | " | 79.5 | 1.2 | 33.4 |
| 70 | " | 79.0 | 1.2 | 48.1 |
| 40 | " | 79.0 | 1.2 | 84.3 |
| 10 | " | 76.8 | 1.8 | 347.2 |
| 7 | 6.00 - 30.00 | 77.5 | 1.7 | 491.5 |
| 4 | " | 79.7 | 1.8 | 836.2 |

Mean value of $\varepsilon' = 78.6 \pm 0.9$: Value according to 'Handbook of Physics and Chemistry' $= 78.4$

In the measurements we changed the frequency at a fixed spacing of the electrodes. This was followed by a change of this spacing and measurements at the same frequencies. This sequence in measuring eliminates in combination with the use of the last equation errors due to conductivity drift and reproducibility of the adjustment of the distance between the electrodes.

To determine the change of $k_\nu$ with $\nu$ due to the relaxation requires a precise measurement of k. This will be treated in detail in a publication which is in preparation.

In figs. 4 and 5 the relaxation data at 25°C are shown of 9.5% (w/w) polystyrene latex with a particle diameter of 559 $\pm$ 36 nm determined by electron microscopy. The specific conductivity at 100 Hz amounts to 0.08653 mho.m$^{-1}$, which has been taken as the static value. The latex has been dialysed against 1 mM $HNO_3$. The charge density of the particles prepared in the same way was -15 $\mu$C.cm$^{-2}$ according to Norde (6).

## NOMENCLATURE

| | |
|---|---|
| A | electrode area |
| $C_m$ | measured parallel capacitance |
| $C_{str}$ | stray capacitance |
| d | separation of the electrodes |
| $G_m$ | measured parallel conductance |
| j | $\sqrt{-1}$ |
| $L_c$ | inductance of the leads connecting the cell to the bridge |
| $L_s$ | inductance of the leads at the standard side of the bridge |

$Y_s$ admittance of the solution
$Z_{el}$ electrode impedance of the lower electrode
$Z_{eu}$ electrode impedance of the upper electrode
$Z_m$ impedance calculated from $G_m$ and $C_m$
$\varepsilon_o$ dielectric constant of the vacuum
$\varepsilon'_\nu$ relative dielectric constant of the solution at frequency $\nu$
$\varepsilon''_\nu$ loss factor at frequency $\nu$
$k_\nu$ specific conductivity of the solution at frequency $\nu$
$\nu$ frequency (Hz)
$\omega$ $2\pi\nu$

## REFERENCES

1 Schwan H P, Schwarz G, Maczuk J and Pauly H, J. Phys. Chem., 66 (1962) 2626.

2 Ballario C, Bonincontro A and Cametti C, J. Coll. Interface Sci., 54 (1976) 415.

3 Dukhin S S and Shilov V N, "Dielectric phenomena and the double layer in disperse systems and polyelectrolytes", John Wiley & Sons, New York (1974).

4 Rosen D, Bignall R, Wisse J D M and Van der Drift A C M, J. Phys. E., 2 (1969) 22.

5 Schwan H P and Sittel K, Trans. Am. Inst. Electr. Eng., 72 (1953) 114.

6 Norde W, Thesis Agricultural University, Wageningen 1976, The Netherlands.

# PHOTOCONDUCTIVITY AND DIELECTRIC PROPERTIES OF POLYHEXAMETHYLENE ADIPAMIDE

G. Guillaud and M. Maitrot

Laboratoire d'Electronique, Université Lyon I,
43 Bd du 11 Novembre 1918, 69621 Villeurbanne, France

The photoelectric properties of polyhexamethylene adipamide thin samples have been studied in near UV and near IR wavelength range. Two effects have been pointed out, with UV light, the first one being sensitive to the total integrated flux, the second varying as $t^{-1/2}$. In IR range a photo dielectric effect is seen.

## INTRODUCTION

The photoelectric properties of polymers synthesised in a variety of ways were recently studied by many authors (1) but among these polymers few have a large dark ionic conductivity.

In ionic polymers, the interface or subelectrode layers play a very important part on DC or low AC conductivity because they can act as sources or sinks of electronic or ionic carriers. So it is interesting to test the photoelectric properties of ionic polymers and see if the general non ionic "trend" is followed. Among polar polymers, polyhexamethylene adipamide was studied in our laboratory many years ago and we well know its dark behaviour (2).

Two main wavelength ranges seem interesting: the near ultraviolet and the near infrared part of the spectrum. In the first range, light assisted electrode injection was the proposed model for non polar polymers. In the second, a photodielectric effect (detrapping of charges stored in shallow traps near the electrodes) was pointed out.

Our work is divided into two parts. First DC and AC relatively "high" voltages were used to test the photoinjection contact properties in the ultraviolet range. On some samples a photocapacitive effect was found, but in all the studied samples electron injection was obtained.

In the second part we tested transport properties in visible and IR range. The photodielectric effect was present and some variations in the AC voltage dielectric spectrum were available (using low voltage).

## EXPERIMENTAL

### Samples

"Commercial" nylon samples were used initially. Later "pure" samples were obtained from a very pure nylon powder provided by the Macromolecular Chemistry Laboratory of the Claude Bernard University. The general preparation procedure follows the following lines. This powder is dissolved in formic acid. Then a drop of this solution is put on a glass slab, calibrated by a roller and evaporated. Variable thickness sheets (in the 10-100 $\mu$ range) are achieved with a crystallinity depending on the evaporating rate. The crystallinity can be rather low.

Formic acid is eliminated by numerous washings with pure acetone and the samples are heated near $100^{o}$ C in a vacuum ($10^{-6}$ Torr) for a complete drying. Dry samples are kept in an Argon atmosphere.

Semi-transparent metallic electrodes (Au, Ag, Al) are deposited by vacuum evaporation ($10^{-6}$ Torr) on the two faces of the sheets. Photometric measurements afford a control for the optical electrode properties.

### Measuring cell

The samples are set in a small regulated oven. The pressing electrical contacts are made with gold or silver or Al metal covered rings to avoid contact voltages. The incident light crosses spectrosil glass slabs (transparent up to 1600 A). The cell can be used in a vacuum ($10^{-6}$ Torr) or in a gaseous controlled atmosphere. Electric leads are insulated by teflon.

Photoemission from metallic parts of the cell to the lightened electrode (or from the lightened electrode to the metallic parts of the cell) can give rise to wrong currents. These photoemission currents disappear by coating the inner surface of the cell by a thin teflon layer.

A high pressure Mercury lamp (HPK 125 Philips burner), a Xenon lamp and a deuterium with some optic filters lamp were used.

Temperature measurements

The true sample temperature is difficult to ascertain: the oven is not really a black body and the sample is heated by the light. Hence the control thermocouple temperature is not the sample temperature.

The best measurements can be made by testing the electrical resistance of the semitransparent electrodes versus temperature in a fully closed oven: temperatures of control thermocouple and sample are then identical. The oven can be "opened" or "closed" from an external device. The true temperature of the lightened sample can then be given; the temperature gradient between the faces is evaluated. The samples are very thin, so we can expect the temperature difference between the electrodes and sample to be low. These measurements are sometimes difficult; so a modulated light can also be used to afford a smaller temperature effect.

Measuring circuits

(a) DC measurements. The electric fields can be high, and the illuminated electrode either positive or negative. From the possible measuring circuits (fig. 1) we used only those on figures 1 a, b and c. With the others the measurements are disturbed by high wrong currents, due to the illuminated face of the sample. The currents enter a Keithley 410 A picoammeter and are recorded with a memory Tektronix oscillograph. Using a second picoammeter the two measured currents are compared: they are identical if there is no wrong parasitic effect.

(b) AC measurements. We studied the variations of the parallel equivalent conductance Gp and parallel equivalent capacity Cp versus AC frequency. We used a small amplitude signal (50 mV) from a generator HP 203 A, in the range $10^{-4}$ to 5 x $10^{4}$ Hz. Various techniques are used in this range: between 20 Hz and 50 KHz we used a conventional General Radio 1615 A bridge. Between $10^{-4}$ and 20 Hz, we recorded the in phase and $90^{o}$ out of the phase components of the current.

## RESULTS

"Pure" samples

(a) Voltage and light steps. The dark conductivity of the samples is far from negligible and dark transients are very high, so it is generally more interesting to use light steps with a constant applied voltage than voltage steps with a constant light.

Two types of decay have been found, as in other polymers already studied (fig. 2). A fast transient, occurring at all wavelengths and a slower decay chiefly occurring in the shorter wavelength region are observed in polyethylene. The wavelength behaviour for nylon, although probably similar, has not been fully studied. These transients,

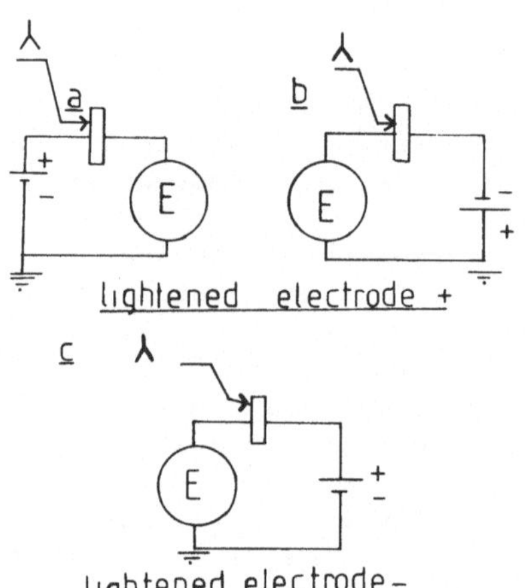


Fig. 1. Measuring DC circuit for a and b, the illuminated electrode is positive, whilst for c the illuminated electrode is negative.

and especially the fast one have a far larger amplitude when the illuminated electrode is negative and no correlation is found between the magnitude of the dark absorption current and the magnitude of these transients. When several following light steps are used in a constant light, a few minutes apart, the fast transient does not appear. A complete recovery is found after an hour in the dark followed by a few minutes in the light, when the sample is short circuited.

The temperature rise can be measured in a permanent state and can be as high as 30 or 40 degrees. When the temperature reaches a constant value with a DC light, the corresponding dark current is far less than the measured current.

(b) Modulated light. With a modulated light the temperature rise can be really weak ( ~ a few degrees) and some interesting and signifi-

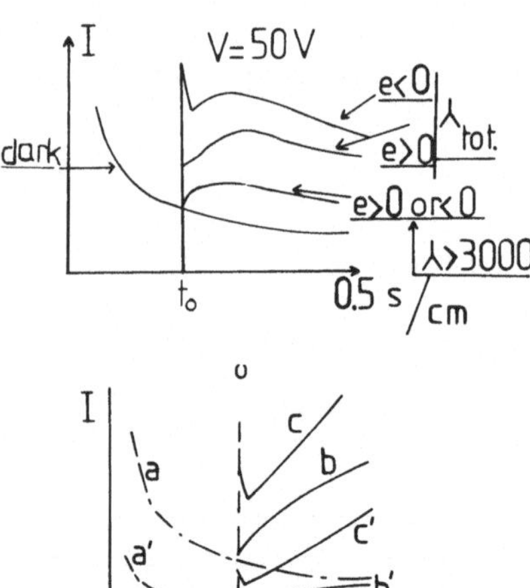


Fig. 2a. Intensity versus time for dark and photocurrent (20 μ Argon). An apparent transit time can sometimes be seen due to a warming effect (in dark we have room temperature).

Fig. 2b. a,a': dark current following a voltage step, a : 18 V to a' : 4,5 V. The voltage is on at t = 0 and the light at $t = t_o$. Curves c and c' illuminated electrode negative: curves b and b' illuminated electrode positive ($10^{-6}$ Torr pressure)

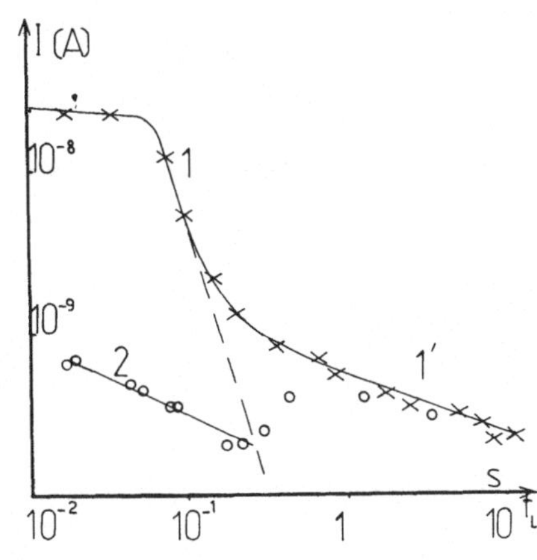


Fig. 3. Negative illuminated electrode - we have i versus $t_L$, where $t_L$ is the total illumination time. The curves are obtained with a modulated light in vacuum (warming effect is small). (1) first illumination we can see that the intensity of the fast transient is constant $i = i_o$ during a time t depending on voltage and light flux. (1') first illumination, slow transient $I \sim t_L^{-1/2}$. (2) second illumination (after several seconds) we have first a curve $I \sim t_L^{-1/2}$ then a weak warming transient effect and then the i value is approximately the same as in (1').

cant remarks can be made on the fast transient.

If the modulation is fast enough, the transient is decomposed into successive and equal pulses. The total charge across the external recorder does not depend on the modulation velocity.

The curves on fig. 3 give $I(t_L)$, $t_L$ is the total illumination time. An intensity plateau is seen for $V > 15$ V. The initial $I_o$ value and the total recorded charge versus V in this case are given in fig. 4. After a time $t_{Lo}$ depending on voltage V, the current decreases sharply in the following light pulses, then the slow transient can be seen. If the light is set off, and then put on again, no fast transient appears.

These results are likely of the same nature as those related to some other polymers (2-6). Several interpretations have been suggested for the polyethylene case: the first is the destruction of the impurity centers and charge release in a thin layer near the electrode (but the results are sometimes the same whatever the sample purity). An alternative is due to space charge accumulation, although dark currents are insensitive to the illumination duration or intensity. It is the

Fig. 4. $I_o$ versus V for a constant light flux and $Q_o = \int i_o dt$ for the fast transient - $i_2$ is (at another scale) the intensity of the slow transient versus V with $t_L \sim 20$ seconds.

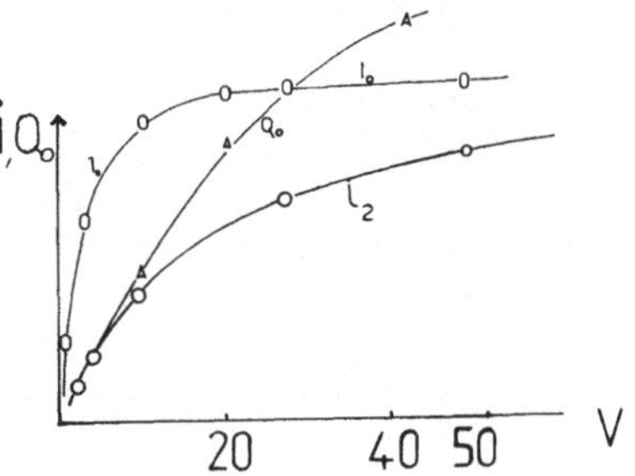

reason why the generated currents are likely crossing the whole sample: with a space charge accumulation the photocarriers would modify the dark conduction.

If the total collected energy $E = F\ t_L$, F is the flux and $t_L$ the illumination time, we have (from Wintle (1)):

$$\frac{I}{F} \not\parallel \frac{C}{E + E_o} \quad \text{C and } E_o \text{ are constant}$$

and when $E \ll E_o$ for short $t_L$ values,

$$I_o = \frac{C\,F}{E_o} \quad \text{the intensity plateau is in direct ratio to the light flux.}$$

This expression can be developed theoretically for different processes. For instance, in a very recent paper (7) Wintle points out the effect of the trapped space charge on the injected Schottly current: the trapping of charge in the insulator results in a decreasing current of the hyperbolic form: $j \sim \frac{1}{t + \tau}$ where $\tau$ is a constant, whether the charges are trapped in a thin sheet or are spatially distributed.

However the proportionality of I versus F is non rigorous for high intensity flux and other processes (excitonic bimolecular interaction) can perhaps better fill the experimental data (7).

The slow transient current varies as $t_L^{-1/2}$ with the illumination time. The corresponding photocarriers likely transit across the whole sample, but the generation mechanism is different (perhaps a bulk mechanism). It is relatively greater for low voltages and thermally activated. An ambient gas decreases this effect, possibly due to a free volume effect. In fact some conductive channels can be supressed by gas molecules.

## Impure samples

Some commercial nylon 66 samples were studied with AC and ramp voltages (with compensation for dark conductivity). Some interesting results were obtained:

A fast capacitive effect is seen when the illuminated electrode is positive (silver electrode) - (cf. fig 5 with a ramp voltage). When a second ramp is used, the capacitive effect is less significant. When the sample is short circuited in the light, a fast recovery is obtained. Some electrons are injected by the negative electrode but holes injected at the positive one seem to be trapped immediately. The effect is hindered by a gaseous atmosphere.

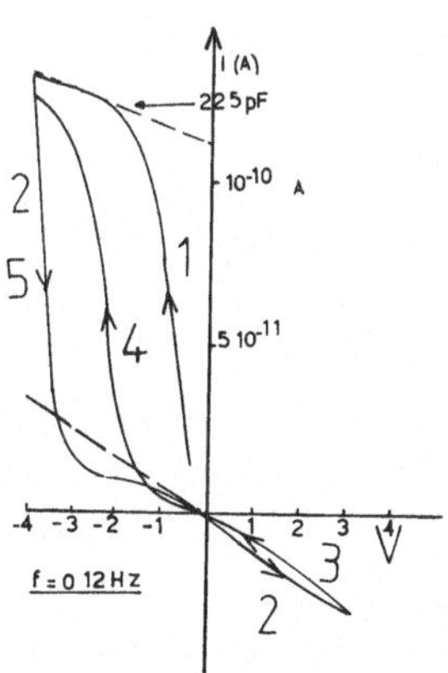


Fig. 5. Capacitive effect with an impure sample. A ramp voltage with compensation for dark current is used. This effect is obtained when the positive electrode is illuminated.

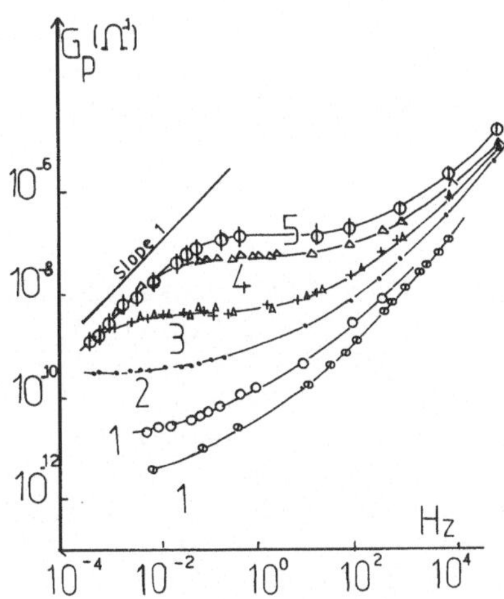


Fig. 6. $G_p$ versus $\omega$ in dark and in infrared light.

Red and near IR range

In this wavelength range transient photocurrents have been reported in the literature and ascribed to a thermal expansion of subelectrode material. This expansion results in motion of charges accumulated in the boundary layers adjacent to the electrodes. On the other hand, a large warming effect of the sample is always seen. In fact, electrodes play an important part in the AC dark conductivity as sources or sinks of carriers and the AC spectra measured with red or near infrared light are very similar to dark spectra (fig. 6). The illuminated sample temperature can be measured from the previously tested electrode resistance. The mechanical stresses following the sample illumination can only raise this resistance, giving an erroneously high temperature value. The true light AC spectrum can be measured. We can see that the AC conductivity plateau is raised by IR light; hence IR light and temperature have a similar effect. This is the detrapping of ionic carriers from shallow traps near the electrodes and perhaps ionisation of carriers from bulk "shallow" traps, raising the mobility value.

## CONCLUSION

Photoconductive properties of nylon 66 provide important and powerful means of characterising surface and bulk properties of this polymer. Much work is yet to be done to fully understand the mechanisms of these effects, as functions of purity, crystallinity, orientation, gazeous atmosphere and free volume of the material.

## REFERENCES


1 Wintle H J, IEEE Transactions on Electrical Insulation, Vol. E 1.12 No. 2 (April 1977) p.97-133.

2 Rosenberg N, Guillaud G, Michel R and Maitrot M, J. Phys. Chem. Solids, (1976) 5-15.

3 Wintle H J and Tibensky G M, J. Polym. Science Pol. Phys. 11 (1973) 25-30.

4 Wintle H J, J. Polym. Science Pol. Phys. 12 (1974) 2135-51.
Cohins J D and Wintle H J, J. Polym. Science Polym. Phys. 10 (1972) 2259-2280.

5 Mizutani T, Takai Y and Ieda M, Japan Journal Appl. Physics, 11 (1972) 597.

6 Chan G Y C and Wintle H J, J. Polym. Sci. Polym. Phys. 13 (1975) 1187-1199.

7 Wintle H J and Sapreka H, Journal of Electrostatics, 3 (1977) 195-202.

# *Colloidal Systems*

'Happy he
With such a mother! faith in womankind
Beats with his blood, and trust in all things high
Comes easy to him, and tho' he trip and fall
He shall not blind his soul with clay.'

ALFRED, LORD TENNYSON (1809-1892)

The Princess: Intro. Song

# ELECTRO-OPTIC STUDIES OF COLLOIDS AND THEIR STABILITY

S. P. Stoylov

Institute of Physical Chemistry, Bulgarian Academy of Sciences, 1000 Sofia, Bulgaria

On the basis of the general ideas of Lebedev for intermolecular forces, and the general theory of Lifshitz for electrodynamic interactions and its recent modifications and extensions on the one side, and the experiments and theories of the interfacial electric polarizability of O'Konski, Schwarz, Stoylov, Jennings, Shurr, Takashima, Dukhin, Shilov and Oosawa on the other, a connection is revealed between electro-optically and dielectrically measured high values of the electric polarizability of some colloid particles and polymer molecules and the stability of their aqueous solutions or suspensions. Conditions are suggested where this connection is believed to be the most pronounced. These are supported by the theoretical expectations and the experimental results of other authors.

## INTRODUCTION

One of the main achievements of electro-optical studies in the last 20 years (1-3) is the clear establishment of the existence of an important interfacial electric polarizability of big molecules and of colloidal particles. In the case of strongly anisodiametric molecules or particles of longest dimension of the order of the wavelength of light, in aqueous media of low ionic strength ($10^{-3}$, $10^{-4}$), the value of the interfacial electric polarizability could reach a value 2 or 3 orders of magnitudes higher than the value of the electric polarizability connected with the characteristics of the volume of the particle. The relaxation of this interfacial polarizability takes place at frequencies which could be 7 to 12 orders of magnitude lower than that known for the volume polarizability. Both great differences from the values which could be predicted following only the electrical properties of the volume of the particles have been explained with the polarization of the interface. In many cases studied, the polarization of the interface is mainly connected with the redistribution under the action of the applied electric field of the ions outside the particle's volume either on the particle's surface or near it. In most cases it is accepted that these are the ions which participate in the compensation of the particle's charge, i.e. in the terms of colloid science these are ions from the particle's double electric layer.

The large values found for the electric polarizability and for its relaxation time are related to the large distances on which the ions are redistributed and correlated in their movements under the action of the external electric field. It deserves mention that electro-optic measurements show this polarizability and relaxation. They are provoked practically by "external" charges with respect to the particle's volume and are strongly connected with the mechanical movement of the particles. In most cases that is in the form of orientation of the particles, i.e. particle rotations.

It is rather astonishing that till now cases (3,4-8) in which the effect of these strongly pronounced particularities on other essential properties of the macromolecular or colloidal systems have been rarely searched for. Firstly, one could study the question in terms of the stability of the respective systems. Till now almost all problems (here interactions and stability being included) in macromolecular and colloid sciences where the electric properties of the big molecules or particles are involved have been considered without paying attention to the interfacial electric polarizability.

When speaking of the connection between electric polarizability of big macromolecules or colloid particles and the interactions between them one can not jump over the fundamental ideas of Lebedev (9) who was the first to point out in a very general and clear way, the far reaching analogy between the behaviour of molecules in external electric fields and in the electric fields of the neighbouring molecules. In both cases the fluctuating electromagnetic fields at all frequencies are the basic factor.

In the 60's the ideas of Lebedev saw a precise theoretical realization in the work of Lifshitz (10), who derived rigorous expressions for the interactions of macroscopic condensed bodies in vacuum. The consequent generalization of the Lifshitz theory for interactions in liquids (11) and in electrolytes (12) as well as its further application to colloid problems (13) did not take into account in a detailed way the frequency interval below 1 MHz, where the interfacial electric polarizability is acting in full force. In some cases, the contribution of the low frequency polarizability seems to have been taken into account through appropriate consideration of the zero frequency term.

## INTERFACIAL ELECTRIC POLARIZABILITY

It is rather a long story to describe the introduction of the notion for the interfacial electric polarizability in science. An attempt to do this has been made recently by the author (14). There are three aspects of this problem. The experimental evidence for the existence of interfacial electric polarizability, the qualitative physical model explaining the characteristics of the polarizability experimentally observed and the quantitative mathematical treatment of the physical model.

Most convincing and the richest experimental evidence for the existence of the interfacial electric polarizability have come from electro-optic (3) and dielectric (15,16) studies. There are some works which deserve to be mentioned in this connection: the dispersion of benzopurpurine suspension at 300 kHz (17); the high value of the electric polarizability of tobacco mosaic virus, its low frequency dispersion, dependence on electrolyte concentration and on pH (18) all followed by electric birefringence; the high value of electric polarizability of polyphosphates (19), the high value of electric polarizability of polystyrene lattices, low frequency dispersion and the square dependence of the critical frequency on particle's radius followed dielectrically (20), the high value of electric polarizability of palygorskite clay, its low frequency dispersion (21), effect of surface active substances on it (4-8) as studied by electric light scattering.

Whilst from the experimental evidence it seems that there is an excellent accordance between the results of almost all authors working in the field, for the qualitative physical model of the interfacial electric polarization the situation is totally different. Hence Mandel (22), Schwarz (23), O'Konski (24), Schurr (25) and others considered that interfacial electric polarization is connected only with the redistribution of counter ions located at the particle surface (bound ions). Some of these authors (24,25) took into account the exchange of ions with the diffuse double layer. On the other hand Dukhin and Shilov (3) considered that the polarizability is connected with the redistribution of ions in the diffuse electric layer, the exchange of ions with the volume of the electrolyte being taken into account. This difference in the physical models leads to some controversy concerning the relation between diffuse and dense electric layer in the particle's interfacial electric polarizability (14).

When the question comes to the quantitative mathematical treatment of the different models (2,22-25) it seems that again there are no great differences among the separate authors either in the mathematical approach to the problem or in the type of relations finally obtained. This is also true to a great extent for the relaxation times of the electric polarizability.

It seems that except for the theory of Dukhin-Shilov (3) no other theory has such a generality as that of Oosawa (26). What is particularly useful for the connection, which is emphasized in this article, is the fluctuational approach of Oosawa to the description of the interfacial electric polarizability. Even in the absence of externally applied electric field the counter ions (which at equilibrium are distributed with a spherical symmetry around the macromolecule or colloidal particle so that they have no net electric dipole moment) under the action of their thermal motion, the macromolecule or the particle has a fluctuating dipole moment. For a rod-like polyion Oosawa (26), expanding the fluctuations of the counter ion concentration along the rod as a Fourier series derived for the j-th Fourier component of the electric polarizability $\gamma_j$:

$$\gamma_j = 2\left(\frac{\ell}{2\pi j}\right)^2 \cdot \frac{n_o e^2}{kT} \cdot \frac{1}{(1 + n_o \phi_j / 2kT)}, \qquad (1)$$

where $n_o$ is the number of counter ions per rod-like polyion, $\ell$ is the polyion length, e is the elementary charge, k is the Boltzmann constant, T the absolute temperature, and $\phi_j$ is the j-th Fourier component of counterion - counterion interaction energy. The relaxation time of the electric polarizability is given by the Fourier transform of the diffusion of time dependent fluctuation of counter ion concentration:

$$\tau_j = \frac{1}{ukT}\left(\frac{\ell}{2\pi j}\right)^2 \frac{1}{(1 + n_o \phi_j / 2kT)}, \qquad (2)$$

where u is the counter ion mobility. For the lowest mode (j = 1) and no counterion - counterion interaction ($\phi_j = 0$) one obtains

$$\tau = \ell^2 / (2\pi)^2 D \qquad (3)$$

where D = u k T is the translational diffusion constant of the ions.

## ELECTRODYNAMIC INTERACTIONS

One of the fields where all the story of the electrodynamic interactions could be found in quite a detailed way is the interaction between colloidal particles. A shortened variant of this story, starting with the microscopic approach to this problem, is attempted in reference (14). Here the Lebedev-Lifshitz approach will be briefly summarised.

According to Lifshitz (10) at every point of a system of material bodies, there exists fluctuational electric and magnetic fields due to fluctuations of electric charges. The charge fluctuations are due both to thermal agitation and to natural uncertainties in the positions and momenta of electrons and atomic nuclei (27). Because of the interactions of these fluctuational electric and magnetic fields at the different points of the system, at every point of the system the fluctuating electric and magnetic fields are the sum of their own fields and those which are the response of the fields elsewhere. Thus the charge fluctuations throughout the system are correlated and the electrodynamic energy of the system is composed of the electrodynamic energies of the separate parts of the system and the exchange of electrodynamic energy among them at all possible frequencies. Physically significant for the interaction is this part of the electrodynamic energy which changes with the variations of the distances among the different parts of the system at frequencies at which there is absorption of electrodynamic energy in any part of the system.

As an illustration, following references (13) and (27), one can write for the electrodynamic free energy of interaction G between two particles L and R through a planar slab m of thickness $\ell$:

$$G(\ell) = \frac{kT}{8\pi\ell^2} \sum_{n=o}^{\infty\,\prime} x \ln\left(1 - \Delta_{L_m}^{(n)} \Delta_{R_m}^{(n)} e^{-x}\right) dx$$

$$\approx \frac{kT}{8\pi\ell^2} \left\{ \frac{\Delta_{L_m}^{(o)} \Delta_{R_m}^{(o)}}{2} + \sum_{n=1}^{\infty} \Delta_{L_m}^{(n)} \Delta_{R_m}^{(n)} (1 - r_n)\, e^{-r_n} \right\}, \tag{4}$$

where $\Delta_{L_m}$ and $\Delta_{R_m}$ are the differences in the dielectric susceptibilities

$$\Delta_{L_m}^{(n)} = \frac{\varepsilon_L - \varepsilon_m}{\varepsilon_L + \varepsilon_m} ; \quad \Delta_{R_m}^{(n)} = \frac{\varepsilon_R - \varepsilon_m}{\varepsilon_R + \varepsilon_m}$$

the susceptibilities being evaluated at frequencies $i\xi_n$

$$\xi_n = \frac{2nkT}{\hbar} \qquad n = 0, 1, 2 \ldots,$$

$2\pi\hbar$ is the Planck constant, $r_n$ is the ratio of the travel time of an electromagnetic signal across the gap $\ell$ and back $\{2\ell\varepsilon_m^{\frac{1}{2}}(i\xi_n)/c\}$ to the characteristic period $(1/\xi_n)$ of the frequency:

$$r_n = (2\ell\varepsilon_m^{\frac{1}{2}}\xi_n / c)$$

and c is the velocity of light in vacuum. The prime in the summation indicates that the n = o term is multiplied by 1/2.

In the last ten years there have been considerable advances in both the simplification of calculation procedures (13) and the importance of temperature dependent electrodynamic forces (13,27). In both cases an important role has been played by Parsegian and co-workers (13,27). Discussing the Kirkwood-Shumaker's proton fluctuation theory (28) it is suggested (13) that for protein molecules with high dielectric increments, the interaction forces could dominate the interaction forces at higher (optical) frequencies by one or two orders of magnitude. So it seems that the zero frequency contribution to the electrodynamic forces could become dominant in highly polarizable colloid and macromolecular disperse systems. When the particles are charged there could arise strong screening effects and additional long range forces due to current - current fluctuation correlations (13). Extremely pronounced effects are expected for highly asymmetric particles (13). Some experimental evidence is appearing for the existence of very long range strong attraction forces. So for example in the case of interactions of red cells with a hydrocarbon surface a range of attraction as large as 2500 Å has been detected (29).

## ELECTRO-OPTIC STUDIES AND COLLOID STABILITY

It is rather a rare case when researchers studying colloid stability make use of electro-optic and dielectric measurements. So for example on the last Discussion of the Faraday Society on Colloid Stability held in Lunteren in April 1978, where one of the principal problems was the effect of polymer adsorption on colloid stability, none of the 20 presented papers used electro-optic or dielectric methods. Furthermore, from the approximately 100 comments, only one word was spoken for dielectric studies.

Against this "unhappy" background the correlation found between colloid stability for some clays and the extremum in the value of the electric polarizability (4-8) seems to show a promising way of going deeper in the understanding of colloid stability. Broad investigations of the interactions and electric polarizability studied dielectrically and electro-optically of colloid systems, stable at the isoelectric point, seem to be one of the recommended ways for the future work in this field. One can find examples for such systems in references 4-8, 30 and 31.

## CONCLUSION

Although very important, the bridge "interfacial electric polarizability - colloid stability" is not the only broad field of application of electro-optical and dielectric information for the electric properties of dispersed particles. Probably this is not the most important application of what we know for the electric polarizability at present. New exciting fields of application are to be expected in the near future. As an example one could think of the intriguing question of the mechanism of the sensitivity of biological systems to weak electromagnetic fields. It is not excluded that a part of the answer sought at present at high frequencies ($\approx 10^{11}$ Hz) (32,33) could be found in the low frequency range ($\approx 10^{6}$ Hz).

## REFERENCES


1 Fredericq E and Houssier C, "Electric Dichroism and Electric Birefringence", Oxford University Press, London (1973).

2 O'Konski C T, Ed. "Molecular Electro-Optics", Marcel Dekker, New York (1976).

3 Stoylov S P, Shilov V N, Dukhin S S, Sokerov S Kh and Petkanchin I B, "Electro-Optics of Colloids", Naukova Dumka, Kiev (1977).

4 Stoylov S P and Petkanchin I B, CR Acad. Bulg. Sci., 24 (1971) 487.

5 Stoylov S P and Petkanchin I B, J. Coll. Interface Sci., 40 (1972) 159.

6 Petkanchin I B, Sokerov S Kh and Stoylov S P, Proc. VI Congress on Surface Active Substances, Zurich, Carl Hanser Verlag, München (1973).

7 Stoylov S P, "Proc. Intern. Conference on Colloid Surface Science", Publishing House Hung. Acad. Sci., Budapest, p.379 (1975).

8 Petkanchin I B and Bruckner R, Coll. Polym. Sci., 254 (1976) 596.

9 Lebedev P N, "Selected Works", GITTL, Moscow, Leningrad (1949).

10 Lifshitz E M, Zh. Exp. Theoret. Fiz., 29 (1955) 94; Sov. Phys. J.E.T.P., 2 (1956) 73.

11 Dzyaloshinskii I E , Lifshitz E M and Pitaevskii L P, Zh. Exp. Theoret. Fiz., 37 (1959) 229; Advan. Phys., 10 (1961) 165.

12 Barash Yu S and Ginzburg V L, Uspekhi Fiz. Nauk, 116 (1975) 5; Sov. Phys. Uspekhi, 18 (1975) 305.

13 Mahanty J and Ninham B W, "Dispersion Forces", Academic Press, London (1976).

14 Stoylov S P, Commun. Dept. Chem., (1978) (Izv. Khim.) in press.

15 Chelidze T L, Deryavenko A I and Kurilenko O D, "Electric Spectroscopy of Heterogenious Systems", Naukova Dumka, Kiev (1977).

16 Takashima S and Minakata A, Digest of Literature on Dielectrics, 37 (1973) 602.

17 Errera J, Overbeek J Th G and Sack H, J. Chim. Phys., 32 (1935) 681.

18 O'Konski C T and Haltner A J, J. Am. Chem. Soc., 79 (1957) 5634.
19 Eigen M and Schwarz G, J. Coll. Sci., 12 (1957) 181.
20 Schwan H P, Schwarz G, Maczuk J and Pauly H, J. Phys. Chem., 66 (1962) 2626.
21 Stoylov S P, Proc. IV Congress on Surface Active Substances, vol. 2, Gordon and Breach, New York, p.171 (1967).
22 Mandel M, Mol. Phys., 4 (1961) 489.
23 Schwarz G, J. Phys. Chem., 66 (1962) 2636.
24 O'Konski C T, J. Phys. Chem., 64 (1960) 605.
25 Schurr J M, J. Phys. Chem., 68 (1964) 2407.
26 Oosawa F, Polyelectrolytes, Marcel Dekker, New York (1971).
27 Parsegian V A, Ann. Rev. Biophys. Bioeng., 2 (1973) 221.
28 Kirkwood J G and Shumaker J B, Proc. Natl. Acad. Sci., USA, 38 (1952) 855.
29 Gingel D, Todd I and Parsegian V A, Nature, 268 (1977) 767.
30 Stoylov S P and Petkanchin I B, God. Sof. Univ., Khim. Fak., 61 (1966/67) 227.
31 Ottewill R H and Watanabe A, Koll. Zs., 170 (1960) 38, 132.
32 Fröhlich H, Proc. Nat. Acad. Sci., USA, 72 (1975) 4211.
33 Grundler W, Keilmann F and Fröhlich H, Phys. Letters 62A (1977) 463.

# SIZE DISTRIBUTIONS OF RIGID COLLOIDS FROM TRANSIENT BIREFRINGENCE DATA

A. R. Foweraker, V. J. Morris and B. R. Jennings

Electro-Optics Group, Physics Department, Brunel University, Uxbridge, Middlesex, U.K.

A log-linear analysis of the transient electric briefringence relaxation data for a polydisperse colloidal dispersion yields a curved plot which is due to contributions from the individual relaxation processes of each size species within the continuous size distribution. The initial slopes of such plots under two specific experimental conditions, namely high frequency, low field and saturation orientation, have been shown to give rise to discrete average rotational diffusion coefficients for the size distribution. Thus by measuring these two average parameters, it is possible to interpret them in terms of the two coefficients of standard two-parameter distribution functions thus characterising the full size distribution.
Illustrative experimental data obtained on dispersions of sepiolite and copper phthalocyanine are presented and used to obtain representative distributions of particle sizes within the sol.

## INTRODUCTION

The industrial use of colloidal dispersions is widespread. In many cases, the processes and products involving such suspensions depend strongly upon the size distribution of the colloidal particles. Knowledge of this distribution is thus of great importance.

In three previous communications we have developed the theory of a method for determining the size distributions of rods of constant diameter (1), ellipsoids of constant axial ratio (2) and discs of constant thickness (3), in terms of monomodel, two-parameter distribution functions. The object of this study is to test the former theory experimentally and to demonstrate the ease of the method.

Consider a dispersion of rigid colloidal particles. At low concentrations, for which there are no interparticle interactions, each particle can be considered as isolated and in some random orientation relative to its neighbours. Any optical anisotropy of the individual particles is averaged over all orientations for the bulk dispersion which thus appears isotropic. If some degree of orientational order is imposed upon the particle array, the optical properties of the bulk dispersion will exhibit to some degree the anisotropy of the individual particles.

In this work the colloidal particles are ordered by the application of a pulsed external electric field which interacts with any induced or inherent electrical dipole moments of the particles. The resultant orientation is monitored by measuring the induced birefringence. As the orientation process takes a finite time, one observes a steady build up in the birefringence of the bulk dispersion. At some stage the electrical orienting forces are balanced by the Brownian disorienting forces and the birefringence reaches a saturation value ($\Delta n_o$) dependent upon the magnitude of the applied field. On cessation of the pulsed electric field the particles revert to a random orientation and the birefringence decays back to zero. The observed decay is dependent upon the size distribution of the dispersed particles. If $\Delta n_t$ is the magnitude of the birefringence at some time t after cessation (for which t = 0), then conventionally one plots the logarithm of the normalised birefringence $\ln(\Delta n_t/\Delta n_o)$ against time (4). For a monodisperse sample, this results in a straight line whose negative slope is 6D where D is the rotational diffusion coefficient of the particles. For a polydisperse sample the semi-logarithmic plot is curved due to a distribution of D values owing to the distribution of particle sizes.

The most accurately determined part of the logarithmic plot is the initial slope $S_o = \frac{\partial}{\partial t} \ln(\Delta n_t/\Delta n_o)$ which yields a discrete average value for D, the nature of which is dependent upon the degree of particle orientation (1-3).

The majority of practical systems may be analysed in terms of a monomodal distribution which can be described by two-parameter functions. One of the dependent parameters defines the positioning of the peak and the other its breadth. A common two-parameter function which is adopted in this study is the log normal distribution. For a sample of particles of major dimension $\ell$, this is described by

$$f(\ell) = \frac{1}{\sigma \ell \sqrt{2\pi}} \exp\left\{-\frac{1}{2}\left[\frac{\ln(1/m)}{\sigma}\right]^2\right\}$$

where $\int_0^\infty f(\ell)\, d\ell = 1$ and m and $\sigma$ are the characteristic parameters of the distribution. Experimentally it is necessary to determine two discrete average values of D from two distinct orientation conditions and use these to obtain m and $\sigma$. The electrical orienting conditions that we have chosen are

(i) High frequency low field orientation such that the birefringence depends upon the square of the applied field strength (4), as for Kerr law behaviour. Here the electrical orienting forces are very small compared to the thermal disorienting forces. At high frequencies any permanent dipole moment orientation has relaxed out. We designate the rotational diffusion coefficient obtained under these conditions as $\langle D \rangle_{f\infty}$.

(ii) Saturation orientation in which the electrical orienting force dominates the Brownian disorienting forces and the particles attain a condition of complete orientation. We designate the rotational diffusion coefficient obtained under these conditions as $\langle D \rangle_{E\infty}$.

The utilisation of these two experimental diffusion coefficients to obtain the parameters m and $\sigma$ has been outlined elsewhere (1-3). It should be noted however that the theories have assumed that the particles are rigid, independent and that the geometrical axes are coincident with the principal axes of the optical and electrical properties. For rod particles, the axial ratio must be in excess of 10 and the diameter must remain constant for a given sample. Ellipsoidal particles have been treated as having a constant axial ratio for a given sample. In each case it has been assumed that the excess polarisability ($\Delta\alpha$) is proportional to the square of the major particle dimension.

## EXPERIMENTAL

### Electric Birefringence

Measurements were made using a conventional apparatus (4). Data were recorded as a function of the applied field strength. Care was taken to ensure that true Kerr law or saturation conditions were obtained, by recording the field amplitude dependence of the birefringence $\Delta n_o$. The parameters $\langle D \rangle_{f\infty}$ and $\langle D \rangle_{E\infty}$ were obtained from the transient decays under the relevant orientation conditions. These were then used to fit a log-normal size distribution for each sample using the relevant model for the shape of the particles.

(a) Sepiolite. This mineral occurs as long, thin, lathe-like particles and is a member of the attapulgite family of clay minerals. The sample used in this work was kindly supplied by Dr. B. Neumann of Laporte Industries, Redhill, Surrey, U.K. and was dispersed in deionised water by continuous stirring for 24 hours. This dispersion was then left to stand for 48 hours to ensure that any large particle aggregates sedimented out. Electric birefringence data were recorded for a pulsed a.c. field of 6 kHz frequency. The required average rotational diffusion coefficients were evaluated from analysis of the

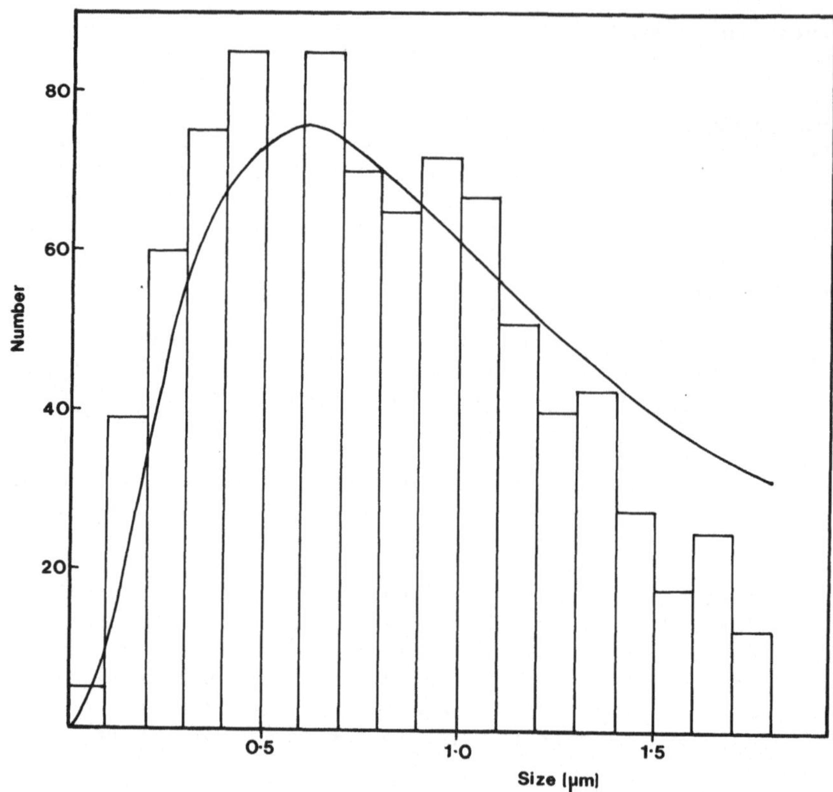


Fig. 1. Comparison of the log normal distribution obtained from the birefringence data (full curve) with the electron microscopy data for a dispersion of the clay mineral sepiolite. The parameters of the log normal distribution curve are m = 1.21 μm and $\sigma$ = 0.85.

initial slopes of the relaxation curves under the relevant orientation conditions. The data obtained were $\langle D \rangle_{E\infty}$ = 23.0 $s^{-1}$ and $\langle D \rangle_{f\infty}$ = 0.3 $s^{-1}$. Using the relevant equations from the theory for rods (1) a value of $\sigma$ = 0.85 was obtained which together with an axial ratio ($\rho$) of $\rho$ = 30 led to m = 1.21 μm. It is worth noting that if values of $\rho$ = 15 or 45 were used, values for m of 1.12 μm and 1.26 μm respectively would have been obtained. This shows the relatively small influence of the choice of $\rho$ on the value of m. The full distribution curve for m = 1.21 μm and $\sigma$ = 0.85 is shown in fig. 1.

(b) Copper phthalocyanine. This is a blue pigment of great commercial importance. The pigment particles are generally rod or ellipsoidal in shape depending upon their length which varies from about 50 nm to approximately 1 μm with their method of preparation. In this work, a sample of $\beta$-copper phthalocyanine was dispersed in an aqueous-surfactant solution, a Mullard ultrasonic drill type E7680/3 at 20 kHz and 50 w power was used as an ultrasonic dispersant for 1 hour. This method of dispersion had the effect of breaking up the majority of large pigment aggregates so as to leave a sol containing mainly individual ellipsoidal particles. The resulting suspensions were centrifuged to further eliminate any persistent particle aggregates. Electric birefringence measurements were made on the dispersion at a wavelength of 500 nm which was well removed from any optical absorption band for this material. In this case square wave d.c. pulses were used since we have previously shown (5) that the sample has no permanent electrical dipole moment. The rotational diffusion coefficients obtained from the initial slope analysis were $\langle D \rangle_{E\infty} = 98.0\ s^{-1}$ and $\langle D \rangle_{f\infty} = 17.1\ s^{-1}$. Inserting these data into the theoretical equations for ellipsoids (3) a value of $\sigma = 0.54$ was obtained. Using this value and an axial ratio

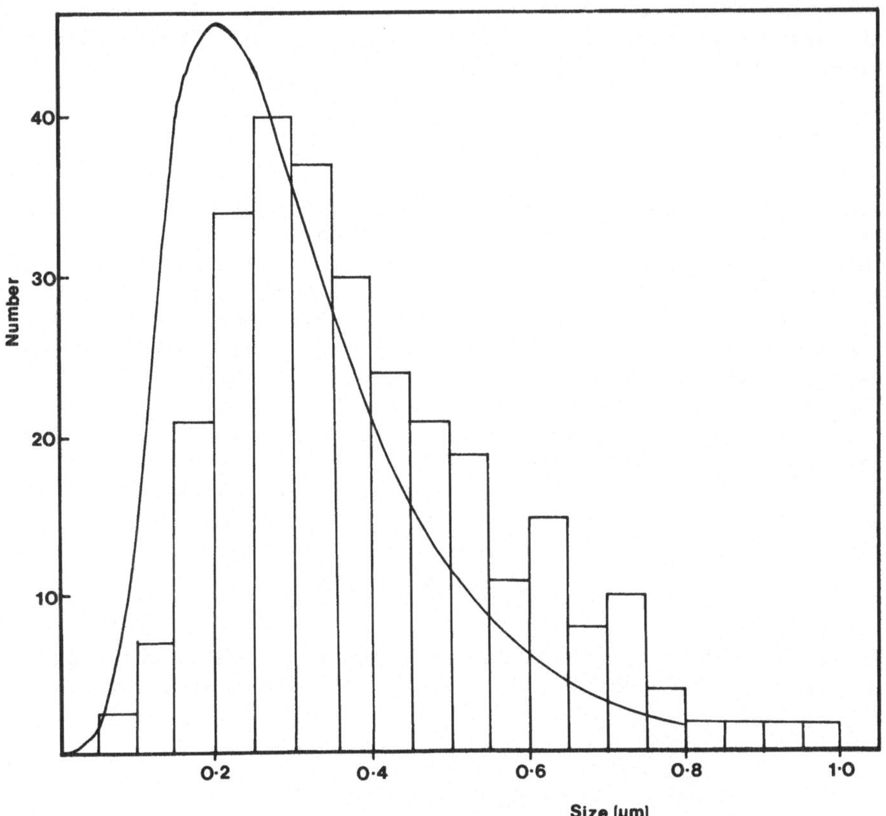


Fig. 2. Comparison of the log normal distribution obtained from the electric birefringence data (full curve) with the electron microscopy data for a dispersion of the pigment copper phthalocyanine. The log normal distribution has $m = 0.26\mu m$ and $\sigma = 0.54$.

of $\rho$ = 4, a value of m = 0.26 $\mu$m was calculated. Again it is worth noting that if axial ratios of 3 or 5 had been used one would have obtained values of m = 0.25 $\mu$m and m = 0.27 $\mu$m respectively. The full distribution curve is shown in fig. 2.

Electron Microscopy

After preparation, an aliquote of each sample was taken and deposited on carbon film substrates for electron micrograph studies. A JEM7 electron microscope was used which had previously been calibrated using a diffraction grating of 2160 lines/mm. Each preparation was inspected to ensure a workable number of particles for efficient counting and measuring. Rejection was made if the number of particles per unit area was too great. Micrographs of acceptable specimens were taken randomly from the whole grid area in order to obtain representative data. For each sample some 200 particles were measured and histograms computed. These are displayed along with the electric birefringence data in figs. 1 and 2.

## CONCLUSIONS

Figs. 1 and 2 show that for both samples the electric birefringence data lead to a well fitting distribution of particle sizes in both positioning of the peak and breadth of the distribution. In the case of the Sepiolite, the deviation of the distribution obtained from the birefringence data with that obtained from the electron micrographs for larger particles is due to counting errors. This occurs because only particles which were clearly visible were counted. With longer particles it was less likely that the complete particle was visible. Hence, the histogram appears to be relatively truncated. The method and analysis is quick and, as can be seen, leads to the full distribution of particle sizes. It thus lends itself as an extremely fast method for the sizing of industrial colloidal dispersions.

## ACKNOWLEDGEMENTS

Grants from Messrs. Ciba-Geigy Ltd. and I.C.I. Ltd. which provided fellowships for A.R.F. and V.J.M. respectively are gratefully acknowledged.

## REFERENCES

1 Morris V J, Foweraker A R and Jennings B R, Adv. Mol. Relxn. & Intn. Proc., 12, (1978) 65.

2 Morris V J, Foweraker A R and Jennings B R, Adv. Mol. Relxn. & Intn. Proc., 12, (1978) 201.

3 Morris V J, Foweraker A R and Jennings B R, Adv. Mol. Relxn. & Intn. Proc., 12, (1978) 211.

4 Fredericq E and Houssier C, 'Electric Dichroism and Electric Birefringence', Clarendon Press, Oxford (1973).

5 Foweraker A R and Jennings B R, Spectrochim. Acta, 31A, (1975) 1075.

# LENGTH DEPENDENCE OF THE IONIC CONTRIBUTION TO THE ANISOTROPY OF THE ELECTRICAL POLARISABILITY FOR RIGID RODS

V. J. Morris, G. J. Brownsey and B. R. Jennings

Electro-Optics Group, Physics Department, Brunel University, Uxbridge, Middlesex, U.K.

A method is developed, based on transient electric birefringence, for evaluating the length ($\ell$) dependence of the ionic component of the anisotropy of the electrical polarisability ($\Delta\alpha$), for polydisperse, dilute suspensions of rod-like particles. Two average values of the rotary diffusion coefficient, obtained from the birefringence decay under specific experimental conditions, are combined with an average translational diffusion coefficient, from photon correlation data, to generate the best log-normal function to fit the particle size distribution and to estimate the length dependence of $\Delta\alpha$ These factors are employed to analyse the field dependence of the birefringence and to obtain the size independent part of $\Delta\alpha$. Illustrative data are given for aqueous suspensions of sepiolite. A value of $\Delta\alpha$ proportional to $\ell^{1.8}$ was obtained.

## INTRODUCTION

In aqueous or ionic media the predominant contribution to $\Delta\alpha$ (the anisotropy of the electrical polarisability) for large colloidal particles is generally believed to arise from interfacial polarisation (1). Thus $\Delta\alpha$ should provide a useful parameter for characterising particle surfaces, investigating particle electrical double layer structures, studying surface interactions and monitoring the stability of colloidal suspensions. An important problem in electro-optics is the determination of the dependence of $\Delta\alpha$ on particle size. Most practical systems studied are polydisperse in particle size. The measured electrical parameters are then average values which depend on the mean particle size and the breadth of the particle size distribution. The interpretation of these averages to yield reduced or size independent parameters, characteristic of the material or particle-medium interface, requires knowledge of the dependence of the electrical parameters on particle size. Further, knowledge of the dependence of $\Delta\alpha$ on particle size is essential for testing theoretical models for $\Delta\alpha$ .

Theoretical analyses for rod-shaped particles predict a dependence of $\alpha$ on length ($\ell$) of the form $\ell^{\gamma}$ with $\gamma$ ranging from 1 to 3 (2-8). The value of $\gamma$ depends on particle size and whether tightly bound or diffuse ions dominate the polarisation process. The extent of interaction between the ions is also of importance. As far as the authors are aware, there have been only two experimental determinations of $\gamma$ . Käs and Brückner (9) have determined $\gamma = 2$ for platelets (bentonite, kaolinite) and $\gamma = 1$ for rods (halloysite) using electric birefringence studies on fractionated suspensions. Takashima (10) has found $\gamma = 2.2$ for sonicated DNA samples using dielectric studies. The latter results are well approximated by the model of Dukhin and Shilov (8) invoking polarisation of the diffuse double layer.

In this report we describe a practical method for evaluating $\gamma$ for polydisperse suspensions which eliminates the need to fractionate the sample. Measurements are made of two averages of the rotary diffusion coefficient $\langle D_R \rangle$ under different experimental conditions using electric birefringence. When combined with a measured average $\langle D_T \rangle$ of the translational diffusion coefficient from photon correlation data, one can determine $\gamma$ , generate a particle size distribution in terms of a simple two parameter function, and evaluate a reduced value of $\Delta\alpha$ . The method has been tested by measurements on aqueous suspensions of the clay mineral sepiolite.

## METHOD

For suspensions of particles $\Delta\alpha$ may be determined by electric birefringence measurements (11).

### Monodisperse Suspensions

Consider a suspension of particles subjected to a burst of alternating electric field, with the pulse length of sufficient length for the induced birefringence to reach an equilibrium value $\Delta n(o)$. The decay following termination of the pulse is given by (11).

$$\Delta n(t) = \Delta n(o).\exp(-6 D_R t) \qquad (1)$$

Here $D_R$ is the rotary diffusion coefficient. It yields information on particle size if the particle shape is known. At sufficiently large amplitude electric fields, all the particles are fully orientated and

$$\Delta n_f (0) = \frac{2\pi}{n} C \Delta g \tag{2}$$

where the subscript f denotes full orientation and n is the refractive index of the bulk medium. Also C represents the volume fraction of the particles and $\Delta g$ the anisotropy of the intrinsic optical factor. At low field amplitudes, as in the Kerr region, and at sufficiently high frequencies such that (a) the permanent dipole moments of the particles are incapable of contributing to particle orientation and (b) the amplitude of the double frequency component of the induced birefringence has become zero, then (11)

$$\Delta n_K (0) = \frac{2\pi}{n} C \Delta g \frac{\Delta\alpha E^2}{15 kT} \tag{3}$$

with the subscript K denoting Kerr region conditions. Also E is the r.m.s. amplitude of the applied field, k the Boltzmann constant and T the absolute temperature. $\Delta\alpha$ is evaluated from a plot of $\Delta n_K (0)$ versus $E^2$ using the value of $\Delta n_f (0)$.

## Polydisperse Suspensions

For polydisperse suspensions the decay process becomes multi-exponential due to the distribution of $D_R$ values, and the form of the decay is dependent on the amplitude of the applied field. Further the equilibrium values $\Delta n (0)$ are average values. To interpret the experimental data it is necessary to know the form of the particle size distribution function and the dependence of $D_R$, $\Delta\alpha$ and particle volume V on particle size. Picture the particles as cylindrical rods, of constant radius r, and variable length $\ell$. Let the particle size distribution function be $f(\ell)$ and be normalised

$$\int_{L_1}^{L_2} f(\ell)\, d\ell = 1 \tag{4}$$

where $L_1$ and $L_2$ are the lower and upper limits of particle length present. It is convenient to consider the average lengths defined by the integrals

$$I(j) = \int_{L_1}^{L_2} \ell^j f(\ell)\, d\ell \tag{5}$$

For rods we have

$$\Delta\alpha = \Delta\alpha_* \ell^\gamma \tag{6}$$

$$V = \pi r^2 \ell \tag{7}$$

and $$D_R = A F(\rho) \ell^{-3} \tag{8}$$

with $$A = \frac{3kT}{\pi\eta} \quad \text{and} \quad F(\rho) = \{\ln(2\rho) - 0.8\} \tag{9}$$

$\Delta\alpha_*$ is the reduced or length independent part of $\Delta\alpha$. The expression for $D_R$ is that due to Burgers (12) with $\rho$ the axial ratio ($\rho = \ell/2r$) and $\eta$ the viscosity of the bulk medium. Consider a continuous distribution of particle lengths for which

$$dC = V dN = N f(\ell)\, d\ell \tag{10}$$

where N is the total number concentration of particles per unit volume.

(a) Electric Birefringence Amplitudes. For a polydisperse system from eqs. 2-7 and 10 one obtains

$$\Delta n_f(o) = \frac{2\pi^2 r^2 N \Delta g}{n} I(1) \tag{11}$$

and $$\Delta n_K(o) = \frac{2\pi^2 r^2 N \Delta g}{n} \left\{ \frac{1}{15} \frac{\Delta\alpha_*}{kT} \right\} E^2 I(\gamma + 1) \tag{12}$$

In the case of $\Delta n_f(o)$ the length dependence arises solely from the quantity C whereas in the case of $\Delta n_K(o)$ there is an additional dependence on length due to the term $\Delta\alpha$.

(b) Electric Birefringence Decay. A polydisperse average value of $D_R$ may be obtained from the initial slope of a semilog plot of $\ln(\Delta n(t))$ versus t, namely (13,14).

$$\langle D_R \rangle = -\frac{1}{6} \operatorname*{Lt}_{t \to 0} \frac{\partial}{\partial t} \{\ln(\Delta n(t))\} \tag{13}$$

Due to the dependence of $\langle D_R \rangle$ on $\Delta n(o)$ the average obtained will be dependent on field amplitude (15,16). Two discrete averages can be defined, involving decay from full orientation $\langle D_R \rangle_f$, and decay from partial orientation in the Kerr region $\langle D_R \rangle_K$ (17). From eqs. 1-10 and 13 one has

$$\langle D_R \rangle_f = A\,F\,(\bar{\rho}) \cdot \frac{I\,(-2)}{I\,(1)} \tag{14}$$

$$\langle D_R \rangle_K = A\,F\,(\bar{\rho}) \cdot \frac{I\,(\gamma-2)}{I\,(\gamma+1)} \tag{15}$$

In deriving eqs. 14 and 15 the term $F\,(\rho)$, which is weakly dependent on $\ell$, has been taken outside the integral over particle length and, as a first approximation, replaced by the mean axial ratio $\bar{\rho}$ ( $= I\,(1)/2r$).

For clay suspensions two facts simplify analysis of the experimental data. Firstly, it is found experimentally that the distribution of particle lengths $f\,(\ell)$ can be modelled by simple two parameter distribution functions such as the log-normal distribution (18). Secondly, for the particle sizes studied in the experimental section of this account it is a reasonable approximation to take $L_2 = \infty$ and $L_1 = 0$ (18).

## Log-Normal Distribution

The log-normal distribution function, when $L_1 = 0$ and $L_2 = \infty$, has the simple form (19)

$$f\,(\ell) = \frac{1}{\sqrt{2\pi}} \cdot \frac{1}{\sigma\ell} \exp\left\{ -\frac{1}{2} \left(\frac{\ln\,(\ell/m)}{\sigma}\right)^2 \right\} \tag{16}$$

where $\sigma$ and m are the two characteristic parameters. The integrals $I\,(j)$ become moments of the distribution function given by (17).

$$I\,(j) = m^j \exp\left(\frac{j^2\,\sigma^2}{2}\right) \tag{17}$$

To obtain m and $\sigma$ it is necessary to measure two experimental averages which depend solely on m and $\sigma$. $\langle D_R \rangle_f$ provides one such parameter and a second may be obtained by measuring an average value of the translational diffusion coefficient $\langle D_T \rangle$ from photon correlation spectroscopy.

## Photon Correlation Spectroscopy

Coherent optical radiation of a fixed wavelength is scattered by a suspension of particles. The random motion of the particles leads to a broadening of the wavelength of the scattered light. This broadening may be observed in the time domain by measuring the auto-correlation function (20). Both rotational and translational motion of the particles occurs. At sufficiently low scattering angles the particles become effectively Rayleigh scatterers and only translational motion need be considered. From the initial slope of the autocorrelation function at a series of scattering angles and an extrapolation of this slope to small scattering angles one obtains a polydisperse average of the translational diffusion coefficient $D_T$ given by (20,21).

$$\langle D_T \rangle = K(\bar{\rho}) \cdot \frac{I(1)}{I(2)} \tag{18}$$

with

$$K(\bar{\rho}) = \frac{kT}{3\pi\eta} \left\{ \ln(2\bar{\rho}) - 0.11 \right\} \tag{19}$$

This equation has been derived by Burgers' method (see for example (21,22)) in order to relate to Burgers' expression for $D_R$. Once again in integrating over particle length the term $K(\rho)$, which is weakly dependent on particle length, has been taken outside the integral and $\rho$ replaced by $\bar{\rho}$.

## Analysis of Data

Measured values of $\langle D_T \rangle$ and $\langle D_R \rangle_f$ provide sufficient information to fit a log-normal distribution function. From eqs. 14,17 and 18

$$\langle D_T \rangle = \frac{K(\bar{\rho})}{m} \exp\left(-\frac{3}{2}\sigma^2\right) \tag{20}$$

$$\langle D_R \rangle_f = \frac{AF(\bar{\rho})}{m^3} \exp\left(\frac{3}{2}\sigma^2\right) \tag{21}$$

Hence

$$\sigma = \left\{ \frac{1}{6} \cdot \ln \left[ \frac{K(\bar{\rho})^3}{AF(\bar{\rho})} \; \frac{\langle D_R \rangle_f}{\langle D_T \rangle^3} \right] \right\}^{1/2} \tag{22}$$

and

$$m = \left\{ \frac{AF(\bar{\rho})}{\langle D_R \rangle_f} \exp\left(\frac{3}{2}\sigma^2\right) \right\}^{1/3} \tag{23}$$

Knowing $\sigma$ and m the average $\langle D_R \rangle_K$ may be employed to obtain a value for $\gamma$. From eqs. 15 and 17

$$\langle D_R \rangle_K = \frac{A F(\bar{\rho})}{m^3} \exp\left\{ \frac{3(1-2\gamma)}{2} \sigma^2 \right\} \qquad (24)$$

Hence

$$\gamma = \frac{1}{3\sigma^2} \cdot \ln\left\{ \frac{A F(\bar{\rho})}{m^3 \langle D_R \rangle_K} \right\} + \frac{1}{2} \qquad (25)$$

Finally, knowledge of $\sigma$, m and $\gamma$ permit an analysis of a plot of $\Delta n_K(o)$ versus $E^2$, in the Kerr region, to obtain the reduced parameter $\Delta\alpha_*$. From eqs. 11, 12 and 17

$$\Delta\alpha_* = \frac{15\,kT}{\Delta n_f(o)} \left\{ \frac{d\,\Delta n_K(o)}{d\,E^2} \right\} \frac{1}{m^{\gamma}} \exp\left\{ - \frac{\gamma(\gamma+2)}{2} \sigma^2 \right\} \qquad (26)$$

A value for $\Delta g$ may be obtained from the measured value of $\Delta n_f(o)$ knowing the concentration of the suspension and the density of the material.

## ILLUSTRATIVE DATA FOR SEPIOLITE SUSPENSIONS

Experimental data obtained for aqueous suspensions of the needle shaped clay mineral sepiolite are displayed in table I. Details of the preparation of the suspensions and the experimental measurement techniques are given elsewhere (21). The pH of the suspension was 6.8 and its specific conductivity $11.8 \times 10^{-6}$ $(\text{ohms cm})^{-1}$. Suspension concentrations were determined using a Rayleigh refractometer and a measured value of the specific refractive index increment ($dn/dc$) of $4.6 \times 10^{-4}$ $cm^3\ g^{-1}$ at a wavelength of 540 nm (21). Photon correlation measurements used single clipped homodyne detection (20) at a wavelength of 441 nm. Electric birefringence measurements were made at a frequency of 400 Hz, using linear detection (23) with the analyser offset by $6^o$, at a wavelength of 633 nm. Weight concentrations for the birefringence and laser line broadening studies were $200\,\mu g\ ml^{-1}$ and $60\,\mu g\ ml^{-1}$ respectively. The value of $\bar{\rho} = 30$ was obtained from electron microscopic data.

The data shown in table I have been analysed by the method outlined above and the parameters obtained are given in table II. The value for $\gamma$ ($= 1.8 \pm 0.3$) suggests an approximate quadratic dependence on rod length. This result is similar to that obtained by Takashima ($\gamma = 2.2$) (10) but differs from that obtained by Käs and Brückner (9) for rod-shaped clays ($\gamma = 1$). However, these results may not be incompatible. The sonicated DNA studied by Takashima was similar in size to the particles studied here. The

Table I. Experimental data for aqueous suspensions of the needle-shaped clay mineral sepiolite at pH 6-8.

| Parameter | Experimental Value |
|---|---|
| $\Delta n_f(o)$ | $(6.6 \pm 0.3) \times 10^{-7}$ |
| $\frac{d \Delta n_K(o)}{dE^2}$ | $(8.8 \pm 0.6) \times 10^{-7} cms^2$ stat volts$^{-2}$ |
| $\langle D_R \rangle_f$ | $(2.90 \pm 0.15)\ s^{-1}$ |
| $\langle D_R \rangle_K$ | $(0.92 \pm 0.09)\ s^{-1}$ |
| $\langle D_T \rangle$ | $(7.0 \pm 0.2) \times 10^{-9} cms^2\ s^{-1}$ |

Table II. Parameters derived from the experimental data presented in table I by the method of analysis suggested in the text.

| Parameter | Calculated Value |
|---|---|
| $\sigma$ | $0.46 \pm 0.01$ |
| m | $(1.90 \pm 0.02)\ \mu m$ |
| $\gamma$ | $1.8 \pm 0.3$ |
| $\Delta\alpha_*$ | $(2.0 \pm 0.8) \times 10^{-6}\ cms^{12/10}$ |
| $\Delta\alpha$ (for a $1 \mu m$ rod) | $(1.3 \pm 0.5) \times 10^{-13}\ cms^3$ |
| $\Delta g$ | $(1.60 \pm 0.08) \times 10^{-3}$ |

clays studied by Käs and Brückner were an order of magnitude smaller in size. It is not inconceivable that $\gamma$ is itself length dependent. In fact the work of McTague and Gibbs (3) on the polarisation of bound ions suggests this is likely.

The units of the reduced parameter $\Delta\alpha_*$ are clumsy owing to the fractional character of $\gamma$. For convenience we quote a value of $\Delta\alpha (= \Delta\alpha_* l^{\gamma})$ of $(1.3 \pm 0.5) \times 10^{-13}$ $cm^3$ for a 1 $\mu$m rod as a representative value. This quantity is characteristic of the particle-medium interface and independent of sample polydispersity. In quoting this value it is necessary to note the pH of 6.8 and specific conductivity of $11.8 \times 10^{-6}$ $(ohm\ cm)^{-1}$ of the suspending medium. The value of $\Delta g$ was calculated using a density of 2.6 g $cms^{-3}$ for sepiolite. This value is for South African sepiolite. The present sample is of Spanish origin and is purer in form. Values of $\Delta g$ may need to be corrected for a lower density value.

ACKNOWLEDGEMENTS

I.C.I. Ltd and S.R.C. are thanked for the financial support provided for V.J.M. and G.J.B. respectively. The facilities provided by the Physics Department are greatly appreciated.

REFERENCES

1 Stoylov S P, Advances Coll. Interf. Sci., 3, (1971) 45.
2 Mandel M, Mol. Phys., 4, (1961) 489.
3 McTague J P and Gibbs J H, J. Chem. Phys., 44, (1966) 4295.
4 Hornick C and Weill G, Biopolymers, 10, (1971) 2345.
5 Oosawa F, "Polyelectrolytes", Dekker, New York (1971).
6 Schurr M, Biopolymers, 10, (1971) 1371.
7 Weill G and Hornick C, "Polyelectrolytes" (Ed. E. Selegny), Reidel Publishing Corp., Dordrecht-Holland (1974) 277.
8 Dukhin S S and Shilov V N, "Dielectric phenomena and the double layer in disperse systems and polyelectrolytes", J. Wiley, New York (1974).
9 Käs H and Brückner R, Berichte der Deutshen Keramischen Gesellschaft, 47 (1970) 550.
10 Takashima S, J. Phys. Chem. 70, (1966) 1372.
11 Fredericq E and Houssier C, "Electric Dichroism and Electric Birefringence", Oxford University Press, London (1973).
12 Burgers J M, Verh. Kon. Ned. Acad. Weten. Afdel. Nat., 16, (1938) Sect.1, No. 4, 113.
13 Boeckel G, Genzling J C, Weill G and Benoit H, J. Chim. Phys., 59 (1962) 999.
14 Matsumoto M, Watanabe H and Yoshioka K, Biopolymers, 12 (1973) 1729.
15 Schweitzer J F and Jennings B R, Biopolymers, 11 (1972) 1077.
16 Schweitzer J F and Jennings B R, Biopolymers, 12 (1973) 2439.

17 Morris V J, Foweraker A R and Jennings B R, Adv. Mol. Relax. Int. Processes, 12, (1978) 65.
18 Foweraker A R, Morris V J and Jennings B R, this volume p. 303.
19 Hastings N A J and Peacock J B, "Statistical distributions", Butterworths (1975).
20 Berne B and Pecora R, "Dynamic Light Scattering", J. Wiley, New York (1976).
21 Morris V J, Brownsey G J and Jennings B R, Mol. Phys.(in press) (1978).
22 Happel J and Brenner H, "Low Reynolds number hydrodynamics", Nordhoff International Publishing, Leyden (1965), Chap.5, 159.
23 Badoz J, J. Phys. Radium, 17, (1956) Suppl. 11, 143.

# FLOW ALIGNMENT OF A COLLOIDAL SOLUTION WHICH CAN UNDERGO A TRANSITION FROM THE ISOTROPIC TO THE NEMATIC PHASE (LIQUID CRYSTAL)

S. Hess

Institut für Theoretische Physik, Universität Erlangen-Nürnberg, D-8520 Erlangen, Germany

The flow alignment and the resulting flow birefringence of a colloidal solution of nonspherical particles is studied theoretically. Point of departure is a Fokker-Planck equation (FPE) for the orientational distribution function where torques exerted by the gradient of the flow velocity field and by an internal (molecular) field are taken into account. It is due to this internal field that a transition into an ordered liquid crystal phase of nematic type can occur if the concentration of the molecules exceeds a certain critical value. From the FPE a nonlinear inhomogeneous relaxation equation for the alignment tensor is derived which can be applied to the isotropic and nematic phases. Thus a unified theory is obtained for the pre-transitional behaviour of the flow birefringence in the isotropic phase and for the flow alignment in the nematic phase as well as in the transition region between both phases.

## INTRODUCTION

Nonspherical particles in a streaming viscous fluid are (partially) aligned. The flow alignment can be detected optically with the help of the ensuing birefringence. The first experimental observation of flow birefringence was reported by Maxwell (1) over a century ago. Theoretical and extensive experimental investigations (2,3) have been performed on flow birefringence in liquids, colloidal and macromolecular solutions, and more recently, in gases (4,5).

Liquid crystals, in particular in the nematic phase, possess a spontaneous alignment. This fact modifies the flow birefringence in a twofold way. Firstly, in the isotropic phase, where the magnitude of the birefringence is proportional to the velocity gradient, the flow birefringence exhibits a pretransitional behaviour (6,7) in the vicinity of the transition point. Secondly, in the nematic phase where the magnitude of the alignment is practically independent of the velocity gradient, the velocity field determines the direction of the spontaneous alignment, in particular the "flow alignment angle". This is the angle between the "director" and the flow direction. Flow alignment in the nematic phase is described by the phenomenological theory of Ericksen and Leslie, cf. refs. 8, 9.

For thermotropic liquid crystals, a unified theory of non-equilibrium alignment phenomena applicable to the isotropic and nematic phases has been developed by the author within the framework of irreversible thermodynamics (10,11). A first step towards a microscopic theory is the use of a Fokker-Planck equation as indicated in ref. 12.

In this article, the Fokker-Planck equation approach is applied to lyotropic liquid crystals, in particular to the flow alignment in colloidal solution which can undergo a transition into a nematic phase if the concentration exceeds a certain critical value. Here, the aim is to point out the basic ideas and to present the essential results rather than to dwell on details of calculation.

## DESCRIPTION OF THE ALIGNMENT

A solution of (axisymmetric) colloidal particles is considered. The unit vector parallel to the figure axis of a particle is denoted by $\underline{u}$. The type of alignment which is of interest here is characterized by the alignment tensor

$$\underline{\underline{a}} = \langle \overline{\underline{u}\underline{u}} \rangle .$$

The symbol $\overline{\cdots}$ refers to the symmetric traceless part of a tensor, i.e. $\overline{\underline{u}\underline{u}} = \underline{u}\underline{u} - \frac{1}{3}\underline{\underline{\delta}}$ where $\underline{\underline{\delta}}$ is the unit tensor. The bracket $\langle \dots \rangle$ indicates an average to be evaluated with the one-particle distribution function.

Provided that the optical anisotropy of the solvent molecules can be disregarded, the anisotropic part of the dielectric tensor which is responsible for the birefringence is proportional to the alignment tensor $\underline{\underline{a}}$. Hence an alignment characterized by $\underline{\underline{a}} \neq 0$ can be detected optically via the ensuing birefringence.

For the special case of a uniaxial alignment, the tensor $\underline{\underline{a}}$ is proportional to $S\,\overline{\underline{n}\underline{n}}$ where $\underline{n}$, referred to as director, is a spacefixed unit vector and S is the scalar order parameter of Maier and Saupe (13), viz. $S = \langle P_2(\underline{u} \cdot \underline{n}) \rangle$. Here $P_2$ is the 2nd Legendre polynomial depending on the angle between the figure axis of a colloid particle and the director. The spontaneous alignment in the nematic phase is of this uniaxial type. It must be stressed that the flow alignment is biaxial, in general.

## MOLECULAR THEORY OF FLOW ALIGNMENT

Point of departure for the theory of the alignment caused by the viscous flow of a colloidal solution are Langevin equations for the orientational motion of the colloidal particles. If the angular velocity of a particle is written as $\underline{\omega} + \underline{\Omega}$ with

$$\underline{\omega} = \frac{1}{2} \text{ rot } \underline{v} \tag{1}$$

where $\underline{v}$ is the flow velocity field, these equations are

$$\dot{\underline{u}} = \underline{\omega} \times \underline{u} + \underline{\Omega} \times \underline{u} , \tag{2}$$

$$\dot{\underline{\Omega}} = - w_r \underline{\Omega} + \Theta^{-1} \underline{T}^{sys} + \Theta^{-1} \underline{T}^{fluc} \tag{3}$$

The dot refers to differentiation with respect to time. The quantity $w_r$ is the rotational friction coefficient which is proportional to $\eta R^3$ where $\eta$ is the viscosity of the fluid and R a characteristic length of a colloidal particle. In eq. 3, $\Theta$ is a moment of inertia, $\underline{T}^{sys}$ and $\underline{T}^{fluc}$ stand for the systematic and fluctuating torques acting on a colloidal particle. Two contributions to $\underline{T}^{sys}$ are considered:

(i) The hydrodynamical torque caused by the gradient of the flow field which is proportional to (2,14)

$$\eta \kappa \underline{u} \times (\overline{\underline{\nabla} \underline{v}} \cdot \underline{u} ). \tag{4}$$

The coefficient $\kappa$ depends on the shape of a particle; in particular for rigid ellipsoids of revolution with semiaxes $R_1$ and $R_2$ one has $\kappa \sim (R_1^2 - R_2^2)(R_1^2 + R_2^2)^{-1}$.

(ii) Due to the interaction between the colloidal particles, a test particle will experience a torque which depends on the orientation of the surrounding particles. In the vein of the molecular field theory, this orientation is approximated by the average alignment $\underline{\underline{a}}$. This contribution to the torque which eventually is responsible for the occurrence of a spontaneous alignment is written as

$$- \underline{\mathcal{L}} \mathcal{H} (\underline{u}) \tag{5}$$

with differential operator

$$\underline{\mathcal{L}} = \underline{u} \times \frac{\partial}{\partial \underline{u}} \tag{6}$$

and

$$\mathcal{H} (\underline{u}) = - \varepsilon \underline{\underline{a}} : \overline{\underline{u}\,\underline{u}} . \tag{7}$$

The characteristic energy $\mathcal{E}$ associated with the alignment is assumed to be proportional to the concentration c of the colloidal particles.

## FOKKER-PLANCK-EQUATION

From the Fokker-Planck equation for the distribution function $\rho = \rho(t, \underline{u}, \underline{\Omega})$ pertaining to the Langevin equations (2,3) a simpler equation for the orientational distribution function $f = f(t, \underline{u}) = \int \rho(t, \underline{u}, \underline{\Omega})\, d\Omega$ can be derived (14). The rotational motion is overdamped (rotational diffusion) since one has $k_B T \Theta^{-1} \ll w_r$. Here T is the temperature of the fluid. The average $\langle \Psi \rangle$ of a quantity $\Psi = \Psi(u)$ is given by $\langle \Psi \rangle = \int \Psi \, f\, d^2u$ (normalization $\int f\, d^2 u = 1$). The resulting Fokker-Planck equation for $f = f(t, \underline{u})$ is

$$\frac{\partial f}{\partial t} + \underline{\omega} \cdot \underline{\mathcal{L}} - w \underline{\mathcal{L}} \cdot \left[ \underline{\mathcal{L}}\, f + \underline{u} \times (\underline{\underline{F}} \cdot \underline{u})\, f \right] = 0 \tag{8}$$

The term containing $\underline{\omega} = \frac{1}{2}$ rot $\underline{v}$ describes the rotational motion induced by the flow field $\underline{v}$. The quantity w is the rotational diffusion coefficient which is related to the rotational relaxation coefficient $w_r$ occurring in eq. 3 by $w = k_B T (\Theta w_r)^{-1}$. The 2nd rank tensor $\underline{\underline{F}}$ is given by

$$\underline{\underline{F}} = (6\, w)^{-1}\, \kappa\, \overline{\underline{\nabla}\, \underline{v}} + \frac{c}{c^*}\, \underline{\underline{a}} \, . \tag{9}$$

These terms stem from the torques caused by the gradient of the flow field and by the internal molecular field, respectively. The reference concentration $c^*$ is defined by

$$\mathcal{E}\, (k_B T)^{-1} = c / c^* \tag{10}$$

It is recalled that the energy $\mathcal{E}$ associated with the alignment is assumed to be proportional to the concentration c.

For a dilute solution ($c \to 0$) where the 2nd term in (9) can be disregarded, eq. 8 reduces to the Fokker-Planck equation used by Peterlin and Stuart (2) quite some time ago. In the absence of a gradient of a flow field ($\underline{\nabla}\, \underline{v} = 0$, $\omega = 0$), the stationary solution of eq. (8) is a distribution function essentially of the type used by Maier and Saupe (13) for the treatment of equilibrium properties of liquid crystals.

The derivation of a nonlinear inhomogeneous relaxation equation for the alignment tensor $\underline{\underline{a}}$ from the Fokker-Plank equation (8) is discussed next.

## RELAXATION EQUATION FOR THE ALIGNMENT

Multiplication of eq. 8 by $\overline{\underline{u}\underline{u}}$ and subsequent integration over $d^2u$ leads to an equation for the time change of the alignment tensor

$\underline{\underline{a}} = \langle \overline{\underline{u}\underline{u}} \rangle$. Due to the presence of the term in eq. 8 which involves the tensor $\underline{\underline{F}}$, this resulting equation for $\underline{\underline{a}}$ also contains the 4-th rank tensor $\langle \overline{\underline{u}\underline{u}\underline{u}\underline{u}} \rangle$. In a similar manner, an equation for this 4-rank tensor can be derived from eq. 8 which contains $\underline{\underline{a}}$ and the corresponding alignment tensor of rank 6. If all tensors of rank $\ell \geqslant 6$ are disregarded, elimination of the 4-rank tensor from the first two equations of the set of coupled relaxation equations leads to closed equation for $\underline{\underline{a}}$, viz.

$$\frac{\partial \underline{\underline{a}}}{\partial t} - 2\, \overline{\underline{\omega} \times \underline{\underline{a}}} + \tau_o^{-1} \left\{ \underline{\underline{\Sigma}} + \tau_1\, \overline{\underline{\nabla}\, \underline{v}} \right\} = 0 \tag{11}$$

with the relaxation time coefficients

$$\tau_o = (6\, w)^{-1}, \qquad \tau_1 = -\, \kappa\, \tau_o. \tag{12}$$

The term involving $\underline{\omega} = \frac{1}{2}\,\mathrm{rot}\,\underline{v}$ describes the time change of $\underline{\underline{a}}$ due to the molecular rotation caused by rot $\underline{v}$. The remaining terms in eq. 11 govern the relaxation of the alignment and its coupling with the symmetric traceless part $\overline{\underline{\nabla}\, \underline{v}}$ of the gradient of the velocity field. The quantity $\underline{\underline{\Sigma}}$ is given by

$$\underline{\underline{\Sigma}} = \left(1 - \frac{c}{c^*}\right) \underline{\underline{a}} + \ldots \tag{13}$$

where the dots stand for terms which are of 2nd and 3rd power in $\underline{\underline{a}}$. Notice that the coefficient of the term linear in $\underline{\underline{a}}$ vanishes for $c = c^*$ and becomes negative for $c > c^*$.

In the absence of a gradient of the flow field ( $\underline{\nabla}\, \underline{v} = 0$, $\omega = 0$), the stationary solution of eq. 11 is determined by $\underline{\underline{\Sigma}} = 0$. Due to the nonlinearity of $\underline{\underline{\Sigma}}$, 3 solutions exist. A stability analysis shows that one has $\underline{\underline{a}} = 0$ for $c < c_K$ corresponding to the isotropic phase and $\underline{\underline{a}} \neq 0$ for $c > c_K$ which corresponds to the nematic phase of a lyotropic liquid crystal. The concentration $c_K$ is somewhat smaller than the reference concentration $c^*$. The transition from the isotropic to the nematic phase is first order.

Here, we are more concerned with the results for the nonequilibrium phenomena which can be inferred from eq. 11.

## RESULTS

### Isotropic Phase

For concentrations c below $c_K$ it suffices to approximate $\underline{\underline{\Sigma}}$ by its term linear in $\underline{\underline{a}}$, cf. eq 13. The relevant relaxation time for the alignment tensor is

$$\tau = \tau_o \left(1 - \frac{c}{c^*}\right)^{-1}. \tag{14}$$

For a stationary flow field and if /rot $\underline{v}$/ is small compared with $\tau^{-1}$, the flow induced alignment as inferred from the linearized version of eq. 11 is

$$\underline{\underline{a}} \sim (1 - \frac{c}{c^*})^{-1} \tau_1 \overline{\underline{\nabla}\underline{v}} . \qquad (15)$$

Thus both the orientational relaxation time $\tau$ and the flow birefringence show a pretransitional increase governed by the factor $(1 - \frac{c}{c^*})^{-1}$ if the concentration c approaches the concentration $c_K < c^*$ where the transition into an ordered nematic phase occurs. This behaviour is quite analogous to the pretransitional behaviour discussed by de Gennes for thermotropic liquid crystals (7,8,9).

Nematic phase

In the nematic phase the alignment tensor $\underline{\underline{a}}$ is of the form $S\,\overline{\underline{n}\underline{n}}$ where the scalar order parameter $S = \langle P_2(\underline{u}\cdot\underline{n})\rangle$ can be approximated by its equilibrium value. The flow alignment angle $\chi$ between the director $\underline{n}$ and the flow velocity $\underline{v}$ is determined by

$$\cos 2\chi = -\gamma_1 / \gamma_2 \qquad (16)$$

(provided that $/\gamma_2/ > \gamma_1$). This relation and the Leslie coefficients $\gamma_1$ and $\gamma_2$ can also be inferred from eq. 13, for details see refs. 10 and 12 where the analogous case has been treated for thermotropic liquid crystals. In lowest order in S, one has $\gamma_1 \sim S^2\tau_0, \gamma_2 \sim S\,\tau_1$, thus $\cos 2\chi \sim S\,\kappa^{-1}$. It is recalled that the dimensionless coefficient $\kappa$ determines strength of the orienting torque exerted by the gradient of the velocity field on a colloidal particle.

Transition Regime

At concentrations very close to $c_K$ it is not possible to linearize the eq. 11 around the equilibrium value of the alignment tensor in isotropic or nematic phase as indicated above. An analysis of the full nonlinear equation (11) is required. For thermotropic liquid crystals, this has been done in ref. 11. Analogous results apply to colloidal systems. At concentrations slightly below $c_K$, a velocity gradient may induce a transition into the ordered phase if its magnitude exceeds a certain value order of $\tau_K^{-1}$. Here $\tau_K$ is the value of the orientational relaxation time for $c = c_K$ (notice that the formal divergence of $\tau$ occurs at $c^* > c_K$, cf. eq. 14). This causes a discontinuity both in the magnitude of the birefringence and in the orientation of the principal axes of the dielectric tensor. Since the orientational relaxation times for colloidal particles are much larger than those of most thermotropic nematic liquid crystals, it should be easier to observe these flow induced discontinuities in lyotropic liquid crystals, in particular in colloidal solutions.

REFERENCES

1 Maxwell J C, Proc. Roy. Soc. London, A 22, (1873) 46.
2 Peterlin A and Stuart H A, Hand- und Jahrbuch der Chem. Phys., eds. Eucken and Wolf, 8 I B, Leipzig (1943).
3 Champion J V and Meeten G H,, Trans. Faraday Soc., 64, (1968) 238.
4 Hess S, Phys. Lett. 30A, (1969) 239; Springer Tracts in Mod. Phys., 54, (1970) 136; A. Naturforsch., 29a, (1974) 1121.
5 Baas F, Phys. Lett., 36A, (1971) 107;
Baas F, Breunese J N, Knaap H F P and Beenakker J J M, Physica, 88A, (1971) 1.
6 de Gennes P G, Phys. Lett., 30A, (1969) 454.
7 Martinoty P, Candau S and Debeauvais, Phys. Rev. Lett., 27, (1971) 1123.
8 de Gennes P G, The Physics of Liquid Crystals, Clarendon Press, Oxford (1974).
9 Stephen M J and Straley J P, Rev. Mod. Phys., 46, (1974) 617.
10 Hess S, Z. Naturforsch., 30a, (1975) 728, 1224.
11 Hess S, Z. Naturforsch., 31a, (1976) 1507.
12 Hess S, Z. Naturforsch., 31a, (1976) 1034.
13 Maier W and Saupe A, Z. Naturforsch., 13a, (1958) 564; 14a, (1959) 882.
14 Hess S, Physica, 74, (1974) 277.

# QUASI-CRYSTALS PRODUCED BY COLLOIDAL PHOTOCHROMIC DYES IN AN APPLIED ELECTRIC FIELD

V. A. Krongauz

Department of Structural Chemistry, The Weizmann Institute of Science, Rehovot, Israel

Colloidal solutions are formed during the photochemical transformation of solutions of photochromic spiropyrans in non-polar solvents. The systems are composed of globules ∼ 0.1 - 0.4 μm in diameter. The globules consist of highly dipolar crystalline nuclei enclosed in amorphous envelopes. In a constant external electric field linear quasi-crystalline material is formed composed of globules joined together like straight strings of beads aligned along the electric lines of force. The distinctive features of quasi-crystal formation were studied in the electron microscope.

## INTRODUCTION

The ability of cyanine dyes to form tightly bound molecular aggregates in solutions is a well-known phenomenon (1). These aggregates exhibit a spectral red shift of the order of 100 nm if the dye molecules are packed in so-called J- or Scheibe- stacks with the tilted deck-of-card structure.

Our previous studies (2-5) revealed that a similar effect takes place upon photochemical transformation of photochromic spiropyrans dissolved in saturated hydrocarbon solvents.

$H_3C$ $CH_3$ A ... N $R_1$ O $NO_2$ $R_2$ $\underset{h\nu,\ kT}{\overset{h\nu}{\rightleftharpoons}}$ B $NO_2$ $H_3C$ $CH_3$ $N^+$ $R_1$ $O^-$ $R_2$

In solution the photocoloured merocyanine form B interacts with uncoloured molecules A to give dimers AB and polar complexes $A_nB$ (with n = 2 - 3). The dimers AB and complexes $A_nB$ produce aggregates which form colloidal solutions during UV irradiation. The relative amounts of AB and $A_nB$ in the colloidal particles depend on temperature and light intensity, because $A_nB$ formation involves an activation energy of about 5 kcal/mole which is associated with the interaction of dimers with further A molecules.

During the irradiation of spiropyran solutions in an external constant electric field (5 - 25 kV/cm) coloured threads were formed, extending from one electrode to the other along the lines of force. The absorption spectra of the threads are identical with those of the colloidal solutions formed in the absence of an electric field, indicating that the threads also consist of AB and $A_nB$.

Measurements of linear dichroism showed that the complexes $A_nB$ are oriented along the thread axis, while the dimers AB are placed at random. Electron microscope studies of the precipitated colloidal solutions and of the quasi-crystalline threads showed that both consist of globules of about 0.1 - 0.4 $\mu$m in diameter. In the quasi-crystals the globules are aligned in straight chains oriented along the electric field.

The X-ray diffraction patterns of the coagulated colloidal solutions and of the quasi-crystalline threads are identical and consist of sharp reflections superimposed on a diffuse pattern. The sharp reflections vanish when the concentration of $A_nB$ in the colloidal particles becomes small. We suggested that $A_nB$ complexes form one dimensional dipolar molecular stacks similar to J- or Scheibe- stacks inherent in many cyanine dyes. A number of such stacks are combined in three dimensional dipolar microcrystals. The globules are formed when the microcrystals become coated with a nonpolar amorphous phase composed of dimers AB. The dipole moments of the nuclei could cause the alignment of the globules in an external electric field (fig. 1).

The present paper is devoted to detailed electron microscopic studies of the structure of globules and of the growth of quasi-crystalline threads in an electric field. The major part of the results were obtained with a spiropyran having $R_1 \equiv -C_2H_4.OCO.C(CH_3) = CH_2$ and $R_2 \equiv H$. The quasi-crystals given by this spiropyran are very stable, probably due to the fractional photopolymerization of the amorphous coatings of globules. Methylcyclohexane was used as a solvent. The results described here were presented in part in ref. 3-5 where the other experimental details can be found.

## RESULTS

Fig. 2 shows an electron micrograph of the sediment of a colloidal solution obtained upon UV irradiation of a spiropyran solution in

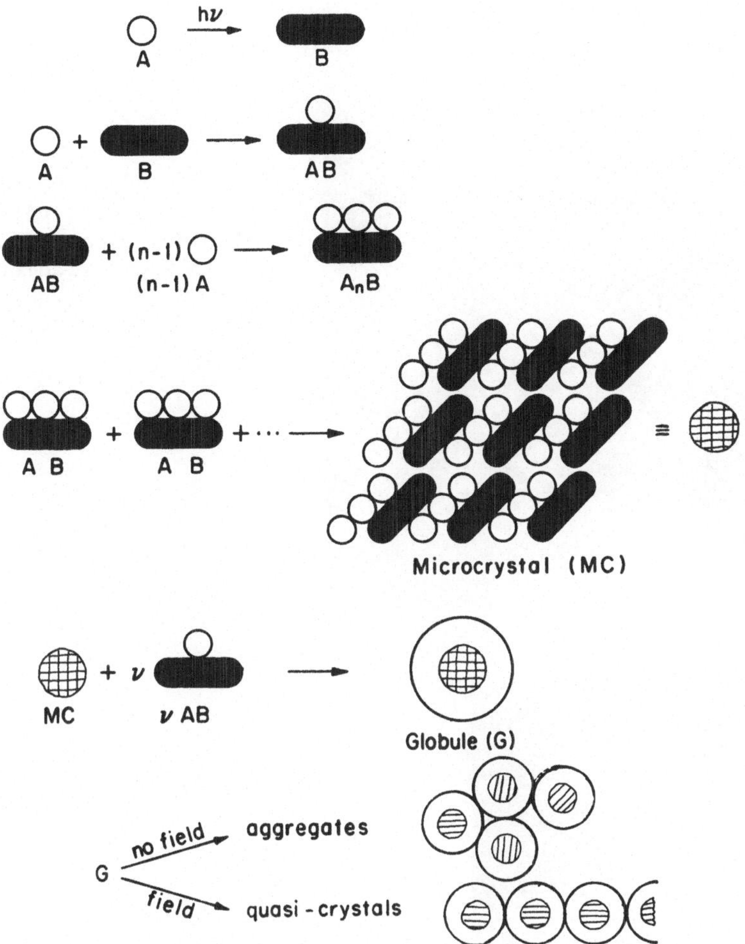


Fig. 1. Scheme of quasi-crystal formation.

the absence of an electric field. In figs. 3 - 5 the quasi-crystals are shown. Macroscopically they look like coloured threads, while in the electron microscope they resemble strings of beads aligned along the lines of force.

A micrograph of an uncoated microcrystal is shown in fig. 6. Such microcrystals are formed after a very short exposure to UV irradiation prior to the formation of globules, and have a streaky texture. The streaks, (probably molecular stacks) ∿ 0.5-1 nm thick are parallel and evenly spaced at a distance of 5-10 nm. The microcrystals exhibit distinct discrete electron diffraction. The globules occurring after coating by amorphous material do not exhibit diffraction.

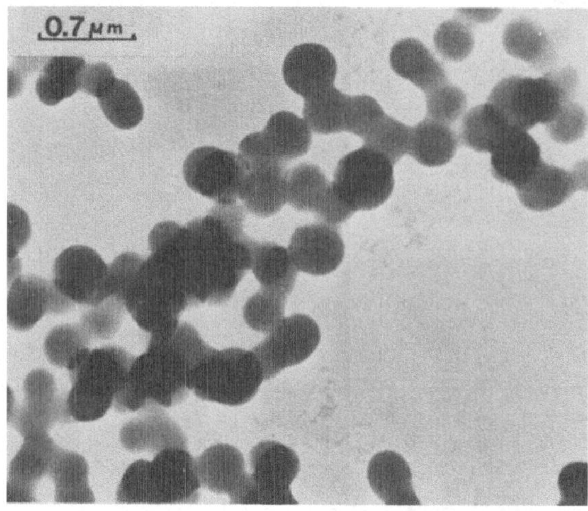


Fig. 2. Electron micrograph of a precipitated colloidal suspension formed upon irradiation of $5.10^{-4}$ M spiropyran solution in methylcyclohexane.

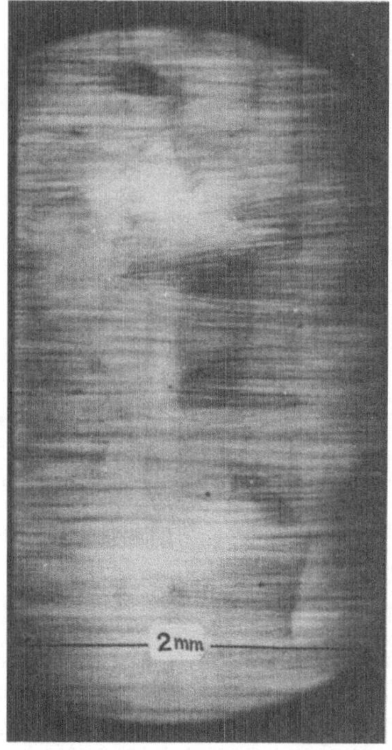


Fig. 3. Photograph of the quasi-crystalline threads. Interelectrode distance is 2 mm.

There are two very important regularities concerning the growth of the quasi-crystals: (a) Every quasi-crystalline thread consists as a rule of globules of equal size. (b) The radii of the globules increase discretely during irradiation, resulting in the appearance of several generations of threads. The first three generations are seen in micrograph 3, the fourth in mocrograph 4. The diameters of the glubules composing the threads of different generations (given in parentheses) are: (1) 0.8, (2) 0.14, (3) 0.23 and (4) 0.36 $\mu$m ($\pm$15%).

To obtain bare nuclei the outer layers of the globules were dissolved by washing for 20 - 30 min. with methylcyclohexane. The amorphous coatings were partly removed and sometimes it was possible to see that globules of the second, third, and fourth generations include a number of smaller units. For instance, the globules of the second generation include 4 - 10 units, 0.03 - 0.04 $\mu$m in diameter (fig. 7).

Irradiation of spiropyran solutions in an alternating electric field (10 kV/cm and 50 Hz) does not produce quasi-crystalline threads, indicating that the dipole moments of the globules are really permanent and they are not induced by the electric field.

## DISCUSSIONS

### Thermodynamics of the Quasi-Crystal Formation

The orientation of single $A_nB$ complexes in weak electric fields (5-25 kV/cm) for any reasonable molecular dipole is very unlikely, in view of the thermal molecular motion at room temperature. We therefore concluded that $A_nB$'s form dipolar molecular stacks, which in turn form highly dipolar crystals and then the nuclei of globules. The set of globules in solution can be regarded as a system of strong dipoles p. These dipoles interact with an external field E and with one another. Thermal motion destroys the orientation caused by electrostatic interactions. If the field is strong enough $pE > kT$, the dipoles become oriented along the field. From this inequality the minimal dipole moment can be estimated which makes dipoles oriented in a given field. For $E = 5$kV/cm, $p > 2.5\ 10^3$D. When the dipole moments are strong enough a transition from the random dipole arrangement to the ordered thread structure takes place. According to a theoretical estimation (4) the condition of the transition is $(2p^2/r_o^3) > -kT\,\ell nC$ where $r_o$ is the distance between the centers of dipoles in the threads, C is the dimensionless concentration of globules in the solution. For the system under investigation we obtained $p > 10^4$D as a condition of transition to the thread structure.

### Kinetic Aspects of Phenomenon

As stated above the microcrystals are highly polar particles. Obviously they can exist as a more or less stable dispersion in a non-

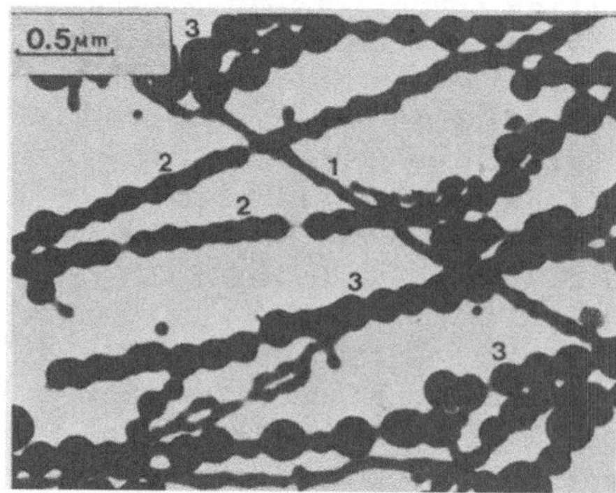


Fig. 4. Electron micrograph of quasi-crystals. 1, 2 and 3 represent threads of the first, second and third generations respectively.

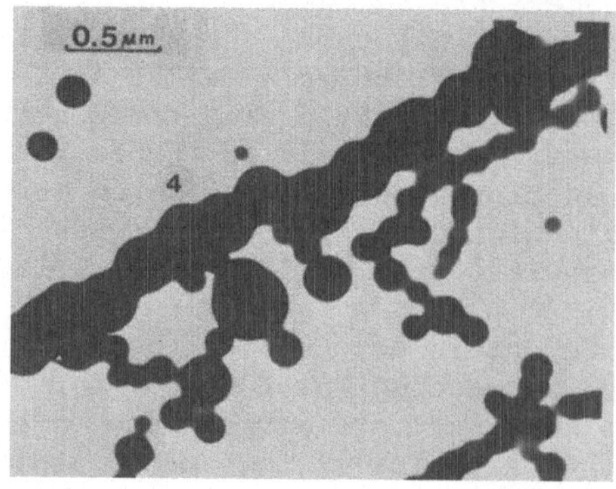


Fig. 5. Electron micrograph of quasi-crystals. 4 - the fourth generation.

polar solvent only if they are separated from it by a less polar amorphous phase. The experiment shows that the life-time of the microcrystals is much shorter than the characteristic time of thread formation. So the microcrystals are enveloped by an amorphous phase before they are involved in a thread.

The nonpolar amorphous envelope must be of some definite thickness, $\lambda$ , to make the colloidal solution of globules sufficiently stable. If the envelope around the nucleus is thinner than $\lambda$ , several such nuclei could be imagined occasionally to join and then to be coated by one envelope, forming in this way a larger globule. If there is an excess of matter forming amorphous envelopes the probability of joining is small and the primary microcrystals are covered by layers of thickness $\geqslant \lambda$ . If there is a deficiency of amorphous matter in the system, the joining of crystalline nuclei occurs, followed by covering. Thus the microcrystals are involved in two competitive processes. The preferred direction of the reaction is governed by the amount of amorphous matter. If the matter responsible for the formation of the amorphous envelopes becomes exhausted during the formation of globules of the first generation, the formation of globules of the second generation can be sharply separated in time. Comparison of electron micrographs of threads obtained at different light exposures confirms this conclusion.

The size of the globules of the second generation can be predicted from rather simple considerations. For simplicity let us suppose that the microcrystals have a spherical form. In fact the primary nuclei do have a form close to spherical. If n of them join, the minimal radius of globules of the next generation will be $R_1 = 2R + \lambda$, where R is the

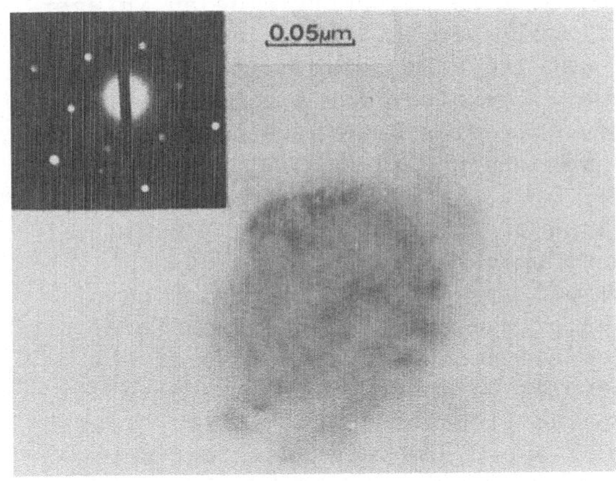


Fig. 6. Electron micrograph of a micro-crystal and the electron diffraction pattern.

radius of the primary nucleus and $\lambda$ is the thickness of the envelope, which is supposed to be equal for globules of all generations. The volume of amorphous matter in a globule of the next generation is

$$V = \frac{4}{3}\pi\left\{(2R+\lambda)^3 - nR^3\right\} \qquad (1)$$

The volume of amorphous phase per n globules of the first generation is

$$nv = \frac{4}{3}\pi n\left\{(R+\lambda)^3 - R^3\right\} \qquad (2)$$

If $V < nv$, the joining is favoured from the point of view of saving amorphous material. Using eqs. 1 and 2 this condition can be expressed as

$$\frac{\lambda}{R} > \frac{2 - n^{1/3}}{n^{1/3} - 1} \qquad (3)$$

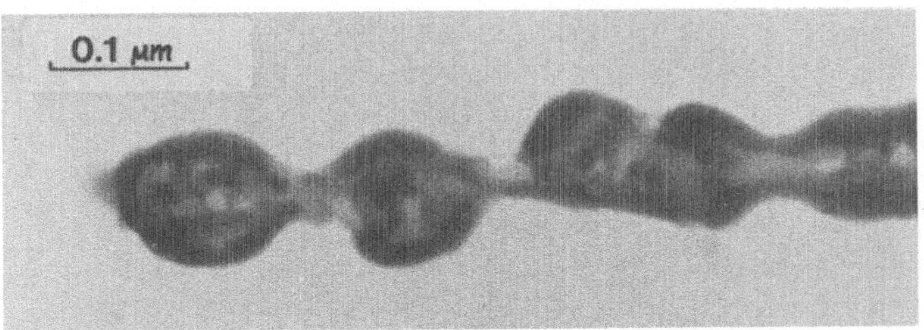


Fig. 7. A quasi-crystalline thread with dissolved amorphous coatings.

Obviously the most favourable situation will occur when the sphere of radius 2R (this is the minimal radius for a sphere containing an integer number of primary nuclei $n > 1$) is closely packed with n nuclei of radius R. The closest face-centered packing of a volume with spheres fills approximately 0.75 of the volume. Therefore for a volume $\frac{4}{3}\pi(2R)^3$ closely packed with spheres of radius R we have $n \simeq 6$. Inserting $n = 6$ into inequality (5) we obtain $\lambda/R > 0.2$. The ratio $\lambda/R$ in the experimental system ($\lambda \simeq 0.02$, $R \simeq 0.02$) fits into the interval restricted by the previous inequality. The value $n = 6$ is also in good agreement with experiment, which gives for n values between 4 to 10. A similar mechanism was assumed for the formation of following generations, i.e. the radius of the nucleus of a globule of the third generation is equal to twice the radius of a nucleus of the second generation and so on. This permits us to calculate the size of the globules of each generation. For the globules of the first - fourth generations we obtained 0.08, 0.12, 0.2 and 0.36 μm in excellent agreement with the experimental data.

## REFERENCES


1 Mees C E and James T H, Theory of the Photographic Process, Ch. 11 and 12, Macmillan, N.Y. (1966).

2 Krongauz V A and Parshutkin A A, Photochem. Photobiol. 15, (1972) 503.

3 Parshutkin A A and Krongauz V A, Molec. Photochem., 6, (1974) 437.

4 Krongauz V A and Goldburt E S, Nature, 271, (1978) 43.

5 Krongauz V A, Fishman S N and Goldburt E S, J. Phys. Chem., in press.

# EFFECT OF CTAB ON COLLOIDAL SUSPENSIONS OF SEPIOLITE - A STUDY BY LIGHT SCATTERING, ELECTRIC BIREFRINGENCE, LASER LINE BROADENING AND ELECTROPHORETIC LIGHT SCATTERING

G. J. Brownsey, V. J. Morris and B. R. Jennings

Electro-Optics Group, Physics Department, Brunel University, Uxbridge, Middlesex, U.K.

A combination of laser line broadening and electro-optic techniques has been used to investigate the stability of suspensions of the rigid needle-shaped clay sepiolite. In distilled water, sepiolite is negatively charged. Addition of the surfactant CTAB was used to take the clay through its isoelectric point. The minimum stability of the suspension occurred at the surfactant concentration at which (a) the anisotropy ($\Delta\alpha$) of the electrical polarisability ($\alpha$), obtained from electric birefringence data, showed a minimum value, and (b) the electrophoretic mobility, obtained from laser line broadening in electric fields, was zero. The minimum in $\Delta\alpha$ was attributed to the isolation of the bulk polarisation contribution from the otherwise predominant interfacial polarisation contribution.

## INTRODUCTION

An important physical characteristic of colloidal dispersions is the tendency of the particles to aggregate. In liquid media, Brownian motion ensures that frequent collisions occur between particles. The stability of a suspension depends on its interparticle interaction which is, in turn, determined by the surface charge of the particles and the properties of their associated electric double layers. Electrophoresis provides a well established method of monitoring the net charge density at the shear plane through measurements of the electrophoretic mobility (u) (ref. 1). Values of u are obtained by observing the translational motion of particles in response to an applied dc electric field.

With electro-optical phenomena, the anisotropy ($\Delta\alpha$) of the electrical polarisability ($\alpha$) is a readily evaluated parameter. For non-spherical particles, $\Delta\alpha$ may be measured by recording the rotational response of the particles to an applied alternating electric field of sufficiently high frequency (2). In ionic media, it appears that the main contribution to $\Delta\alpha$ arises from surface polarisation (3-4).

This possible surface origin of the main contribution to $\Delta\alpha$ has led to comparative studies of u and $\Delta\alpha$ under various experimental conditions. In particular, studies of the effect of surfactants added to clay suspensions (4-8) have suggested that the minimum stability of the suspension, with respect to added surfactant concentration, is related to a minimum in $\Delta\alpha$ which is displaced from the surfactant concentration at which the zeta-potential ($\xi$) is zero. Such studies have been made on suspensions of the rod-shaped clay mineral palygorskite (5-7) and the plate-shaped clay mineral kaolinite (8), with cetyl-pyridinium chloride (CPC) as the added surfactant.

In this paper we present further experimental data of a comparison of u and $\Delta\alpha$ on a clay suspension in which $\xi$ has been varied from negative values through zero to positive values. In addition the instability of the suspension has been followed by noting the extent of particle aggregation. This has been done in four ways. Firstly, by recording the increase in scattered intensity at a fixed angle of observation. Secondly by measuring an average value of the rotary diffusion coefficient ($D_R$). Thirdly by measuring an average value of the translational diffusion coefficient ($D_T$) and fourthly by visual observation of the suspensions. Suspensions of the needle-shaped clay mineral sepiolite were employed as a model system. When suspended in distilled water, individual sepiolite particles are negatively charged. Originally it was proposed to vary $\xi$ by varying the pH of the suspension. However, sepiolite remained negatively charged at pH's as low as 3. Thus $\xi$ was varied by addition of the surfactant cetyl trimethyl ammonium bromide (CTAB).

## EXPERIMENTAL

Samples of sepiolite, of Spanish origin, were supplied by Dr. B.S. Neumann of Laporte Industries, Redhill, Surrey, U.K. The clay was dusted into freshly distilled water, the suspension stirred for 24 hours, allowed to stand for several days and then the upper layer was removed. This upper layer was filtered under pressure through 5$\mu$m Mitex Millipore filters. As monodisperse a sample as possible was obtained by centrifugation, although a significant degree of polydispersity remained. The resultant suspension was then diluted to the required concentration with clean freshly distilled water or clean solutions of CTAB in distilled water. All suspending media were cleaned by filtration through 0.45 $\mu$m Millipore filters. Clay concentrations were determined using a Rayleigh refractometer and taking a measured value of 4.6 x $10^{-4}$ $cm^3$ $g^{-1}$ for the specific refractive index increment ($dn/dc$) at a wavelength of 540 nm (9).

Conventional light scattering intensity measurements were made on a FICA 5000 photometer at scattering angles of 30°, 90° and 150° and at a wavelength of 436 nm. The clay concentration was 0.44 x $10^{-4}$ g $cm^{-3}$.

Values of u were obtained using the recently developed method of photon correlation spectroscopy in applied dc electric fields (10,11). Single clipped heterodyne detection was employed at a wavelength of 441 nm. Electric fields were applied in the form of pulses. Data were recorded and averaged for twenty pulses, each of 100 V $cm^{-1}$ amplitude and 200 m sec duration. Care was taken to record data only after an initial dead time for each pulse of 40 m sec, in order to eliminate particle rotational responses and to ensure that they had attained a uniform translational velocity. Alternate pulses were of opposite polarity in an attempt to minimise electrode polarisation processes. Experimental average $\langle D_T \rangle$ values of the translational diffusion coefficient were obtained by standard photon correlation spectroscopy using single clipped homodyne detection. The autocorrelation function was evaluated at a number of discrete scattering angles in the range 5 to 120 degrees. Z average values of $D_T$ were calculated from the short time region of the autocorrelation function (12) and extrapolated to zero scattering angle. Clay concentrations for the photon correlation studies were 0.6 x $10^{-4}$ g $cm^{-3}$.

A standard birefringence apparatus was used to evaluate $\Delta\alpha$ and the averages of $D_R$. Linear detection (13) was employed with the analyser offset by 6°. The wavelength of the incident light was 633 nm and the clay concentration 2.05 x $10^{-4}$ g $cm^{-3}$. Single shot bursts of alternating electric field of r.m.s. amplitudes (E) between 27 and 935 V $cm^{-1}$, at a frequency of 400 Hz, and pulse durations up to 5s were applied to the suspensions and the transient birefringence responses recorded. Values of $\Delta\alpha$ were obtained from the slope of the $E^2$ plot of the steady-state equilibrium value of the birefringence ($\Delta n_s$) at low field amplitudes (2). At least four pulses were applied to obtain average values of $\Delta n_s$. Experimental average values $\langle D_R \rangle$ were calculated from the short time region of the field-free decay of the birefringence in the $E^2$ region (14). For all surfactant concentrations the pH was 6.8 and the specific conductivity of the suspension varied from 11.8 x $10^{-6}$ (ohm cm)$^{-1}$ at zero surfactant concentration to 72 x $10^{-6}$ (ohm cm)$^{-1}$ at a surfactant concentration of 1.44 x $10^{-6}$ g $cm^{-3}$.

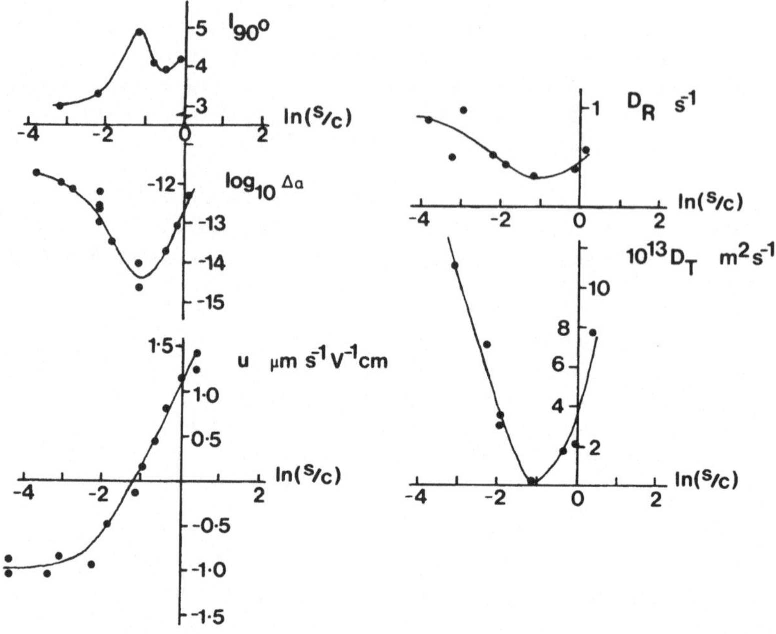


Fig. 1. Plots of $I_{90^o}$, $\log_{10}\Delta\alpha$, u, $D_R$ and $D_T$ as a function of the relative surfactant concentration (s/c). s is the surfactant concentration and c the clay concentration. All results have been normalised to a clay concentration of 2.05 x $10^{-4}$ g/cc. The values of $I_{90^o}$ have arbitrary units. The units of $\Delta\alpha$ are $cm^3$.

## RESULTS AND DISCUSSION

The effect of varying the surfactant concentration (s) on the parameters $\Delta\alpha$, u, $\langle D_R\rangle$, $\langle D_T\rangle$ and the intensity of scattered light ($I_{90^o}$) at $90^o$ is illustrated in fig. 1. The results have been normalised to a clay concentration of 2.05 x $10^{-4}$ g $cm^{-3}$. We note that the maximum in $I_{90^o}$ occurs at the same surfactant concentration as the minima in $\Delta\alpha$, $\langle D_T\rangle$ and $\langle D_R\rangle$ and is coincident with the zero value of the electrophoretic mobility u.

Qualitatively the results may be explained in the following manner. The maximum in $I_{90^o}$ and the minima in $\langle D_T\rangle$ and $\langle D_R\rangle$ are attributed to particle aggregation. Such aggregation was confirmed visibly. It thus appears that maximum particle aggregation occurs at the surfactant concentration where u becomes zero and hence the net charge density at the shear plane of the clay particles is zero. The presence of a minimum value for $\Delta\alpha$ may be explained if $\Delta\alpha$ is considered to consist of at least two components. A realistic situation would be the bulk polarisation of charges within the material and the surface polarisation of ions external to the particle surface but within the double layer structure surrounding the particle. Equations are available for estimating the magnitude of surface (15) and bulk (16) contributions to $\Delta\alpha$ . Such calculations, for large colloidal particles in ionic media, suggest that surface contributions may be as much as two orders of magnitude greater than bulk effects. The surface polarisation process is of correct order of magnitude to account for values of $\Delta\alpha$ measured experimentally for large colloidal particles (3,17). The interrelation between $\Delta\alpha$ and the charge density at the shear plane (18) suggests that the surface contribution should show a minimum when the value of u is zero. We note that the minimum value of $\Delta\alpha$ is about two and a half orders of magnitude lower than the value obtained for zero surfactant concentration. Thus we tentatively suggest that at the point where u is zero we are predominantly observing the bulk contribution to $\Delta\alpha$ .

Ideally one would wish to quantitatively analyse the data obtained. The use of more than one technique plus the measurement of well defined average values of the parameters should facilitate such an analysis. However, for a system such as that studied here, such an analysis is difficult as both the form and size of the particles and the form and shape of the particle size distribution function are changing. This is emphasised by the relative variation of $\langle D_R \rangle$ and $\langle D_T \rangle$ with surfactant concentration. Since $D_R$ is proportional to $\ell^{-3}$ (where $\ell$ is the particle length) and $D_T$ is proportional to $\ell^{-1}$, one would expect a greater variation of $\langle D_R \rangle$ than $\langle D_T \rangle$ with surfactant concentration. The fact that this is not observed experimentally is probably because the different methods view the mixture of particles and aggregates in a different manner. Further both bulk and surface contributions to $\Delta\alpha$ , with their different dependences on particle size and shape, complicate the interpretation of the measured values of $\Delta\alpha$ and $\langle D_R \rangle$.

## CONCLUSIONS

We observe for sepiolite suspensions that maximum aggregation occurs when $\Delta\alpha$ attains a minimum value and u is zero. Previous workers have found the minimum in $\Delta\alpha$ and the isoelectric point displaced by an order of magnitude of surfactant concentration (4-8). These different results may be due to the use of different clays, different surfactants or the fact that the present measurements were not made at constant ionic strength.

Clearly electro-optic measurements are important in studies on colloid stability. Further, in the region where u tends to zero, measurements of u are difficult and $\Delta\alpha$ measurement may provide an alternative probe of the particle-medium interface. In addition this region may assist in the isolation of bulk and surface components of $\Delta\alpha$ and a more complete picture for $\Delta\alpha$ for particles suspended in ionic media.

## ACKNOWLEDGEMENTS

The Science Research Council and Messrs. I.C.I. are thanked for grants to B.R.J. under which G.J.B. and V.J.M. held fellowships. The facilities offered by this physics department are gratefully acknowledged.


## REFERENCES


1 Shaw D J, "Electrophoresis", Academic Press, London (1969).

2 Fredericq E and Houssier C, "Electric Dichroism and Electric Birefringence", Oxford University Press, London (1973).

3 O'Konski C T and Haltner A J, J. Amer. Chem. Soc., 79, (1957) 5634.

4 Stoylov S P, Advances Colloid Interface Sci., 3,(1971) 45.

5 Stoylov S P, Petkanchin I and Sokerov S, Proc. IVth Intern. Congress on Surface Activity, Barcelona, Editiones Unidas, S.A., 2 (1968) 163.

6 Stoylov S P and Petkanchin I, Compt. Rend. de l'Academie Bulg. des Sci., 24, (1971) 487.

7 Stoylov S P and Petkanchin I, J. Coll. Int. Sci., 40, (1972) 159.

8 Petkanchin I and Bruckner R, Colloid and Polymer Sci., 254, (1976) 596.

9 Morris V J and Jennings B R, J. Coll. Interf. Sci., 66,(1978) 313.

10 Ware B J and Flygare W, Chem. Phys. Letts., 12, (1971) 81.

11 Brownsey G J and Jennings B R, J. Chem. Phys., 3, (1978) 926.

12 Berne B J and Pecora R, "Dynamic Light Scattering", John Wiley and Sons Inc., London, Chap. 8, (1976) 164.

13 Badoz J, J. Phys. Radium, 17, Suppl. 11, (1956) 143.

14 Morris V J, Foweraker A R and Jennings B R, Adv. Mol. Relax. Inter. Processes, 12, (1978) 65.

15 Maxwell J C, "A Treatise on Electricity and Magnetism", Oxford University Press, London (1892).

16 Peterlin A and Stuart H A, "Handbuch und Jahrbuch der Chemischen Physik", 8, Section 1B, (1943).

17 Jennings B R and Morris V J, J. Coll. Interf. Sci., 49, (1974) 89.

18 Brownsey G J, Jennings B R and Morris V J, J. Coll. Interf. Sci., 63, (1978) 597.

# Laser and High Field Effects

'Macht doch den zweiten Fensterladen
auch auf, damit mehr Licht hereinkomme':

'Open the second shutter, so that more
light can come in.'

JOHANN WOLFGANG VON GOETHE (1749-1832)

His Dying Words

# SOME EVIDENCES FOR A CONFORMATIONAL CHANGE OF POLYPEPTIDE INDUCED BY STRONG ELECTRIC FIELDS

Hiroshi Watanabe and Koshiro Yoshioka

Department of Chemistry, College of General Education,
University of Tokyo, Komaba, Meguro, Tokyo, Japan

A newly constructed electro-optical apparatus equipped with a 15 kV, 10 ms rectangular or reversing pulse generator, an AD converter and a mini-computor was used for finding a conformational change of polypeptide (PBLG) in the helix-coil transition region induced by strong electric fields. Birefringence measurements were used as the proof. When the pulse width was longer than two milliseconds, the Joule heating effect was more dominant than the effect sought. By use of PBLG with lower molecular weight, Kerr signals free from Joule heating could be observed. Field strength dependencies of steady state birefringence showed clear abnormality in the transition region, strongly suggesting the desired effect. Abnormalities which also suggest the presence of the effect were observed in the initial part of the rise and decay of the signal.

## INTRODUCTION

Since a segment of polypeptide has a larger contribution in helix conformation than in coil conformation to the overall dipole moment of the polymer, a strong electric field might stabilize the helix conformation. Such effect may be rather easily realized in the coil-to-helix transition (CHT) region, because of a cooperativity between segments of helix, as suggested theoretically by Schwarz (1). Experimental facts which were so far thought to be evidences of conformational changes induced by the strong electric field have been mainly concerned with polyelectrolyte systems (2-6) in which a change of external physical environment caused by the external field is a dominant factor for the effect. In the case of non-electrolytic polypeptides, such as poly-$\gamma$-benzyl-L-glutamate (PBLG), the situation is very different, because the cooperative intra-molecular interaction between the segments is the motivating factor in the CHT. From both the theoretical and biophysical point of view, it must be very interesting to reveal the existence of the effect by some direct evidences. Only indirect evidence has been obtained by electric birefringence measurements (7).

The purpose of the present work is, using PBLG as the material, to bring out some direct evidences of CHT induced by strong electric fields. The measurements which have been used to identify the effect are electric birefringence (2,3,7), change of optical absorption and dichroism (4-6), and change of optical rotation (8). Although the change of optical rotation can be a more direct proof of the effect than the others, it is not easy to eliminate the effect of linear birefringence as discussed in another paper presented in these Proceedings (9). Electric birefringence was therefore used as the probe of the effect in this work.

## EXPERIMENTAL

A newly constructed electro-optical apparatus for birefringence and dichroism measurements was used. Rectangular and reversing pulses adjustable up to 15 kV in amplitude (peak-to-peak) and up to 10 ms in duration were generated by the spark gap method. The transients of the pulses were less than 0.1 $\mu$s. Photo signals were transferred to a AD converter (Riken Denshi Co. Ltd.) which had two channels of 1K words, 10 bits memories. The quickest sampling time was 2 $\mu$s. The photo signals recorded in the AD converter were displayed on a pen-recorder, or an oscilloscope, or transferred to a mini-computer (System 70, Union Giken Co. Ltd.) for futher data processing, as required. A schematic diagram of the apparatus is shown in fig. 1. Two samples of PBLG (PBLG-T; Mw = 280,000 and PBLG-7; Mw = 10,000) were used. 1,2-Dichloroethane (DCE), m-cresol and glycerol were used as helicogenic solvents; dichloroacetic acid (DCA) and trifluoroacetic acid (TFA) were used as helix breaking solvents. In the middle of the transition region, breaking down in acid containing solutions did not afford us the highest field strength available when the pulses were applied through 2 mm spaced electrodes. The maximum field strength sustained in the acidic solutions was about 45 kV/cm.

## ABNORMAL KERR SIGNALS IN THE CHT REGIONS

No matter what the combination of helicogenic solvent and acid, an abnormally slow rise always followed the initial quick rise of the Kerr signal, when the strong field was applied on the solution of PBLG-T in the CHT region beyond two milliseconds or so. The relaxation time of the secondary slow rise was of the order of milliseconds. A typical example of such abnormal Kerr signals is shown in fig. 2. When the high molecular weight PBLG was used, the second slow rise usually folds over the first quick rise, concealing the steady state of signals.

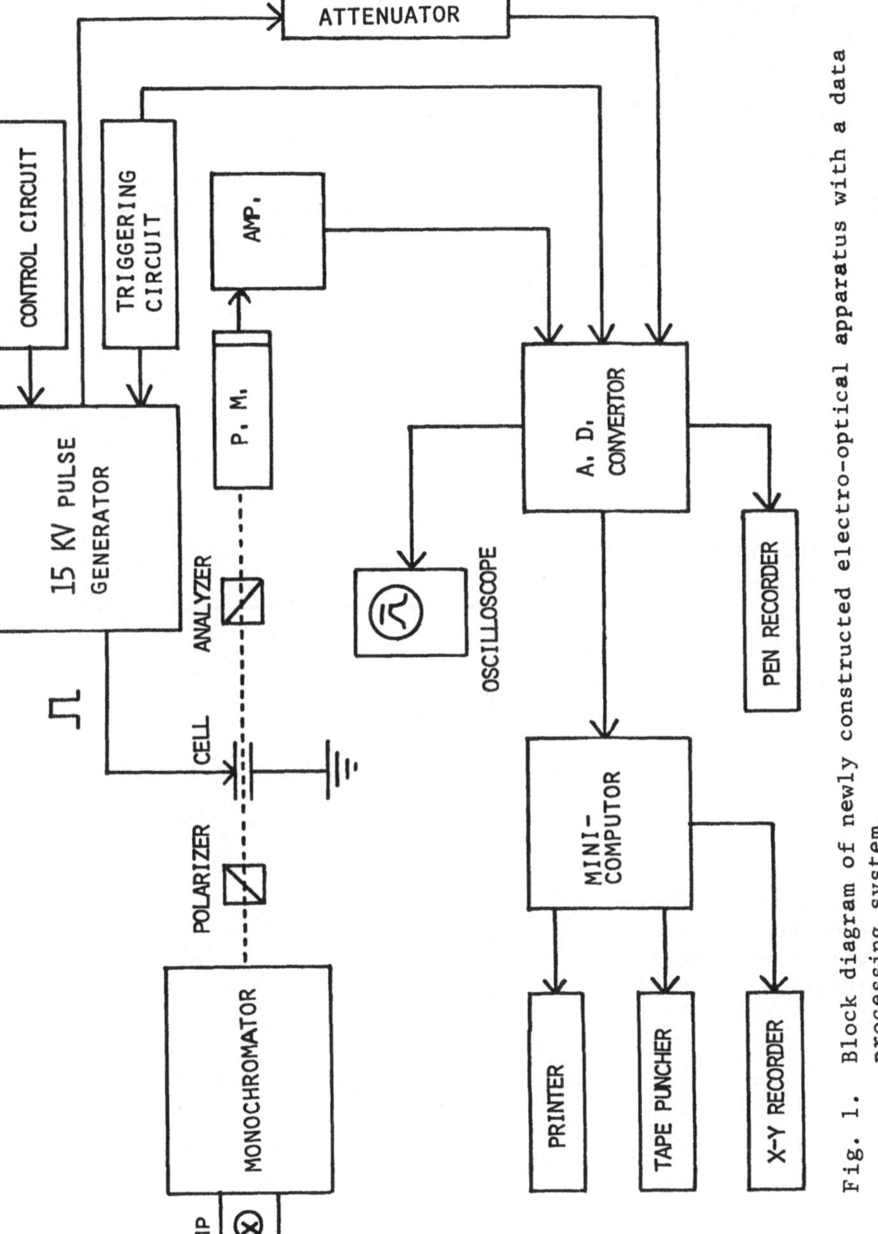


Fig. 1. Block diagram of newly constructed electro-optical apparatus with a data processing system.

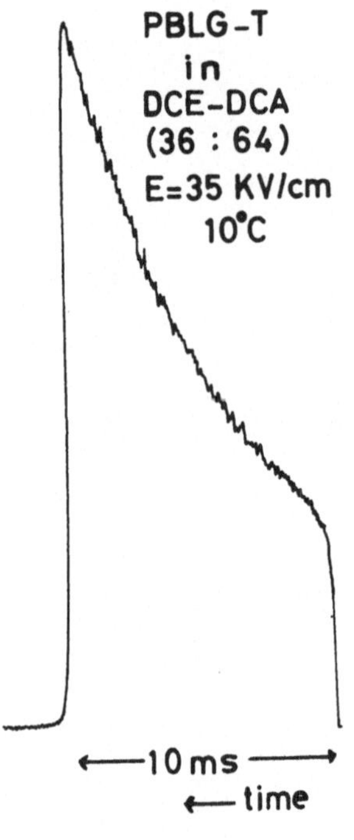


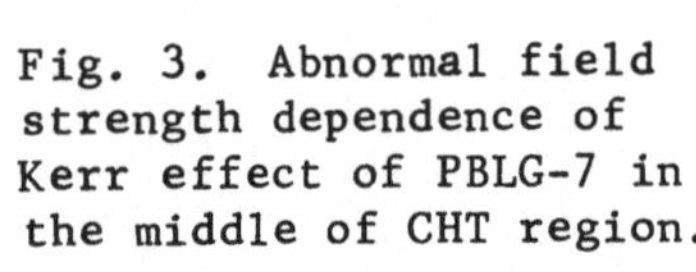


Fig. 2. Typical example of unusual rise of Kerr signal in the middle of CHT region displayed on a penrecorder.

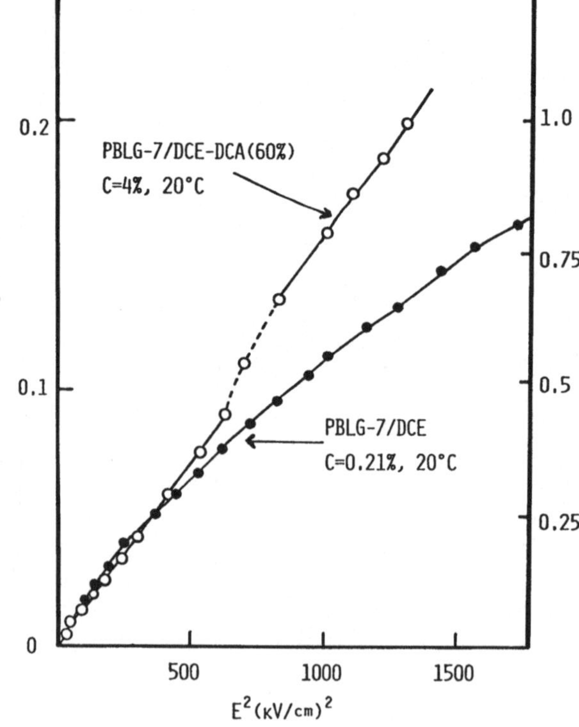


Fig. 3. Abnormal field strength dependence of Kerr effect of PBLG-7 in the middle of CHT region.

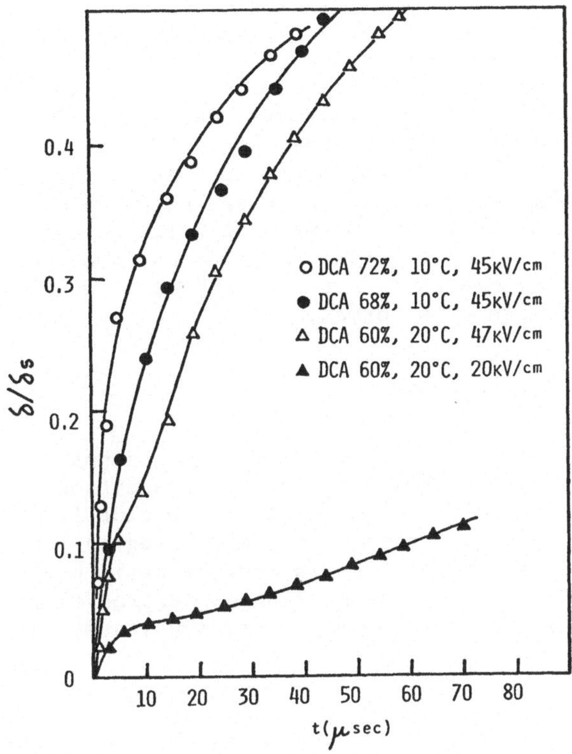


Fig. 4. Very rapid initial rise of Kerr signal of PBLG-T in the middle of CHT region.

For the following reasons, we cannot consider the abnormal secondary rise as an evidence of the field induced CHT. Firstly, the rise time is too slow to be thought of as the transition. Secondly, the temperature rise due to the Joule heating reaches several degrees which can be enough to produce a reasonable degree of the transition in the CHT region, when the high field is applied for a few milliseconds. As the specific resistance of the acidic solutions is of the order of $3 \times 10^6 \Omega$ cm, if we apply a field of 30 kV/cm for 10 ms, the heat generated per solution of 1 cm x 1 cm x 0.2 cm amounts to $(6 \times 10^3)^2 \times 10^{-2} \div (6 \times 10^5) = 0.6$ J. The temperature increment can be about 2 K, enough to give rise to a detectable amount of transition from coil to helix in the CHT region. The fact that the initial slope of relaxation from any part of the rising signal was essentially the same could be further support for the above discussion. Other physico-chemical processes such as the electrophoretic drift of protonated polymer to the cathode resulting in a highly concentrated region in the neighbourhood of the cathode or the space charge effect which may cause an apparent excess dipole moment on the polymer (10) cannot be ruled out.

## FIELD STRENGTH DEPENDENCE OF THE STEADY STATE BIREFRINGENCE

In order to obtain the steady state signals which are free from heating, we use PBLG-7. Pulses of $2 \times 10^{-4}$ sec duration were enough to give steady state signals. The temperature increment is less than 0.1 K even if the maximum field of 45 kV/cm is applied. This is an overestimated value because the pulse width could be reduced to less than $10^{-4}$ seconds when the highest field was used. A field strength dependence of steady state birefringence in the CHT region expressed in optical retardation $\delta$ is shown in fig. 3. As can be seen, the steady state birefringence abruptly increases beyond a threshold field strength, indicating an enhancement of helix content due to the strong field applied to the solution. For comparison, the field strength dependence of the same sample in pure DCE is shown in the fig. In pure DCE, the dependence is normal.

## THE TRANSIENTS

We may also expect very fast rises and decays at the beginning of transients as evidence of field induced CHT, for the transition rate was reported as the order of $10^{-8}$ - $10^{-6}$s (11,14). We closely examined the initial parts of rises and decays in the CHT region for PBLG-T. DCE and DCA were used as the helicogenic and the helix breaking solvents, respectively. The rise of the Kerr signal expressed in $\delta/\delta_s$, where $\delta_s$ is the optical retardation at the steady state, for

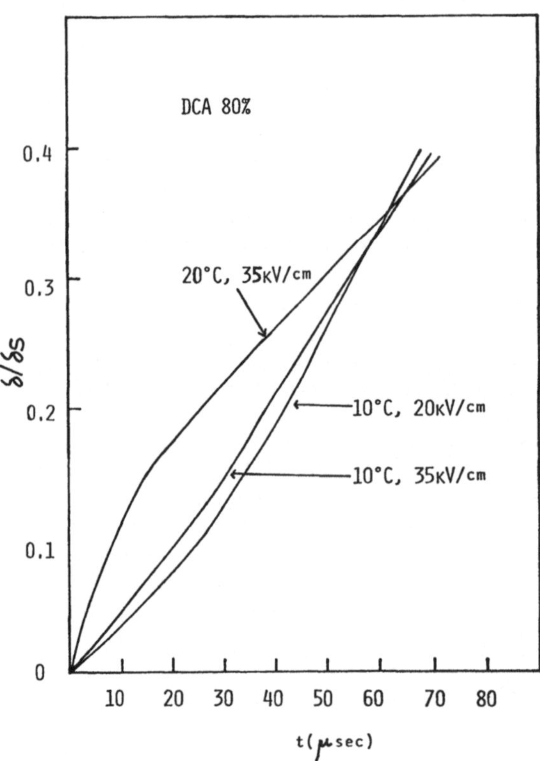


Fig. 5. Effect of temperature on the initial rise of Kerr signal of PBLG-T in DCA 80% solutions.

different conditions is shown in fig. 4. Doubtless the initial slopes of the rises are abnormally larger than we expect for the rise due to permanent dipole orientation where the initial slope should be very small even for high fields. Since the sampling time of the AD converter used in this work is 2 $\mu$ s per word, the largest slope observed may be determined by the sampling time. At the border between the helix and the transition region (DCA = 60%, 20°C), we still observed very quick initial rise under a field of 47 kV/cm, and an appreciable degree of step-up of signal under a moderate field of 20 kV/cm. In fig. 5, an effect of temperature on the initial rise in DCA 80% solutions is presented. The solution lies in the border between the coil and the transition region at 20°C, hence PBLG is expected in the random coil conformation at 10°C. At 10°C, the initial rise depends little on the field strength, as is expected. With increase of temperature to 20°C, the initial rise became faster, suggesting a beginning of the field effect.

Fig. 6. shows the initial decay of Kerr signal expressed in $\delta / \delta_s$. The field strength, the DCA content in the solvent and the temperature are written in the fig. At the instant the external field (which displaced the helix-coil equilibrium towards the helix and the higher degree of alignment subsequently) is removed suddenly, we may observe a very

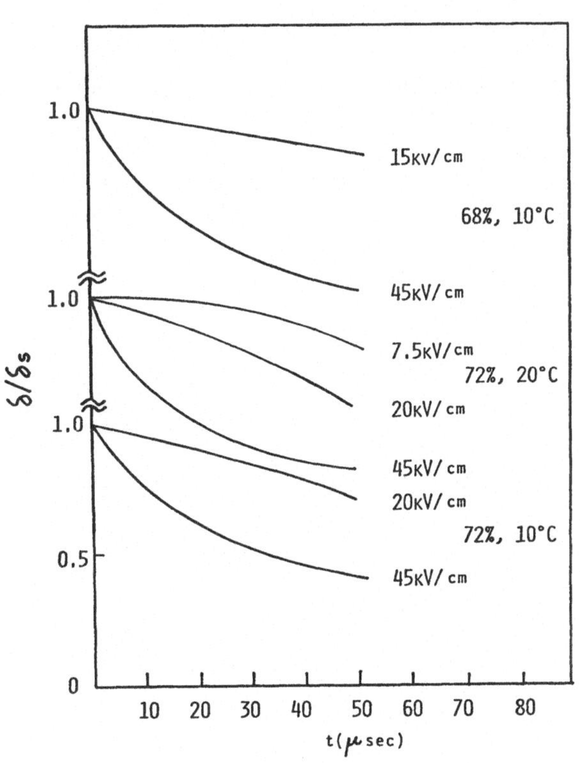


Fig. 6. Initial decay of Kerr signal of PBLG-T in the middle of CHT region.

rapid decay which indicates a collapse of helix to some extent. Since the initial slope of the decay is the larger the stronger the field is, our expectation is partly satisfied. The values of the relaxation rate are, however, much smaller than expected. In some cases unusually slow decays were observed when the signals were relaxed from low fields, as shown in the fig. Closer examination of the phenomena is necessary.

## REFERENCES


1 Schwarz G, Biopolymers, 14 , (1975) 1173.
2 Kikuchi K and Yoshioka K, Biopolymers, 12, (1973) 2667.
3 Kikuchi K and Yoshioka K, Biopolymers, 15, (1976) 1669,
4 Neuman E and Katchalsky A, Proc. Nat. Acad. Sci. U.S., 69 (1972) 993.
5 Pörschke K, Biopolymers, 15, (1976) 1917.
6 Pollak M and Glik H A, Biopolymers, 16,(1977) 1007.
7 Schwarz G and Schrader U, Biopolymers, 14,(1975) 1181.
8 Cummings A and Eyring E M, Biopolymers, 14, (1975) 2107.
9 Watanabe H, this volume p.43.
10 Tsuji K and Watanabe H, J. Chem. Phys., 66, (1977) 1343.
11 Lumry R, Legare R and Miller W G, Biopolymers, 2, (1964) 489.
12 Burke J J, Hammes G G and Lewis T B, J. Chem. Phys., 42, (1965) 3520.
13 Saksena T K, Michels B and Zana R, J. Chem. Phys., 65, (1968) 695.
14 Barksdale A D and Stuehr J E, J. Am. Chem. Soc., 94, (1972) 3334.

# NONLINEAR DIELECTRIC EFFECT OF PBLG IN NONPOLAR MEDIUM

R. Ooms and L. Hellemans

Department of Chemistry, University of Leuven,
Celestijnenlaan 200 D, 3030 Heverlee, Belgium

Measurements of the nonlinear dielectric effect in solutions of helical PBLG in benzene and in trans-dichloroethylene are reported. Molecular weights vary from 30000 to 150000. The observations extend over the frequency range of 50 kHz to 50 MHz. The dependence of the effects on the field strength up to nearly complete saturation yields values of the effective dipole moment and the anisotropy of polarizability. The moment reveals various types of aggregation, the ansiotropy is linear with molecular weight. The variation of the field-induced changes of permittivity and loss tangent with frequency are discussed in terms of orientational relaxation of the permanent moment.

## INTRODUCTION

Dissolved in nonpolar solvents the synthetic polypeptide poly-$\gamma$-benzyl-L-glutamate (PBLG) is known to be in the $\alpha$-helix conformation, forming long thin cylinders with a diameter of 18.3 A (1). Each turn of the helix counts 3.6 aminoacid residues, each residue adding 1.5 A to the total length of the macromolecule. In the $\alpha$-helix conformation the dipole moments of the individual peptide bonds are directed nearly parallel to the molecular axis, so that resultant dipole moments of 1000 D or more are not uncommon. In view of the large moment of the helical state, PBLG will readily show saturation behaviour in electric fields. In our laboratory a sensitive method has been developed to measure the field-induced changes of both real and imaginary parts of the permittivity ( $\varepsilon = \varepsilon' - j\varepsilon''$) over a wide range of frequencies (2).

In this paper we report on the nonlinear dielectric effect measured for solutions of PBLG in benzene and in trans-dichloroethylene (t-DCE). The nonlinear contributions $\Delta\varepsilon'$ and $\Delta\varepsilon''$ are analyzed for their dependence on the frequency of the measuring field, and their variation with high field intensity is considered. Nearly complete saturation has been observed. The electric parameters of the molecule are determined as functions of the molecular weight of the polypeptide.

This work constitutes a preliminary to the kinetic study of field-perturbed helix-coil transitions.

## THEORY

The alignment of molecules in large electric fields depends both on their permanent moment and on the anisotropy of their induced moment. Significant orientation of the molecules by a steady field alters the degree of polarization effected by small time-dependent fields. The permittivity observed with a small alternating field superposed on the high field varies as the field strength is increased. Such behaviour in general constitutes a nonlinear dielectric effect (3). (Another type of nonlinearity arises when the chemical composition of the dielectric is influenced by the field). The effects are characterized by $\Delta\varepsilon$ , the difference between the permittivity measured at high and at zero field strength.

Kielich and Przeniczny have given elaborate expressions for the various contributions to the nonlinear effect to be expected for non-interacting axially symmetric macromolecules (4). In such particles the permanent moment $\mu$ lies along the axis of symmetry, the anisotropy of polarizability is defined as $\Delta\alpha = \alpha_{\parallel} - \alpha_{\perp}$, the difference between the polarizabilities parallel and perpendicular to the axis.

When the electrostatic interaction of the field E with the permanent moment is much more important than with the induced moment ($\mu E \gg \Delta\alpha E^2/2$), the net effect $\Delta\varepsilon$ is negative. This remains true as long as the frequency f of the probing field is low enough for the dipole to rotate in phase, while it is seriously hindered in its movements by the pull of the steady field. Strictly speaking the argument bears only on the <u>average</u> degree of orientation of the particles. The negative change $\Delta\varepsilon$ in this part of the spectrum represents the well-known Langevin saturation. The effect is given by:

$$\Delta\varepsilon(f\to 0) = -(4\pi N\mu^2/3kT)(3L_1^2 - 3L_2 + 1) \qquad (1)$$

where we have assumed the external and directing fields to be equal. N is the number density of polar particles, kT is the thermal quantum and $L_n$ is a generalized Langevin function related to the average angle of orientation $\theta$ between the molecular axis and the field direction by:

$$L_n = \langle \cos^n \theta \rangle \tag{2}$$

It can be indicated that $L_1(a) = \coth a - 1/a$ and that $L_2(a) = 1 - 2L_1(a)/a$ with $a = \mu E/kT$.

At higher measuring frequencies the contribution of eq. 1 disappears gradually when the region of orientation relaxation is entered. The remaining nonlinear effect is now due exclusively to the anisotropy of the induced moment in molecules oriented by the steady field thanks to their permanent moment:

$$\Delta\varepsilon(f \rightarrow \infty) = (8\pi N/3)\ (\alpha_{\parallel} - \alpha_{\perp})\ (1 - 3L_1/a) \tag{3}$$

In this instance the sign of $\Delta\varepsilon$ depends on the type of anisotropy, which is often characteristic of the shape of the molecule.

The orientation function in eq. 1 reaches its limiting value of 1 more rapidly with increasing field strength than the function in eq. 3. For the limiting case of nearly complete saturation when coth a $\rightarrow$ 1, eq. 3 can be approximated by:

$$\Delta\varepsilon = (8\pi N/3)\ (\alpha_{\parallel} - \alpha_{\perp})\ (1 - 3/a + 3/a^2) \tag{4}$$

One recognizes in eq. 3 the orientation factor first given by O'Konski and coworkers to describe saturation of the Kerr effect at the visible end of the electromagnetic spectrum (5). Finally, it is of interest here to write the approximate relation between $\Delta n_s$, the birefringence at infinite field strength and $\Delta\alpha_{opt}$, the optical anisotropy on the basis of eq. 3:

$$2n\Delta n_s \simeq 8\pi N\ \Delta\alpha_{opt}/3 \tag{5}$$

wherein n is the index of refraction.

## EXPERIMENTAL

### Apparatus

The measuring technique has been described earlier (2,6). It will be reviewed here in brief. The apparatus has been built to detect nonlinear field effects and their evolution in the frequency domain. A diagram illustrating the basic operation is given in fig. 1.

A high electric field at 85 Hz is applied to a symmetric cell, filled with the sample. By choosing an inductor from a home-made set and connecting it in parallel to the cell, a resonant network is obtained. The characteristic frequency of the circuit covers the range of about 0.05 to 100 MHz depending on the inductor chosen. In this work the coils are grounded at their midpoint to ensure complete symmetry of the voltages.

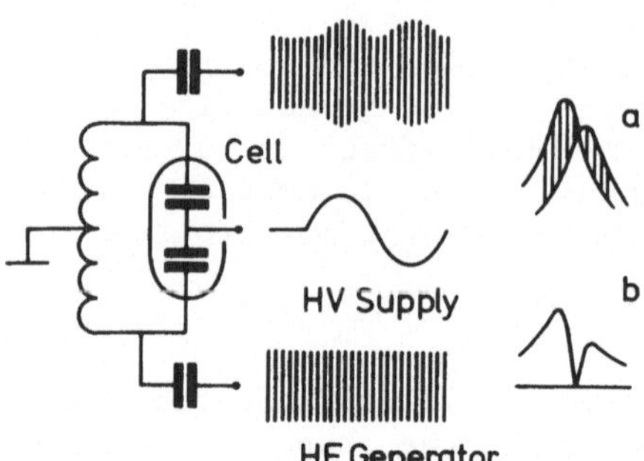


Fig. 1. Diagram of the resonance circuit, excited by the high frequency generator-sweeper and modulated by high voltage at 85 Hz. Both cell capacitors contain the sample. The inductor is exchangeable. The hatched area of inset (a) corresponds to the modulation depth at any point of the sweep and provides the signal (b) upon demodulation.

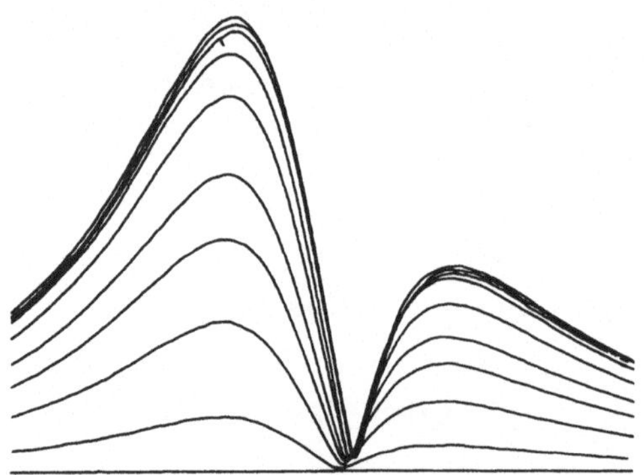

Fig. 2. Difference signals recorded at increasing field strength for a solution of 0.74 g/l PBLG (MW = 150000) in benzene at 25°C. Measuring frequency 25.372 MHz, resonant voltage 1.82 V, circuit quality 412, sweep width 127 kHz, field strength 7, 17, 26, 35, 54, 72, 92, 111 and 128 kV/cm. The maximum degree of modulation is about 0.9%.

The circuit is excited by a tuned generator, whose output frequency is slowly swept through resonance. The resonant voltage is always much smaller than the high voltage applied to the sample. As soon as the impedance of the solution is modified by the action of the periodic high field, amplitude modulation of the high frequency resonant voltage occurs. The modulation depth reflects the difference between the resonance curves at peak field strength and at zero field. Demodulation and consecutive amplification provide the signals which are recorded as functions of the swept frequency. An example of such traces obtained for a solution of PBLG in benzene at different field strengths in shown in fig. 2. Saturation of the nonlinear effects is obvious.

The signals reveal a shift of the resonant frequency together with changes of bandwidth and resonant voltage. A fit of the data to the derivative of the expression relating the voltage to the resonance parameters yields the field-induced changes of the complex impedance, namely $\Delta\varepsilon'/\varepsilon$ and $\Delta\tan\delta$. For measurements in the frequency range studied, $\varepsilon$ can be taken as the permittivity of the solvent. The change of the loss tangent ($\tan\delta$) can be taken as dielectric loss ($\Delta\varepsilon''/\varepsilon$) exclusively, since no field-dependent changes of the conductivity appear. The small contribution $\Delta\varepsilon'/\varepsilon$ noted for the neat solvents is neglected.

## Materials

PBLG was purchased from Sigma Chemical Co. The mean molecular weight is viscosity determined by the manufacturer. The peptide is used without further purification. Benzene is dried with Na-K alloy and fractionally distilled under reduced pressure, while t-DCE is dried over $CaH_2$ before fractional distillation. All solutions are made up gravimetrically in a drybox.

# RESULTS AND DISCUSSION

The real part of the complex permittivity of solutions of PBLG in either benzene or t-DCE varies under high field conditions with the frequency of the measuring field as illustrated in fig. 3. At low frequencies the tail of a dispersion region can be discerned. Here the value of $\Delta\varepsilon'$ is relatively large and negative. At about 1 MHz sign reversal of $\Delta\varepsilon'$ occurs and above 5 MHz the nonlinear effect remains essentially constant in the frequency range covered.

Clearly, at the highest frequencies any contribution from dipolar orientation with the measuring field has ceased. The positive effect $\Delta\varepsilon'$ corresponds to that described by eq. 3. One immediately concludes that $\alpha_{\parallel} - \alpha_{\perp} > 0$. The observed dependence on the field strength at fixed frequency is represented in fig. 4. Saturation is complete for about 85%.

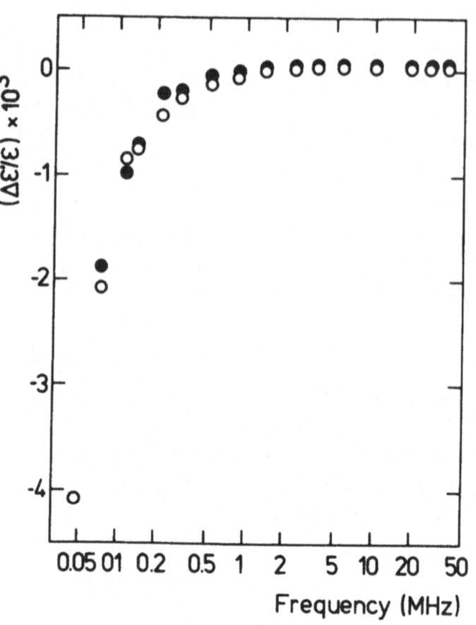


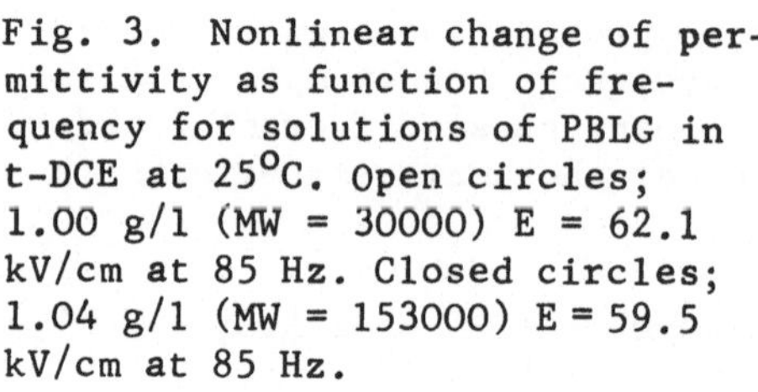
Fig. 3. Nonlinear change of permittivity as function of frequency for solutions of PBLG in t-DCE at 25°C. Open circles; 1.00 g/l (MW = 30000) E = 62.1 kV/cm at 85 Hz. Closed circles; 1.04 g/l (MW = 153000) E = 59.5 kV/cm at 85 Hz.

**Table I. Dielectric parameters of PBLG in nonpolar solvents.**

| Solvent | | MW | $C_o$ (g/l) | T (K) | μ (D) | μ/N (D) | Δα ($10^{21}$ cm³) | Ref. |
|---|---|---|---|---|---|---|---|---|
| Benzene | 2.27 | 150000 | 1.56 | 298 | 1490 (1520) | 2.2 | 5.8 (5.7) | |
| Benzene | 2.27 | 150000 | 0.74 | 298 | 2010 (2050) | 2.9 | 6.6 (6.6) | |
| Benzene | 2.27 | 150000 | 0.74 | 311 | 1890 (1890) | 2.8 | 6.2 (6.2) | |
| t-DCE | 2.14 | 153000 | 1.04 | 298 | 2470 (2280) | 3.5 | 5.3 (5.5) | |
| t-DCE | 2.14 | 70000 | 1.02 | 298 | 1620 (1640) | 5.1 | 2.9 (2.9) | |
| t-DCE | 2.14 | 30000 | 1.00 | 298 | 905 (1130) | 6.6 | 0.8 (0.7) | |
| DCE | 9.9 | 195000 | 0.2 | 302 | 2700 | 3.0 | 1.5 | 5 |
| t-DCE | 2.14 | 220000 | 1.16 | 293 | 1930 | 1.9 | 8.3 | 7 |
| Dioxane | 2.21 | 220000 | 1.16 | 293 | 1410 | 1.4 | 10.5 | 7 |
| | | | | | | | 12.6 | 8 |

MW molecular weight, $C_o$ concentration, μ/N dipole moment per residue. t-DCE (trans-dichloroethylene), DCE (dichloroethane)

The high-field data have been fitted by a least-squares procedure to the polynomial expression in eq. 4. This allows calculation of the dielectric parameters $\mu$ and $\Delta\alpha$ , which are given in table I. The validity of the approximation made is tested by plotting $\Delta\varepsilon'$ against $L_1/E$, calculated on the basis of the former result for $\mu$ . A good linear relationship exists as shown in fig. 5. The new parameters $\mu$ and $\Delta\alpha$ found from the least-squares line through the points are collected in table I between brackets. The agreement between both sets of values is quite good; it is not surprising that it is least satisfactory for the low molecular weight experiment, since a comparable degree of saturation will only be achieved at higher field intensity.

The solid lines in fig. 4 are then calculated with the first set of parameters according to eq. 3. One notices the quadratic region at low field strength. It is found that this type of plot is more critical for checking the agreement between experiment and theory than the plot of fig. 5.

We have included in table I some literature values for the sake of comparison. The data of ref. 5 are derived from birefringence, the other from dielectric measurements.

From the numerical results the assumption $\mu E \gg \Delta\alpha E^2/2$ for helical PBLG is confirmed. The permanent moments are of the order of 1000 D. They do not vary with the field strength. Our results for the benzene solutions show a concentration dependence of the dipole moment per residue. It can be explained by side-by-side (antiparallel) aggregation of the macromolecules in which the dipole moments partially cancel each other (9). A rise in temperature does not seem to affect the aggregation equilibrium, although the temperature effect may be the complex result of various interactions opposite to one another (10). It is of interest to note that the small increment of temperature (13 K) shifts the curves in fig. 4 and 5 considerably.

The measurements in t-DCE reveal the relation of the dielectric parameters with the average molecular weight. The parmanent moment increases monotonously with molecular weight, although the moment per residue decreases. The experiments have been performed at constant number of residues per unit volume. The data of ref. 7 fit with the latter trend, showing that the effective moment has a maximum around molecular weight 150000 at a concentration of about 1 g/l. The effects seem to indicate that several types of aggregation are operative simultaneously: head-to-tail (parallel) aggregation as well as side-by-side aggregation. At constant concentration of residues, it depends on the molecular weight which type of aggregation will predominate. Indeed, the experimental moment per peptide bond becomes either smaller or larger than 3.4 D, the accepted value for the dipolar contribution per residue of PBLG (11). As expected, the head-to-tail association is most important for the low molecular weight sample, where the ratio of polymer ends to the total number of residues is largest.

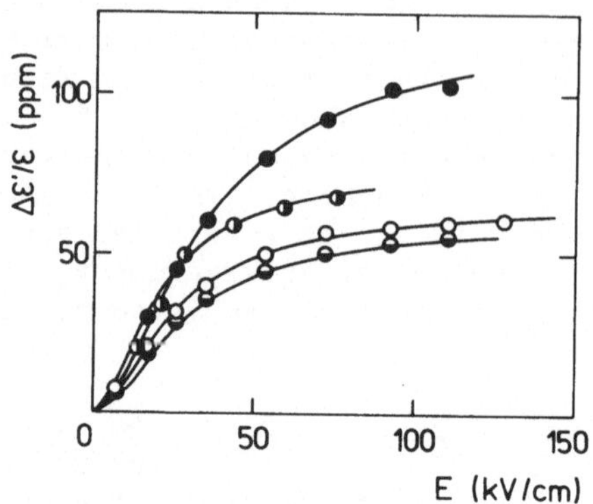


Fig. 4. Nonlinear change of permittivity as function of field strength for solutions of PBLG.
Closed circles: 1.56 g/l (MW = 150000) in benzene, 25°C and 6.5 MHz. Open circles: 0.74 g/l (MW = 150000) in benzene, 25°C and 25.4 MHz. Horizontal segments: 0.74 g/l (MW = 150000) in benzene, 38°C and 25.7 MHz. Vertical segments: 1.04 g/l (MW = 153000) in t-DCE, 25°C and 18.9 MHz.

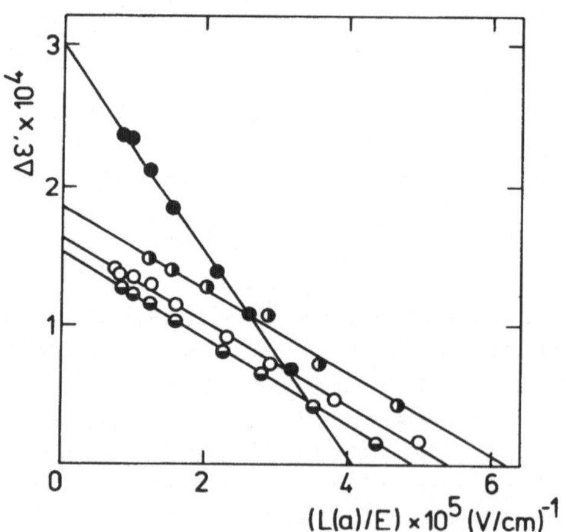


Fig. 5. Nonlinear change of permittivity as function of the ratio L(a)/E with a = $\mu E/kT$ for solutions of PBLG, symbols same as in fig. 4.

One would expect the field to perturb strongly the different aggregation equilibria, to which large changes of electric moment are connected (3,12). We did not find any experimental evidence for this, probably because the processes are slower than the period of the high field (13). It is clear from table I that the solvent determines which interaction will prevail, not so much by its bulk permittivity as by the local polar groups, and their ability to compete for hydrogen bonding.

The anisotropy of the polarizability does not seem to be influenced very much by the nature of the solvents used in this study. It changes less with concentration than does the effective dipole moment. The data including that from ref. 7 for t-DCE solutions show a regular increase of $\Delta\alpha$ with molecular weight. Davies and coworkers have noted that the anisotropy of polarizability obtained by dielectric methods (in the MHz region) is much larger than the value obtained by birefringence studies (7). They indicate that this difference may be due to the spring-like motion of the helix, whose successive loops are held by flexible hydrogen bonds (8). At any rate, the intensity of the microwave absorption which they detect, rises with the degree of polymerization. It is hard to predict how the values of $\Delta\alpha$ would be affected by aggregation.

The conclusions drawn from the results in table I must be handled with some care in view of the uncertainty of the molecular weight and the effect of polydispersity.

In fig. 6 the change of the loss tangent is plotted against frequency for two different molecular weight samples. The fact that the change is negative in high fields means that energy is saved as compared to the dielectric loss at zero field strength. This is in agreement with the Kramers-Kronig relations for any dispersion region where the permittivity decreases by the action of the field. The corresponding decrements $\Delta\varepsilon'$ can be noted in the low frequency region of fig. 3. The slope of the lines in the double-logarithmic plot of fig. 6 is somewhat less than unity, as a Debye-type relaxation would require, possibly reflecting the degree of polydispersity of the samples.

One expects $\Delta\tan\delta$ to depend on the field intensity as predicted by eq. 1. We found that $\Delta\varepsilon'$ and $\Delta\tan\delta$ measured at about 100 kHz varied differently with the field. Such behaviour may indicate that the critical frequency of the orientation process is dependent on the field. Block and Hayes have made such observations (14).

The frequency range investigated at present does not extend far enough to warrant the validity of estimates of the orientation relaxation time from the data of fig. 6. Calculation of the orientation time for the different polymer samples studied shows that in each case most of the dipolar relaxation region is to be expected at frequencies lower than 20 kHz.

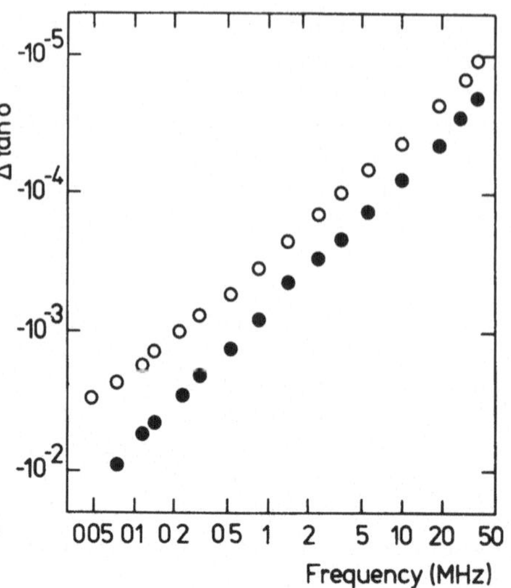


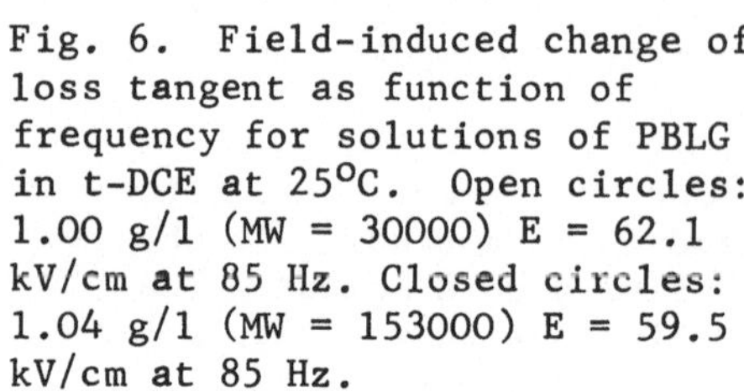
Fig. 6. Field-induced change of loss tangent as function of frequency for solutions of PBLG in t-DCE at 25°C. Open circles: 1.00 g/l (MW = 30000) E = 62.1 kV/cm at 85 Hz. Closed circles: 1.04 g/l (MW = 153000) E = 59.5 kV/cm at 85 Hz.

It would be of interest to analyze by means of high field studies the effect of de-aggregants on the dielectric parameters. Ultimately, the dielectric method will become a valuable tool for studying the dynamics of the unwinding of the helix.

## ACKNOWLEDGEMENT

R.O. is grateful to the Belgian I.W.O.N.L. for the award of a predoctoral fellowship. L.H. is research fellow of the Belgian Nationaal Fonds voor Wetenschappelijk Onderzoek. This project is supported in part by F.K.F.O. grant 2.0051.77N.

## REFERENCES

1 Wada A, Bull. Chem. Soc. Japan, 33, (1960) 822.
2 Hellemans L and de Maeyer L, J. Chem. Phys., 63, (1975) 3490.
3 Böttcher C J F, "Theory of Electric Polarization", Vol. I, 2nd ed., Elsevier, Amsterdam (1973) p.289.
4 Kielich S and Przeniczny Z, Chem. Phys. Letters, 6, (1970) 72.
5 O'Konski C T, Yoshioka K and Orttung W H, J. Phys. Chem., 63, (1959) 1558.
6 Persoons A and Hellemans L, Biophys. J., 23, (1978) 000.
7 Gregson M, Parry Jones G and Davies M, Trans. Faraday Soc., 67, (1971) 1630.
8 Davies M, Maurel P and Price A H, J.C.S. Faraday II, 68, (1972) 1041.
9 Powers J C and Peticolas W L in "Ordered Fluids and Liquid Crystals", Advances in Chemistry Series, No. 63, American Chemical Society, Washington, D.C. (1967) p.217.

10 Tsuji K, Ohe H and Watanabe H, Polymer J., 4, (1973) 553.
11 Yoshioka K in "Molecular Electro-Optics", O'Konski C T ed., Part 2, Marcel Dekker, New York (1978) p.601.
12 Bergmann K, Eigen M and De Maeyer L, Ber. Bunsenges. Phys. Chem., 67, (1963) 819.
13 Parry Jones G in "Dielectric and Related Molecular Processes", Davies M ed., Vol. 2, The Chemical Society, London (1975) p.198.
14 Block H and Hayes E F, Trans. Faraday Soc., 66, (1970) 2512.

# LASER AND ELECTRIC FIELD INDUCED KERR EFFECT STUDIES ON NEMATIC LIQUID CRYSTALS

H. J. Coles* and B. R. Jennings

Electro-Optics Group, Physics Department, Brunel University, Uxbridge, Middlesex, U.K.

The optical and d.c. pulsed field Kerr effects have been used to study the pretransitional behaviour in the isotropic phase for both positive (pentyl cyanobiphenyl) and negative (methoxybenzyladine butyl aniline) dielectric liquid crystals. From the dynamic and static measurements both the relaxation time ($\tau$) and the induced birefringence ($\Delta n$) have been studied as a function of temperature. These results have been examined in terms of the phenomenological Landau - de Gennes model of phase transitions. Finally both 5 CB and MBBA have been studied in solution in $CCl_4$ and the sign changes recorded as a function of field frequency and concentration for MBBA are contrasted with those for 5 CB where no such changes are measured. This paper reviews the more important results obtained with the method and such typical liquid crystalline materials.

## INTRODUCTION

It has been demonstrated recently that both the electrical (1-5) and optical (6-10) Kerr effects may be used to study the isotropic to nematic phase transition in liquid crystals. In the electrical Kerr effect a pulsed d.c. field is used to induce the birefringence whereas in the optical Kerr effect a pulsed laser field is used. This allows us a method of examining the ordering induced by the two different mechanisms, i.e. the interaction with the applied field of the permanent or low frequency induced dipole moment or the polarisability anisotropy.

*Current address: Centre de Recherches sur les Macromolecules, 6 rue Boussingault, 67083 Strasbourg Cedex, France.

Both methods have been used herein in the same experiment to determine the static properties and from the pulsed field free relaxation the dynamic behaviour of the cooperative motions for both a positive (pentyl cyanobiphenyl - 5 CB) and negative (methoxybenzylidene butyl aniline - MBBA) nematic liquid crystal. These materials, which are both of commercial interest, are room temperature nematogens of similar size but different dipolar structure. It is the object of this article to illustrate how their different structure is reflected in their different electro-optical characteristics both as a function of temperature in the isotropic phase and as a function of concentration in solution. For the isotropic phase the results have been examined in terms of the Landau - de Gennes phenomenological model (11), and the parameters of this formalism, which are important in characterising liquid crystals, have been given. Both liquid crystals have been studied in solution in $CCl_4$ and electrical and optical parameters have been given with the concentration dependencies of the Kerr constants. Finally a possible development of the method has been discussed.

## THEORY

### Nematic - Isotropic Phase Transition

In the current work the Landau - de Gennes model (11) has been used to interpret the pretransitional behaviour and as this theory has recently been considered in greater detail elsewhere (12,13) we will list only the main points of the theory below. In the model the excess free energy F is given as a series expansion of the scalar order parameter such that

$$F = \frac{a}{2}(T - T^*)\,Q^2 + \frac{b}{3}Q^3 + \frac{c}{4}Q^4 + \; \ldots\ldots \qquad (1)$$

where a, b, c are the phenomenological constants characteristic of the material and the transition. $T^*$ is the second order transition temperature that the system would have in the absence of the intervening weak first order transition that occurs at the clearing temperature $T_c$. For the case of orientation induced in an electrical or optical field the amount of order produced is small and terms in $Q^3$ and higher order may be neglected, and it has been shown (12) for the electric field case that

$$B_{DC} = \frac{\Delta n_o \; \Delta\varepsilon_o}{3a\;\lambda_o\,(T - T^*)} \qquad (2)$$

where $B_{DC}$ is the d.c. field Kerr constant ($B_{DC} = \frac{\Delta n}{\lambda_o E_{DC}^2}$), $\lambda_o$ is the probe wavelength and $\Delta n_o$ and $\Delta\varepsilon_o$ are the refractive index and dielectric anisotropies respectively. Further for the same model it has been shown (14) that the field free relaxation ($\tau$) is given by

$$\tau = \frac{\nu}{a} (T - T^*)^{-\gamma} \qquad (3)$$

where $\nu$ is a weakly temperature dependent viscosity coefficient and $\gamma$ is an exponent that is unity in any mean field theory.

Solutions

In order to interpret the dilute solution measurements we have used the Peterlin and Stuart formalism (15) to obtain the Specific Kerr constant $B_{sp}$ extrapolated to zero concentration where

$$B_{sp} = \frac{\Delta n}{\lambda_o . c_v . E^2} = \frac{2\pi(g_1 - g_2)}{15n\,\lambda_o} \left[ \left\{ \frac{\mu'}{kT} \right\}^2 + \left\{ \frac{\Delta\alpha}{kT} \right\} \right] =$$

$$\frac{2\pi(g_1 - g_2)(\Theta)}{15n\,\lambda_o} \qquad (4)$$

where $(g_1 - g_2)$ is the optical anisotropy and is related to the optical polarisability by $(\alpha_1 - \alpha_2)^o = 4\pi\,\varepsilon_o'\,v(g_1 - g_2)$, $\Delta\alpha$ = low frequency polarisability, $\mu'$ is the effective permanent dipole moment, v is the particle volume, n is the solvent refractive index, $\varepsilon_o'$ is the permittivity of a vacuum, $c_v$ is the volume fraction and $\lambda_o$ is the detecting wavelength. This model assumes coincidence of the geometric, electrical and optic axis on the axis of rotation, and also that the Lorentz-Lorenz local field factors are equal to unity. We estimate the error to be introduced in this latter assumption to be small (5 - 10%) for the values of $\mu'$ and $\Delta\alpha$ determined for the materials studied herein.

## EXPERIMENTAL

The principles of the apparatus have been described in greater detail elsewhere (10) as have recent improvements relevant to the current 5 CB measurements (16). Both the d.c. and optical Kerr effects were measured using either a low powered Argon-ion laser ( $\lambda_o$ = 488 nm) or a 10 mW HeCd laser ( $\lambda_o$ = 441.6 nm). The HeCd laser has an inherently lower noise ($<$1%) than the Argon-ion laser. The birefringence induced by the electrical (d.c. pulses of duration up to 25 $\mu$s and between 0 and 10 kV) and laser (Q-switched unfocussed $TEM_{oo}$ pulses of duration 50 ns and field strength 0 to 9 $MV.m^{-1}$ from a $Nd^{3+}$ YAG laser) fields was measured between two glan-laser polariser prisms crossed with the azimuth at 45° to the applied field direction. The direction for both the d.c. and polarised laser pulsed fields was horizontal. The beam passing through the analyser as a result of the induced birefringence was detected by a fast blue sensitive photomultiplier (EMI - 9813QKB), the resulting anode pulse amplified by a 100 MHz preamplifier, and then displayed on a fast storage oscilloscope screen. This display could be analysed or photographed for later processing.

At the light levels used in the current measurements the stray or strain birefringence ($\Delta n_{(stray)} \sim 10^{-9}$) was negligible and quadratic detection was often used. The cell, which had an electrode separation of 2mm and an optical path length of 49 mm, was thermostated to $\pm 0.05°C$. It is worth noting that each measurement takes only a few seconds at a given concentration, temperature or field setting and it is possible to change between pulsed laser or d.c. fields by simply changing a triggering switch position. Thus the advantages of each method are available under the same experimental conditions without resorting to separate measurements.

The alkyl cyano biphenyl was a gift from B.D.H. Ltd. (Poole, Dorset, U.K.) which is gratefully acknowledged, and the MBBA was purchased from Messrs. Kodak Ltd. These samples were used as supplied in their sealed containers, and no further purification was attempted. Glassware and cells were rigorously cleaned and thoroughly dried before use.

## RESULTS

### Isotropic Phase

Both the transient optical and d.c. Kerr effects were studied in the two nematogens at temperatures near to and above the nematic-isotropic transition temperatures (i.e. 41.5 and 35.1°C for MBBA and 5 CB respectively) and typical results have been given in figs.1 and 2. Fig. 1 illustrates the birefringence response as a function of temperature for a fixed field strength under optical or d.c. fields, and fig. 2 similarly gives the field free relaxation time $\tau$ as a function of temperature. In the latter case the relaxation times were monoexponential. Both the static ($\Delta n$) and dynamic ($\tau$) properties show strong pretransitional behaviour. For all of the laser induced studies the birefringence was positive whereas in the d.c. field case the birefringence was negative for MBBA and positive for 5 CB.

### Solutions

Below $T_c$ nematogenic liquid crystals are effectively opaque. However Kerr effect studies may be made at lower temperatures in solutions with suitable solvents. In the current work $CCl_4$ has been used as a solvent and concentration dependencies have been established. At the field strengths used solvent effects were generally negligible except for very dilute solutions of MBBA/$CCl_4$ when the solvent effect was subtracted out. For pulsed laser fields the birefringence was positive and as shown in table I concentration independent for low concentrations for both MBBA and 5 CB. At the highest concentration ($0.73\ cm^3/cm^3$) for 5 CB there is a slight dependence on concentration. For pulsed d.c. fields the Kerr constant changes sign for MBBA and shows a marked concentration dependence whereas for 5 CB it is always

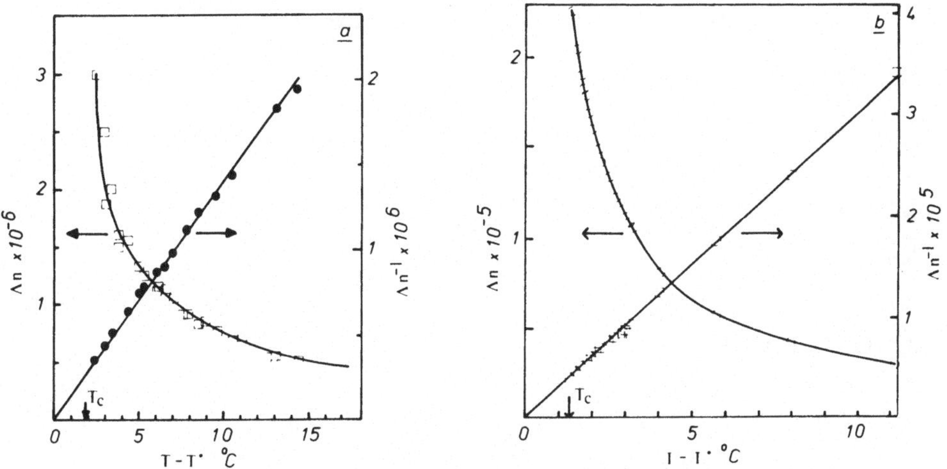


Fig. 1. Birefringence (l.h.s.) and inverse birefringence (r.h.s.) as a function of temperature in the isotropic phase for (a) MBBA under a laser field of $\langle E_0^2 \rangle^{\frac{1}{2}}$ = 2.7 MV.m$^{-1}$ and (b) for 5 CB under a pulsed d.c. field of $E_{DC}$ = 0.58 MV.m$^{-1}$. (a) MBBA = $CH_3O.\emptyset.CHN.\emptyset.C_4H_9$. (b) 5 CB = $CN.\emptyset\emptyset.C_5H_{11}$

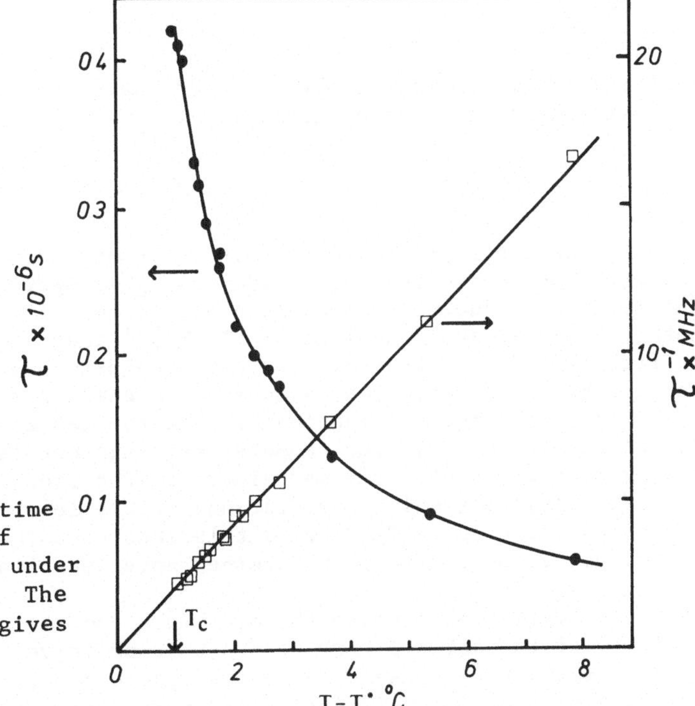


Fig. 2. Relaxation time ($\tau$) as a function of temperature for 5 CB under pulsed laser fields. The right hand ordinate gives $\tau^{-1}$.

Table I. Solution Kerr Constants

(i) M.B.B.A.

| | | | | | | | |
|---|---|---|---|---|---|---|---|
| $c_v$ ($cm^3/cm^3$) | 0.02 | 0.08 | 0.2 | 0.25 | 0.34 | 0.5 | 0.6 |
| Optical $B^o_{sp}$ ($10^{-14} m.V^{-2}$) | - | - | 1.3 | - | - | 1.1 | 1.2 |
| Static $B^{dc}_{sp}$ ($10^{-14} m.V^{-2}$) | 9.6 | 6.2 | - | 2.1 | 1.3 | -3.7 | -6.3 |

(ii) 5 CB

| | | | | |
|---|---|---|---|---|
| $c_v$ ($cm^3/cm^3$) | 0.21 | 0.37 | 0.56 | 0.73 |
| Optical $B^o_{sp}$ ($10^{-14} m.V^{-2}$) | 2.7 | 2.8 | 2.9 | 3.7 |
| Static $B^{dc}_{sp}$ ($10^{-12} m.V^{-2}$) | 2.8 | 2.9 | 2.7 | 4.5 |

positive and is approximately constant until high concentrations when it starts to increase markedly.

## DISCUSSION

In the isotropic phase the birefringence shows strong pretransitional behaviour increasing rapidly as $T_c$ is approached. Following the de Gennes model (11) and using eqs. 1 to 3 it can be seen that both $\Delta n$ and $\tau$ are functions of $(T - T^*)^{-1}$ under the conditions given in figs. 1 and 2. These results are intended to be illustrative of the technique and have been discussed in greater detail elsewhere (10, 11). The important observation is that the Landau - de Gennes model is upheld over the limited temperature range considered herein, and the existence of molecular associates in the isotropic phase is confirmed. These molecular associations become less effective the higher the temperature above $T_c$. Using this model and the above results the phenomenological constants determined have been listed in table II.

In the isotropic phase the d.c. Kerr constant is negative for MBBA whilst the optical Kerr constant is positive. The structure of MBBA is such that it has a permanent dipole moment that is predominantly across the major molecular axis, and this dipole acts in quadrature with the induced axial dipole. Thus in the d.c. case the permanent

Table II. Molecular Parameters

| | 5 CB | M.B.B.A. |
|---|---|---|
| Phenomenological constant a : | $0.15 \times 10^6 J.m^{-3}K^{-1}$ | - |
| Viscosity coefficient : | 0.73P | - |
| Transition temperature $T_c$ : | 35.1 (±0.05) | 41.5(±0.05) °C |
| 2nd order transition temperature $T^*$ : | 34.0 (±.2) | 39.6(±.2) °C |
| Optical polarisability anisotropy : | 30 (± 3) | $32\ (\pm\ 3)\times10^{-40}Fm^2$ |
| Dipole moment : | $13.7(\pm1.5)\times10^{-30}$ | - |

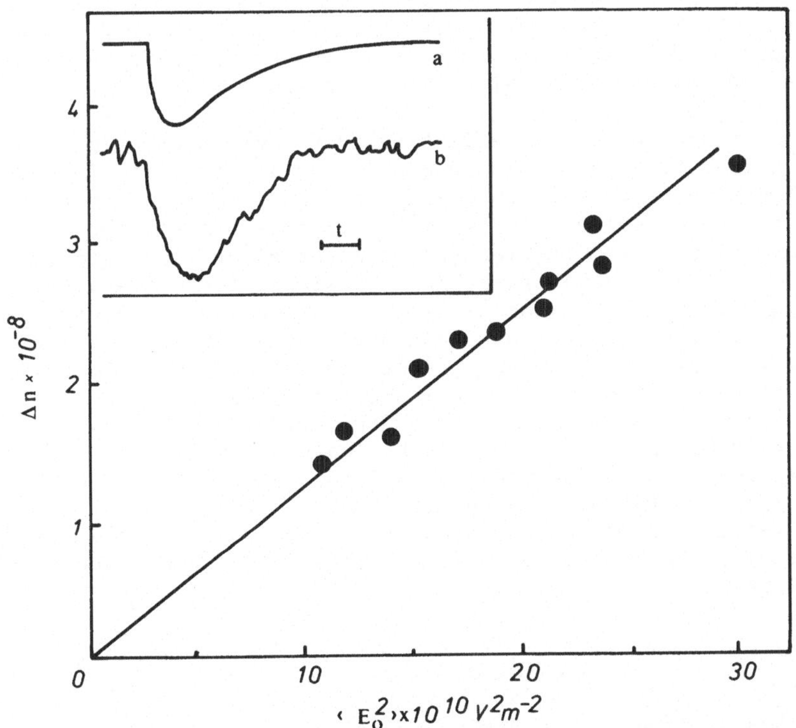


Fig. 3. Laser induced field dependence of the birefringence in MBBA using single $TEM_{00}$ fixed 'Q' pulses, and T = 46.4°C. The inset is a tracing of (a) the orienting pulse for $\langle E_0^2 \rangle^{\frac{1}{2}} = 4.8 \times 10^5\ V.m^{-1}$ and (b) the induced birefringence using quadratic detection where $\Delta n = 3.1 \times 10^{-8}$ and t = 100 μs.

dipole moment orientation dominates giving a negative birefringence whereas in the high frequency or laser field case the axial induced dipole moment dominates giving a positive birefringence. In the latter case the field alternations ($\sim 10^{15}$ Hz) are too fast for the permanent dipole moment to follow the field. In the case of 5 CB the structure is such that both the permanent and induced dipole moment have the same axial direction and this leads to positive birefringences only, the magnitudes of which vary for the two distinct mechanisms. Thus the method allows us to resolve clearly structure-related frequency dispersion effects in liquid crystals.

Further evidence of strongly structure-dependent properties comes from the solution measurements where for MBBA a birefringence sign reversal was recorded as a function of concentration for applied d.c. fields but not laser fields (5,10). No sign reversal was recorded for 5 CB in either case (10,16). These results have been discussed in detail in the references cited. However it is worth noting that at low concentrations the Kerr constants were concentration independent (except for MBBA in pulsed d.c. fields) and this allows us using eq. 4 to calculate the molecular constants also listed in table II.

Finally for the laser Kerr effect we have concentrated in this review on our results obtained using Q-switched laser pulses in which the relaxation time of the medium may in fact be longer than that of the applied orienting pulse. It is worth pointing out however that as is the case with macromolecules (17) the normal or fixed Q pulse may be used to orient liquid crystals in their isotropic phase (18). The preliminary results obtained using this method have been given in fig. 3 where the birefringence has been given as a function of field strength. The inset shows the actual recorded signal as a function of time. Whilst in this time regime it is not possible to determine relaxation times such measurements do allow us to verify that the assumptions made in calculating $\Delta n$ for the Q-switched case are valid. No assumption as to the response of the apparatus to a Dirac pulse (19) has to be made, as the birefringence follows the applied field pulse (the applied laser pulse width being some 200 times longer than the longest field free relaxation time). In the early work we verified (18) that both methods were equivalent.

## CONCLUSION

In summary we have measured the optical and static Kerr constants in the isotropic phase and in $CCl_4$ solutions of both MBBA and 5 CB. Simultaneous measurement has allowed us to determine the constants listed in table II in the same experiment and also to examine the phenomenological Landau - de Gennes model of phase transitions. Both from dynamic Kerr measurements and consideration of the Kerr constants it has been verified that $\Delta n$ or $\tau$ are proportional to $(T - T^*)^{-1}$, and thus a, $T^*$ and $T_c$ have been determined. By using both high and low frequency fields dielectric dispersion effects have been monitored and the birefringence sign changes recorded have been related to structural differences between MBBA and 5 CB.

Finally in the comparison between MBBA and 5 CB we would draw the reader's attention to the similarity of the constants recorded above and yet the large differences recorded for the Kerr constants, table I, for optical and d.c. fields. In both cases the Kerr constants are greater for 5 CB than for MBBA and if we take into account the very good stability of 5 CB in comparison with that of MBBA then we conclude that 5 CB is an excellent choice of material (a) for further study and (b) for use in optical devices.

## ACKNOWLEDGEMENTS

B.R.J. thanks the SRC for an equipment grant with which much of the apparatus was established and H.J.C. thanks the same body for a research fellowship.

## REFERENCES

1 Shadt M and Helfrich W, Mol. Cryst. Liq. Cryst., 17 (1972) 335.
2 Johnston A R, J. Appl. Phys., 44 (1973) 2971.
3 Filippini J C and Poggi Y, J. Phys. Lett., 35 (1974) 99.
4 Bischofberger T, Yu R and Shen Y R, Mol. Cryst. Liq. Cryst., 43, (1977) 287.
5 Coles H J and Jennings B R, Mol. Phys., 31, (1976) 1225.
6 Wong G K L and Shen Y R, Phys. Rev. Lett., 30, (1973) 895; Phys. Rev., A 10, (1974) 1277.
7 Prost J and Lalanne J R, Phys. Rev. A 8, (1973) 2090.
8 Lalanne J R, Phys. Lett. A 51, (1975) 74.
9 Hanson E G, Shen Y R and Wong G K L, Phys. Rev., A 14, (1976) 1281.
10 Coles H J and Jennings B R, Mol. Phys., 31, (1976) 571.
11 De Gennes P G, Mol. Cryst. Liq. Cryst., 12, (1971) 193; and "The Physics of Liquid Crystals", Oxford University Press, London (1974).
12 Poggi Y, Filippini J C and Aleonard R, Phys. Lett., 57A, (1976) 53.
13 Poggi Y, Atten P and Filippini J C, Mol. Cryst. Liq. Cryst., 37, (1976) 1.
14 Lalanne J R and Lefebvre R, J. Chim. Phys., 73, (1976) 3.
15 Peterlin A and Stuart H A, Handb. Jahr. Chem. Phys., 8, (1943) section 1b.
16 Coles H J and Jennings B R, Mol. Phys., in press.
17 Coles H J and Jennings B R, Biopolymers, 14, (1975) 2567. Jennings B R and Coles H J, Proc. Roy. Soc., 348, (1976) 525.
18 Coles H J, Ph.D. thesis, Brunel University (1975).
19 Pouligny B, Sein E and Lalanne J R, this volume, p.375.

# SLOW NON-CRITICAL MOLECULAR REORIENTATION IN THE ISOTROPIC PHASES OF NEMATOGENS

B. Pouligny, E. Sein and J. R. Lalanne

Centre de Recherche Paul Pascal, Domaine Universitaire
33405 Talence, France

We present measurements of the Optical Kerr Effect in the isotropic phase of the nematogen NEMATEL 105 (p-methoxy-benzoate-p'-n-pentyl benzene). In the vicinity of the nematic-isotropic transition temperature, as is well known, the response of the medium is essentially due to the fluctuations of the macroscopic order parameter, in agreement with the Landau-De Gennes theory. However, at high temperatures, the Kerr constant decreases more slowly than predicted by the theory. An explanation for this discrepancy is proposed for the first time involving the very short range <u>cybotactic</u> correlations of liquid crystal molecules in the isotropic phases.

## INTRODUCTION

From a macroscopic point of view, the isotropic phase of a liquid crystal is a disordered medium. However, locally and at a given time, there are intermolecular orientational correlations, prefiguring the nematic state and whose extension and lifetime increase drastically in the vicinity of the isotropic to nematic transition temperature.

Light Scattering and Optical Kerr Effect are very well suited techniques to the study of the amplitude and dynamics of such correlations. They allow one to check the validity of the Landau-De Gennes phenomenological theory (1) of the isotropic to nematic phase transition. This we shall briefly review below.

According to this theory, the intensity and the inverse of the spectral width of the scattered light, as the amplitude and relaxation time of the birefringence induced by an electric field, should depend on temperature as $(T - T^*)^{-\alpha}$:

Here $T^*$ is a virtual second order transition temperature, slightly below the actual temperature $T_K$ of the isotropic to nematic transition, which is weakly first order. The parameter $\alpha$ is a critical exponent, equal to 1 in a mean field theory, such as Maier and Saupe's (2,3).

For the last few years, several publications have reported experimental results in good agreement with the mean field law ($\alpha = 1$) (4-8). However, values of $\alpha$ less than 1 have been measured in some isolated cases: particularly Prost and Lalanne in 1971 (9) reported the value $\alpha = 0.5$ for the optical Kerr Constant of p-methoxybenzilidene-p'-n-butylaniline (MBBA).

This "anomalous behaviour" was never explained satisfactorily, and some people tried to impute it to experimental inadequacies. Thus, it seemed interesting to us to repeat Prost and Lalanne's experiment, improving both the experimental set-up and the data analysis.

Finally, we give our results and discuss the behaviour of the Optical Kerr Constant of our sample as a function of temperature. An interpretation is proposed for the apparent deviation from the mean field law which was sometimes observed.

## THE MAIN FEATURES OF THE LANDAU - DE GENNES THEORY

In the Landau-De Gennes (1) model for the isotropic - nematic transition, it is assumed that the free energy of the isotropic phase per unit volume can be expanded in a power series of an order parameter $Q_{ij}$, namely:

$$F = F_o + \frac{1}{2} a (T-T^*)^{\alpha} Q_{ij}Q_{ji} + \frac{1}{3} bQ_{ij}Q_{jk}Q_{ki} + \frac{1}{4} dQ_{ij}Q_{jk}Q_{k\ell}Q_{\ell i}$$

Any traceless symmetric second rank tensor characteristic of the medium can be chosen as $Q_{ij}$. In the case of the Optical Kerr Effect, the most appropriate is given by the anisotropic part of the dielectric susceptibility:

$$\chi_{ij} - \frac{1}{3} \delta_{ij}\chi_{ii} = \frac{2}{3} \Delta\chi \, Q_{ij}$$

where $\Delta\chi = \chi_{\perp} - \chi_{\parallel}$ for the completely aligned medium.

Then, in the presence of an intense electric field $E_I$ ($\omega_I$) of pulsation $\omega_I$, F becomes:

$$F = F_o + \frac{1}{2} a (T-T^*)^{\alpha} Q_{ij}Q_{ji} + \frac{1}{3} b Q_{ij}Q_{jk}Q_{ki} + \frac{1}{4} d Q_{ij}Q_{jk}Q_{k\ell}Q_{\ell i} - \frac{1}{3} \Delta\chi Q_{ij}E_iE_j$$

If we neglect the coupling of $Q_{ij}$ with the hydrodynamic shear strains, the equations describing the dynamic behaviour of the medium reduce to:

$$\nu \frac{\partial Q_{ij}}{\partial t} = - \frac{\partial F}{\partial Q_{ij}}$$

where $\nu$ is a viscosity coefficient, with a temperature dependence of the form $\nu = \nu_o$ ext $(T_o/T)$.

The values of the order parameter induced in the Optical Kerr Effect are very low ($\lesssim 10^{-3}$). Then, we can limit the development of F to the term of second order in $Q_{ij}$, and the kinetic equation becomes linear. Its solution is given by the convolution product:

$$Q_{ij} = \frac{\Delta\chi}{3a(T-T^*)^{\alpha}} (E_iE_j - \frac{1}{3} E^2 \delta_{ij}). f$$

where f is the response of the medium to a Dirac pulse:

$$f(t) = \frac{a(T-T^*)^{\alpha}}{\nu} y(t) \exp - \left\{ - \left[ \frac{a(T-T^*)^{\alpha}}{\nu} t \right] \right\}$$

Let us assume the inducing field $E(\omega_I)$ to be parallel to the direction x of a (x y z) frame linked to the laboratory and propagating along z. The induced birefringence is analyzed by a probing beam of pulsation $\omega_A$ and wavelength $\lambda_A$: it is given by:

$$\Delta n = B \lambda_A E^2(\omega_I). f$$

where B is the Kerr Constant of the medium:

$$B = \frac{2\pi}{n_A \lambda_A} \frac{\Delta\chi(\omega_A) \Delta\chi(\omega_I)}{9a(T-T^*)^{\alpha}}$$

## EXPERIMENTAL SET-UP AND DATA ANALYSIS

The principle of our set-up (fig. 1) is not different from the one already described in several publications about Optical Kerr Effect in the isotropic phases of liquid crystals (6-9).

We point out some of its main characteristics:

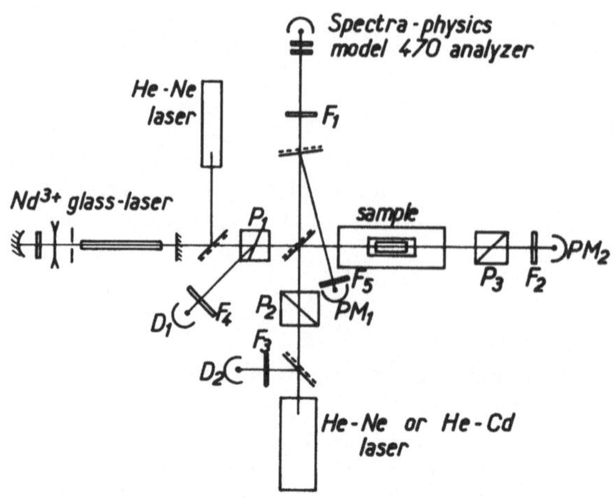


$P_1, P_3$ : Glan polarizers
$P_2$ : Wollaston polarizer
$D_1$ : CSF CPA 1443 cell
$D_2$ : Hewlett-Packard 5082-4207 Pin photodiode
$PM_1$ : La Radiotechnique 150 CVP
$PM_2$ : La Radiotechnique XP 1002 or XP 2020
$F_1, F_3$ : MTO arthervex Ta filters
$F_2$ : Oriel interference filter
$F_4, F_5$ : MTO DIH 70b filters

Fig. 1. Experimental Set-up

Firstly, the inducing field is provided by a $Nd^{3+}$-glass laser, Q-switched by a solution of BDN in chloroform. The pulses are highly reproducible and of nearly gaussian temporal shape, with 30 ns duration. A 1.5 mm pinhole isolates the TEM 00 mode. Moreover, this laser is longitudinal monomode. It's power ($\sim$ 100 kW) is weak enough to avoid any self focusing and stimulated effects inside the sample.

Secondly, a COHERENT RADIATION Model 80 Helium-Neon laser and a SPECTRA PHYSICS Model 185 He-Cd laser were successively used as probing waves. Both were TEM 00 and longitudinal monomode.

Thirdly, the studied compound, the p-methoxy benzoic acid, p'-phenyl n-pentyl ester (NEMATEL 105) was chosen because of its high stability. Moreover, it is colourless and its mesomorphic transition temperature is roughly the same as that of MBBA, i.e. roughly 42°C.

## Data Analysis

Let $\alpha_I$ and $\alpha_A$ the absorption coefficients of the medium at the wavelengths of the inducing and probing waves; $\ell$ the length of the Kerr cell; $I_I$ the intensity of the inducing wave; and $I_A$ the intensity of the probing wave. Then, the probe intensity transmitted through $P_2$ (fig. 1), is given by:

$$S_K(t) = I_A \left\{ \pi B \rho \left( I_I(t) \cdot f(t) \right) \right\}^2$$

where $\rho$ is a factor taking into account the absorption of the medium at the inducing and probing frequencies:

$$\rho = e^{-\frac{\alpha_A}{2}\ell} \left( \frac{1 - e^{-\alpha_I \ell}}{\alpha_I} \right)$$

Let R(t) be the response of the photomultiplier $PM_2$ (fig. 1) to a DIRAC pulse. The actual signal that we deal with is:

$$S(t) = S_K(t) . R(t)$$

It will be shown in a forthcoming publication how neglecting the convolution by R(t) can give reliable results. Let us simply say that the results we report in the following section have been deduced under conditions where this approximation is valid.

## RESULTS

### Relaxation Time

We found our measurements of the relaxation time of the induced birefringence to be in good agreement with the mean-field law ($\alpha = 1$), and to be correctly represented by:

$$\tau = 4.17 \left\{ \frac{e^{1500/T}}{T - 313.69} \right\} \text{ in nanoseconds.}$$

The virtual second-order phase transition temperature $T^*$, (= 313.69 K or 40.54°C) is approximately 2°C below the actual first order phase transition temperature.

### Optical Kerr Constant

Fig. 2 shows the variations of the inverse $\frac{1}{B}$ of the Optical Kerr Constant as a function of temperature. We see that in the vicinity of the clearing point of the material (~42.5°C), B may be considered to obey the mean field law with a value of $T^*$ which is the same as the one deduced from the relaxation study. However, if the temperature is increased above 45°C, B decreases more slowly than predicted by the mean field law. Thus, we give a qualitative confirmation to the "anomalous behaviour" of the Kerr Constant reported a few years ago by Prost and Lalanne in the case of MBBA (9).

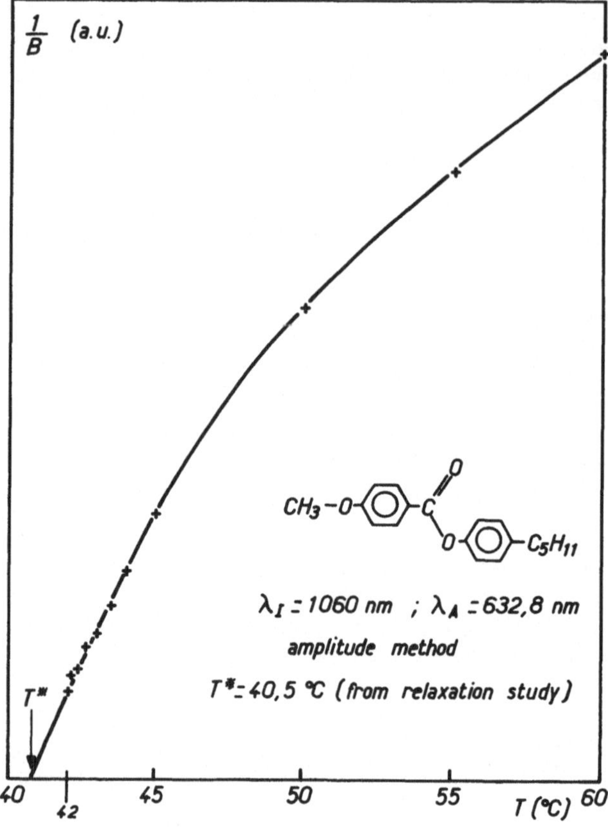


Fig. 2. Inverse of the Optical Kerr Constant of NEMATEL 105 as a function of temperature

The results shown in fig. 2 were deduced from a first experiment performed with the He-Ne laser as a probing beam ( $\lambda_A = 632.8$ nm). We point out that the values of B given by a second experiment, with $\lambda_A = 441.6$ nm, and after a new purification of our sample, show a variation with temperature still in disagreement from the mean-field law at high temperatures, but somewhat less pronounced than before.

## DISCUSSION

We have adopted the Landau - De Gennes model to describe the response of the medium to an intense electric field: this means that we have considered the fluctuations of the order parameter as the only mechanism responsible for the Optical Kerr Effect in the isotropic phase of liquid crystals. This approximation is justified in the vicinity of the transition, where these fluctuations increase drastically and can hide any non critical mechanism which could be coupled to the electric field. However, such a contribution should not decrease significantly when one increases the temperature and thus could explain the abnormally high values of the Kerr Constant at high temperatures.

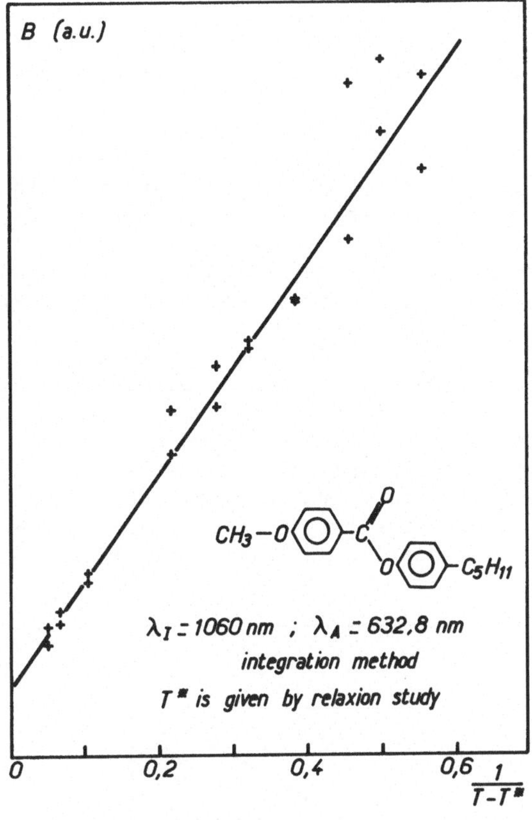


Fig. 3. Optical Kerr Constant of NEMATEL 105 as a function of $\frac{1}{T-T^*}$

$T^*$ is deduced from computer least square method.

We think this point of view is well supported by the curve shown in fig. 3, where the variation of B versus temperature is represented by a function of the form:

$$B = \frac{B_1}{T-T^*} + B_2 \text{ , where } B_2 \text{ is constant} \quad (1)$$

Here $T^*$ has been deduced from a computer calculation giving the best straight line. It is remarkable that its value is the same as the one given by our relaxation study, within the experimental uncertainties. The value of $\frac{B_2}{B_1}$ is $\sim 4.6 \cdot 10^{-2}$ for our first experiment ($\lambda_A = 632.8$ nm) and $\sim 1.8 \cdot 10^{-2}$ for our second experiment ($\lambda_A = 441.6$ nm), that is after the new purification of our sample.

We think that the amplitude of the response due to the fluctuations of the order parameter depends strongly on the purity of the compound, whereas $B_2$ should not be significantly affected by it. This could explain why neither Wong and Shen (6,7) nor Coles and

Jennings (8) did observe any departure from the mean field law as far as ∿ 20°C above the clearing point of MBBA. But we suppose that they should have seen such a departure when increasing the temperature beyond this limit.

A microscopic interpretation of the non critical contribution $B_2$ could be the following: as seen from X-rays studies performed in both the nematic and isotropic phases of non smectogen liquid crystals (10), there exist very short range correlations of molecules, which form what is called CYBOTACTIC (11) groups, whose size is almost independent of temperature. From a dynamic point of view, they correspond to local orientational fluctuations which then must give rise to a non critical component to the Kerr Constant of the medium. Besides, one must keep in mind the fact that the relaxation time of the induced birefringence does not show any departure from the mean field law. We think that the time correlation of the cybotactic fluctuations must be much greater than that of the fluctuations of the macroscopic order parameter. Another argument for our interpretation comes from a recent paper by Philippini and Poggi (12) which suggests the existence of a non critical process at high temperatures. Moreover, a Rayleigh Scattering

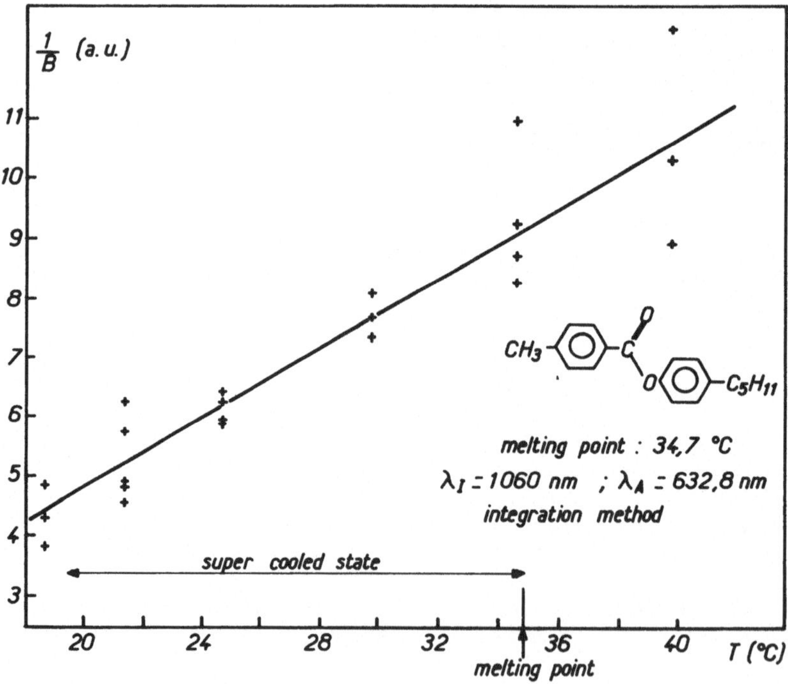


Fig. 4. Inverse of the Optical Kerr Constant of p-methyl benzoate p'-n pentyl benzene versus temperature in the super cooled state.

experiment made in our laboratory (13) on the same sample as the one used in the Optical Kerr Effect has given results still in agreement with eq. (1). Similar conclusions have been drawn by Coles from very recent measurements performed in the isotropic phase of the pentyl-cyano-biphenyl (5CB) (14).

This study must convince the reader of the complexity of the COOPERATIVE molecular orientational phenomena in the isotropic phases of nematogens. Such a slow temperature critical reorientation also exists in supercooled phases of non-nematogens, as revealed by a recent experiment, the results of which are reported in fig. 4. Moreover, although these mechanisms dominate the response of these media to an electric pulse in the nanosecond range, INDIVIDUAL molecular reorientation gives rise to much faster contributions. These can be revealed by picosecond excitations, as was shown by experiments recently performed in our laboratory (15).

## REFERENCES


1 De Gennes P G, Phys. Lett. A, 30, (1969) 454 and Mol. Cryst. Liq. Cryst., 12, (1971) 193.
2 Maier W and Saupe A, Z. Naturforsch., 14a, (1959) 882 and Z. Naturforsch., 15a, (1960) 287.
3 Lefebvre R, Thèse de Docteur Ingénieur, Bordeaux (1975) No. 223.
4 Stinson T W and Litster J D, Phys. Rev. Lett., 25, (1970) 503.
5 Stinson T W, Litster J D and Clark N A, J. de Phys. Coll. $C_1$ Suppl. N 2-3, 33, (1972) C1-69.
6 Wong G K L and Shen Y R, Phys. Rev. A, 10, (1974) 1277.
7 Hanson E G, Shen Y R and Wong G K L, Phys. Rev. A, 14, (1976) 1281.
8 Coles H J and Jennings B R, Mol. Phys., 31, (1976) 571.
9 Prost J and Lalanne J R, Phys. Rev. A, 8, (1973) 2090.
10 Leabetter A J, Richardson R M and Colling C N, J. de Phys. Suppl. N3, 36, (1975) C 1 - 37.
11 Stewart G W and Morrow R M, Phys. Rev., 30, (1927) 232.
12 Philippini J C and Poggi Y, J. de Phys. Lett., 37, (1976) L 17.
13 Bothorel P, Lalanne J R, Maelstaf P and Pouligny B, J. Coll. Interf. Sci., 63, (1978) 178.
14 Coles H J, private communication.
15 Lalanne J R, Martin B, Pouligny B and Kielich S, Opt. Com., 19, (1976) 440.

# A COMPARISON OF OPTICO-OPTICAL SCATTERING AND BIREFRINGENCE MEASUREMENTS ON WYOMING SODIUM BENTONITE SUSPENSIONS

J. W. Parsons, R. L. Rowell and R. S. Farinato

Department of Chemistry, University of Massachusetts,
Amherst, Massachusetts, 01003, U.S.A.

A new light scattering technique for investigating anisotropy is described. Wyoming sodium bentonite particles were suspended in water and were oriented by the electric field of a high energy infrared laser pulse while a 488 nm CW argon laser beam was used to monitor changes in particle orientation by measuring $H_v$ or $V_v$ intensities at a 90° scattering angle. $H_v$ and $V_v$ changes were observed to be in opposite directions. Laser induced birefringence experiments were also made and relaxation times were obtained. The results were used to probe polydispersity and time-dependent phenomena.

## INTRODUCTION

Most macromolecules and many colloidal systems are both geometrically and optically anisotropic. In dilute solution they are randomly oriented and are in continuous random motion. When an orienting force is applied to these molecules via an external field, the resulting torque tends to align them in the direction of the applied field. This results in a change in the properties of the system as a whole. Electric fields from ac and dc sources have been used extensively to orient particles in birefringence experiments (1-4). The material refracts light parallel to the orienting field differently from light perpendicular to the field, an effect first observed by Kerr (5). Recently, Jennings has reported on a new birefringence experiment, optico-optical birefringence, in which the electric field of a 200 $\mu$s pulse from a Nd-YAG laser operating at 1.06 $\mu$m was used to orient various macromolecules in suspension (6,7).

We report new optico-optical measurements which include a comparison of both birefringence and light scattering from the same system, Wyoming sodium bentonite.

## EXPERIMENTAL

Wyoming sodium bentonite was obtained from Fisher Scientific Company and was used without further preparation. Twice distilled water was used to suspend a $4.8 \times 10^{-4}$ g/ml sample of bentonite. The absence of "blips" in the light scattering showed the system to be free of dust.

A Nd-glass pulsed infrared laser (Holobeam Lasers, Inc., Model 811) operating at 1.06 $\mu$m with a variable pulse energy up to about 20 joules and a duration of 1.5 ms was used to orient the sample. Two intra-cavity Brewster polarizer stacks gave 99% vertical plane polarization. Changes in light scattering due to particle orientation were monitored by use of an argon laser (Coherent Radiation, Model 52) operating at 488nm. Fig. 1 shows the experimental configuration for the optico-optical scattering apparatus which was readily converted for the birefringence work.

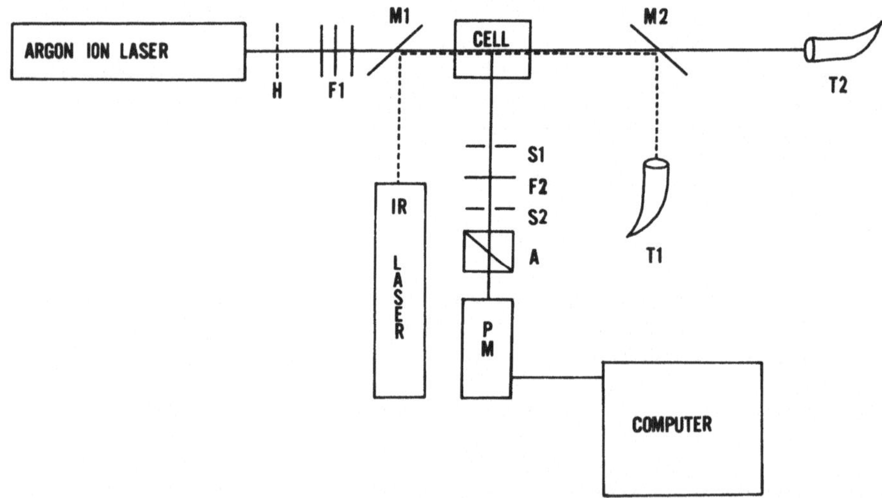


Fig. 1 Optico-optical scattering apparatus. F1 neutral density filters. M1,M2 mirrors, transparent to 488 nm, reflective to 1.06 $\mu$m. S1, S2 slits. F2 neutral density filters and 488 nm bandpass filter, 4 nm half bandwith. A, analyzer. T1,T2 light traps. PM, photomultiplier. Argon ion laser was polarized with E-vector perpendicular to plane of paper. H, halfwave plate used in the measurement of birefringence. For the birefringence experiment, the observation channel S1-computer was rotated to replace T2 and the polarizer and analyzer were crossed at 45$^{\circ}$ to the vertical E-vector of the IR laser.

In a second series of experiments the laser was rebuilt using a second laser rod in an oscillator-amplifier combination (Holobeam Lasers, Inc., Model 3080) with a spatial filter inserted in the oscillator between the laser mirrors to give the $TEM_{00}$ mode. Maximum pulse energy was about one joule with a pulse duration of 0.4 ms and polarization was achieved as before using Brewster stacks.

A standard arrangement for measurement of birefringence was used. A 3 cm path length rectangular cell was placed between a crossed polarizer and analyzer. A half-wave plate was used to rotate the vertically polarized light from the argon beam to an angle of $45^{\circ}$ with respect to the polarization direction of the electric field of the orienting infrared pulse. The transmitted intensity changes were monitored with a photon counting detection system which used a thermoelectrically cooled ($5^{\circ}C$) EMI 6256 SA photomultiplier tube, a SSR Model 1120 amplifier-discriminator, and a Nicolet Model 1074 instrument computer. The argon beam was plane polarized to within one part per thousand. The overall crossed-component artifact introduced by optical components was one part in six hundred. In the later experiments a quarter-wave plate was added to enhance sensitivity.

An important consideration in the birefringence experiment is to have a long path length. This was accomplished by making both the orienting and the monitor beam collinear while going through the sample cell. To do this, two mirrors were manufactured (Valpey Optical Coating Division) with dielectric coatings allowing approximately 70% transmission at 488 nm and nominally 100% reflection at 1.06 $\mu$m. They were placed in appropriate positions in front of and behind the sample cell (see fig. 1). The purpose of the second mirror, behind the sample cell, was to prevent the infrared pulse from entering the detection system. A 4 nm bandpass filter (FWHB) centered at 488 nm was also used in front of the detecting system to further reduce the likelihood of any infrared radiation being detected. Despite these precautions a small signal due to the flashlamp was observed along with two additional sharp spikes which consistently appeared 25 ms and 35 ms after the initial pulse.

The light scattering arrangement used the same equipment, except for the removal of the half-wave plate. This resulted in the polarization direction of both the monitor and orienting electric fields being parallel. The slit system and photomultiplier were rotated $90^{\circ}$ in the scattering plane to measure scattered light instead of transmitted light.

An 11.4 cm focal length double convex lens was used in both configurations with the Model 811 Laser to increase the irradiance of the Holobeam pulse and was placed in front of the cell so that the cross-sectional area of the focused orienting pulse at the center of the cell was reduced to one-fourth of the unfocused pulse area. All area measurements were made from burn patterns on exposed Polaroid film.

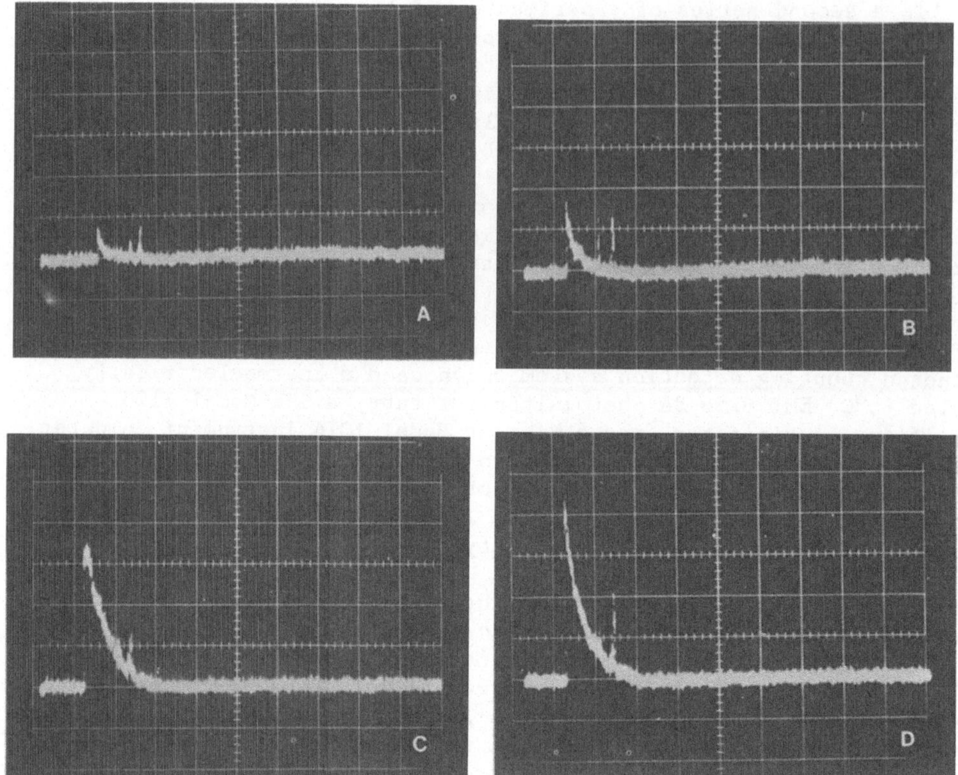


Fig. 2. Orientation of bentonite by the Holobeam 811 laser, 1.5 ms pulse, multimode. Increasing pulse energy (and irradiance) as follows: A, 0.25 joule (0.3 $kW/cm^2$); B, 2.25 joule (2.3 $kW/cm^2$); C, 5.0 joule (5.3 $kW/cm^2$); D, 8.0 joule (8.3 $kW/cm^2$). Time base was 400 ms for the full 10 cm oscilloscope trace shown.

The power output of the Holobeam laser was changed by varying the charge on the capacitors. Nominal pulse power was obtained from a plot of capacitor charge vs. power output provided by the manufacturer.

## SINGLE-ROD LASER RESULTS

Our initial measurements were made on bentonite samples using the Holobeam 811 configuration. The orienting pulse was multimode and the pulse duration was 1.5 ms. The intensity change without a quarter-wave plate and with a crossed polarizer-analyzer combination as des-

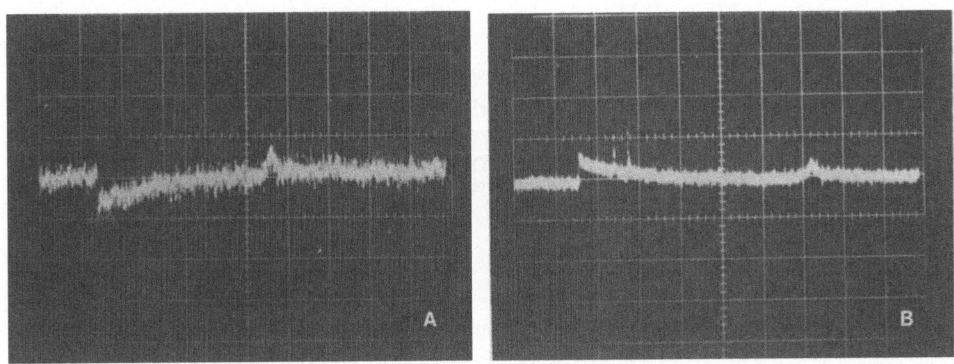


Fig. 3. Optico-Optical scattering with 5.0 J orienting pulse energy. The polarization state of the observing laser was $V_v$ (on left) and $H_v$ (on right).

cribed above resulted in a positive intensity change for a concentration of $4.8 \times 10^{-4}$ g/ml. The signal increased with increasing orienting pulse irradiance up to around a 5 joule orienting pulse where further increase in irradiance broadened the peak. The results are shown in fig. 2 where the orienting pulse varied from 0.25 joule to 8.0 joule giving irradiances of 0.3 kW/cm$^2$ and 8.3 kW/cm$^2$ respectively.

Our first analyses of the data of fig. 2 were done by plotting the logarithm of transmitted intensity change vs. time to obtain an approximate relaxation time. We obtained relaxation times in the range of 5 to 15 ms which were higher than the 1.6 - 1.9 ms reported by Jennings and Coles (6,7). However, in the latter work (8) the sample was allowed to stand for several months and measurements were made on a portion of the supernatant that had been syringed off so that it was likely that a smaller size fraction gave the smaller relaxation time. Sample polydispersity and sedimentation are considered further below. Equally important, however, is that other work in progress (9) has demonstrated the presence of aggregates under our conditions of preparation.

We also carried out some optico-optical scattering measurements using the Model 811 laser. A 5.0 joule orienting pulse was used and the 488 nm monitor laser showed opposite effects for the $V_v$ scattering and the $H_v$ scattering as shown in fig. 3. The data were not quantitatively analyzed because of the undefined contributions arising from multimode operation, sample polydispersity and the possibility of a size distribution extending well into the large size range (beyond Rayleigh-Gans) for which a theoretical interpretation was lacking. However, the relaxation time observed in the scattering experiment was somewhat longer than that observed in the birefringence measurements.

## LASER OSCILLATOR-AMPLIFIER RESULTS

With the results of the single laser rod experiments as a guide we established the requirements for an improved system. The laser was upgraded to a Holobeam Model 3080 Nd: Glass Oscillator/Amplifier with a pulse energy of 1.0 joules $TEM_{00}$, selectable 200 and 400 microsec pulsewidth, 5 kW minimum peak power, diffraction limited divergence, vertical polarization and 1 mm nominal beam diameter.

Measurements with the Model 3080 laser were qualitatively similar to the previous results but continued experiments allowed us to investigate sample polydispersity. We used the optico-optical birefringence technique and carried out two series of experiments. In the first series of measurements, a freshly prepared, unfiltered and freshly stirred bentonite sample was used. In the second series a bentonite sample which passed through a 0.4 $\mu$m Nuclepore filter but not through a 0.2 $\mu$m filter was used. The filtered sample was also freshly stirred before each measurement.

The birefringence of the unfiltered sample increased linearly with $E^2$ up to an orienting field of approximately 4.5 kV/cm. The filtered sample showed a linear change up to approximately 3.9 kV/cm. The field strengths were calculated from the pulse irradiance (9). The birefringence $\Delta n$ at a field strength of 3.8 kV/cm was $9 \times 10^{-9}$ for the filtered sample and $75 \times 10^{-9}$ for the unfiltered sample (9). Our results are consistent with the value of $5 \times 10^{-9}$ at a field of 3.8 kV/cm reported by Jennings and Coles (6,7) on a sample of bentonite which was allowed to settle for several months.

Considering the wide variations in composition for different bentonite samples and the differences in sample treatment (such as stirring versus not stirring prior to a measurement) it is not surprising to observe the wide differences in birefringence. Indeed, the important point to note is that the optico-optical birefringence is a sensitive and rapid probe for the anisotropy of a system.

Another change was observed which is still under study but should be pointed out (9). The irradiance of the orienting pulse was held constant at 22.3 $kW/cm^2$ and the birefringence was observed at intervals over a 3.5 hour period. The signal was observed to reverse direction with increasing time and ultimately to vanish into the noise level. Subsequent re-stirring restored the original signal direction. Aggregation followed by settling would account for the trend. Larger aggregates have a measured birefringence which would result in a negative intensity change upon orientation and could completely dominate the signal. After settling, smaller particles would be left which would achieve a lower orientation at the same field strength and could contribute a positive intensity change. Continued settling would account for the vanishing response at times greater than 5 hours. Subsequent stirring and restoration of the original signal implies a reversible resuspension.

## CONCLUSION

We have reported for the first time the new experimental technique of optico-optical scattering on suspensions in the colloidal domain. The polarized scattering intensities $H_v$ and $V_v$ have been shown to be qualitatively different on a bentonite sample and indicate a longer relaxation time than that obtained in birefringence measurements. Optico-optical birefringence measurements have been shown to be generally consistent with earlier work on bentonite and the approach has been utilized as a tool to examine polydispersity and time-dependent phenomena in bentonite suspensions.

## REFERENCES


1 O'Konski C T, Encyl. Polym. Sci. Technol., 9, (1969) 551.
2 Jennings B R, in Light Scattering from Polymer Solutions (Academic Press, New York (1972), ed. Huglin M B).
3 O'Konski C T, (ed.), Molecular Electro-Optics, 1 (Marcel Dekker, New York (1977)).
4 Fredericq E and Houssier C, Electric Dichroism and Electric Birefringence (Oxford University Press, London (1973)).
5 Kerr J, Phil. Mag., 50, (1875) 337.
6 Jennings B R and Coles H J, Nature, 252, (1974) 33.
7 Jennings B R and Coles H J, Proc. R. Soc. Lond., A 348, (1976) 525.
8 Coles H J, Ph.D. Thesis, Brunel University, (1975).
9 Parsons J, Ph.D. Thesis, University of Massachusetts, Amherst, (1978).

# LASER PHOTOINDUCED CHANGES IN THE HIGH FREQUENCY DIELECTRIC CONSTANT OF CHLOROPLASTS AND DYES

O. E. Anitoff

Laboratoire d'Optique, DRA-SRIRMa, Centre d'études nucléaires de SACLAY, BP No. 2, 21190 Gif sur Yvette, France.

Fast dielectric constant changes, which are recorded with a new and original apparatus, are found to occur in spinach chloroplasts and various dyes, when submitted, between room temperature and liquid nitrogen temperature, to a single light pulse from a ruby laser. These preliminary experiments show some kind of storage effect in chloroplasts at 77°K.

## INTRODUCTION

The conversion of light into chemical energy, as performed by the photosynthetic apparatus of green plants, involves photon harvesting by a pigment antenna and then charge separation and transfer through an electron transport chain. Such a charge carrier (electron or hole, maybe proton) migration should be associated with the exchange of electrical information with suitable probes. However, till now, very few techniques can give such information on intact chloroplasts systems. Photoconductivity and photovoltaic effects have been observed on chlorophyll films deposited on metallic electrodes (1,2) but such systems are very different from the photosynthetic apparatus. Direct measurement of photoinduced potentials into chloroplasts (3) are difficult to interpret, because the required microelectrodes are one order of magnitude larger than the thylakoid membrane, which contains the photoactive part of the photosynthetic system. Hyperfrequency photoconductivity and Faraday rotation measurements on chloroplasts (4) show that the observed charge carriers are positive holes (a well known result for photoconductive pigments such as phtalocyanines and chlorophyll) and that their mobility is of the order of 1 $cm^2/V.s$, but such experiments have not been further developed. The purpose of the present work is to try to develop a new technique for monitoring electrical events in photosynthetic systems.

## EXPERIMENT

A preliminary communication on this subject has already been published (5), and technical details will be found there. The new apparatus, which was designed for these experiments, consists of a highly stable high frequency (12 MHz) MOSFET oscillator. The tank coil is wound on alumina; the tuning capacitor, which contains the sample to be investigated, can be cooled down to liquid nitrogen temperature. A plexiglass window allows the sample to be irradiated with a ruby laser beam, after reflexion on a copper mirror. This laser produces a single flash (duration: 20 ns) at a wavelength of 694,3 nm. So the investigation is restricted to systems which effectively absorb this wavelength (chloroplasts and some dyes: chlorophyll, chloro-aluminium chlorophtalocyanine and dicyanine A). The maximum incident energy on the sample is of the order of 0.5 Joule, and can be reduced with suitable filters. Any change (as small as 10 ppm) in the dielectric constant of the sample leads to a frequency change, and after multiplication by a chain of multipliers-converters, is stored into a memory (with a capacity of 4 kBit in my experiment) and then displayed on a X-Y recorder. The minimum analysis time is 1.2 microsecond, but could be reduced further, if it were useful. Actually, a resolution of 12 microseconds is generally sufficient. All this system is synchronized on the laser flash by means of a fast photodiode. The resulting picture is the record of a time dependent dielectric constant variation of the sample, following the flash.

Experiments performed at room temperature have already been reported (5), but more interesting phenomena are observed at liquid nitrogen temperature. Some records are reproduced on the following figures.

Whereas at room temperature, only one peak is observed, with a rise time of the order of 200 microsecond and a very slow decay (100 ms to 1 second), two peaks can be observed at liquid nitrogen temperature: a fast one, immediately following the flash, with a rise time of the order of 200 microsecond and a decay time of the order of 400 microsecond, sometimes with a multiple structure (in the case of chloro-aluminium chlorophtalocyanine: fig. 3), and a slow peak, which can follow the first one after a delay of about 1 millisecond.

In the case of chloroplasts, this second peak is associated with the storage of some information, as shown on figs. 1a to d and 2. When irradiating the sample (spinach chloroplasts) for the first time (fig. 1a), only the fast first peak is observed. On subsequent shots (with a time interval of about 3 minutes between them), the slow peak appears (fig. 1b) and becomes larger and larger (figs. 1c and 1d), whereas the first peak becomes smaller (fig. 1: shot No. 10). After warming to room temperature and then cooling again down to liquid nitrogen temperature, the first peak has again its full magnitude, and the slow one is no more observed, even on further shots. So, the slow peak may be associated with the storage of electrical charge in

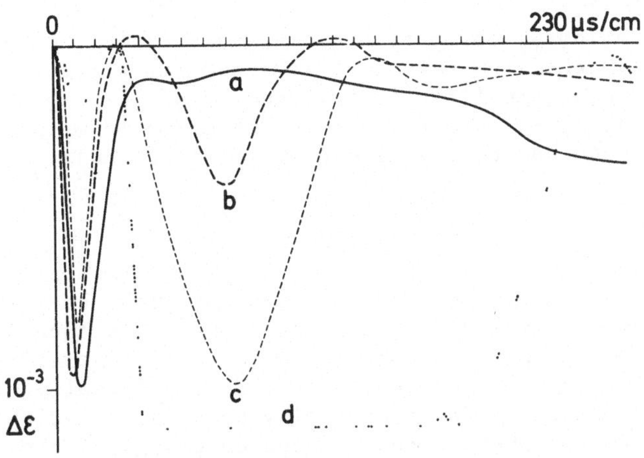


Fig. 1. Dielectric change in chloroplasts. a: 1st shot; b: 2nd shot; c: 3rd shot; d: 10th shot.

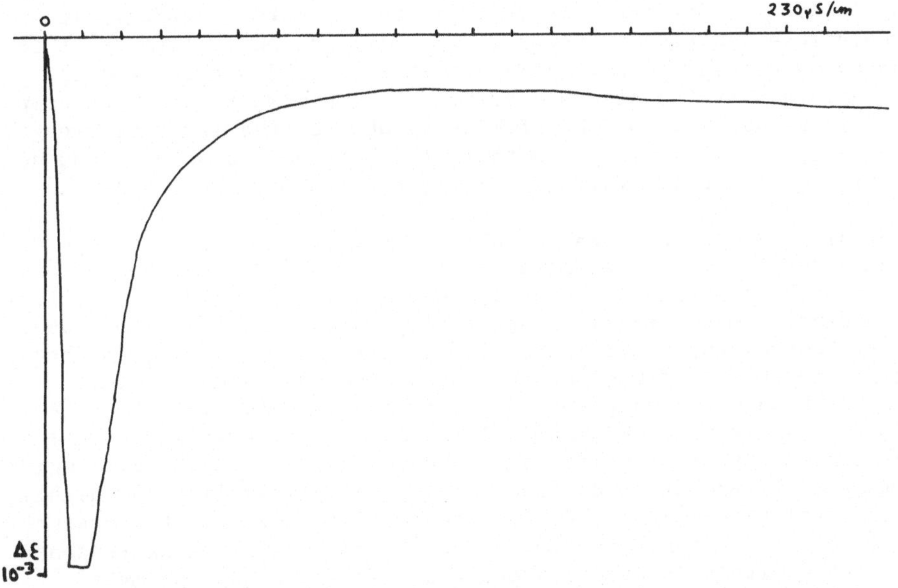


Fig. 2. Dielectric change in chloroplasts after warming to room temperature and then cooling again to 77° K.

the photosynthetic system, this charge being destructively dissipated when warming to room temperature.

In the case of dyes (chloroaluminium phtalocyanine, chlorophyll a and dicyanine A), the slow peak is always and immediately observed. However, when reducing the flash energy, the first peak may disappear (fig. 4b: dicyanine A, reduced energy; to be compared with fig. 4a: full energy. In both cases, the slow peak is saturated).

## INTERPRETATION

The sample area is 0.5 $cm^2$. With a maximum flash energy of 0.5 Joule, the maximum energy density is 1 $Joule.cm^{-2}$. At such high energy densities, biphotonic processes can take place. Moreover, processes involving the absorption of two photons (even if not really biphotonic) must occur if some charge separation really happens, since the photon energy is only 1.79 eV, whereas the ionization energy of the investigated dyes is higher than 2 eV.

One might argüe that such dielectric effects are simply associated with a temperature increase of the sample. Indeed, when used at its maximum sensitivity, my apparatus gives no information on the sign of the dielectric constant change (if it were a thermal effect, it would be always negative). However this explanation is irrelevant for the following reasons: - a temperature effect could not explain the fine structure which is always observed. - experiments on inert samples, such as nickel sulfate, which effectively absorbs the laser wavelength and so should become warmer if a thermal effect had to occur, show no dielectric change at all (within the first three milliseconds).

Actually, it is not clear at all, whether thermal effects are NOT observed after the flash, since an energy pulse of 0.5 Joule should lead to a high temperature increase, very easily detected (the dielectric constant of most materials has a temperature dependence of about -0.5%/$^{o}$K, whereas the sensitivity of my apparatus is 10 ppm, so that a temperature change of 0.01$^{o}$K could be detected). Since thermal effects, with such an equipment, are very easily observed when submitting the sample to a CONTINUOUS light source, irradiating with a single high energy light pulse may produce a thermal shock wave, which needs many milliseconds to diffuse through the pulverulent and highly thermally resistant sample. Other artefacts, coming from the measuring cell itself, are not expected to occur for the following reasons: - there is no direct interaction between the sample and the metal electrodes (aluminium) of the tuning capacitor: this rules out such effects as hole photoinjection. - the radio frequency electric field into this capacitor is very small (about IV/mm). - the rise time of the frequency multipliers chain is very small with respect to the minimal analysis time (1200 nS), the latter being itself very small with respect to the rise time of the observed phenomena.

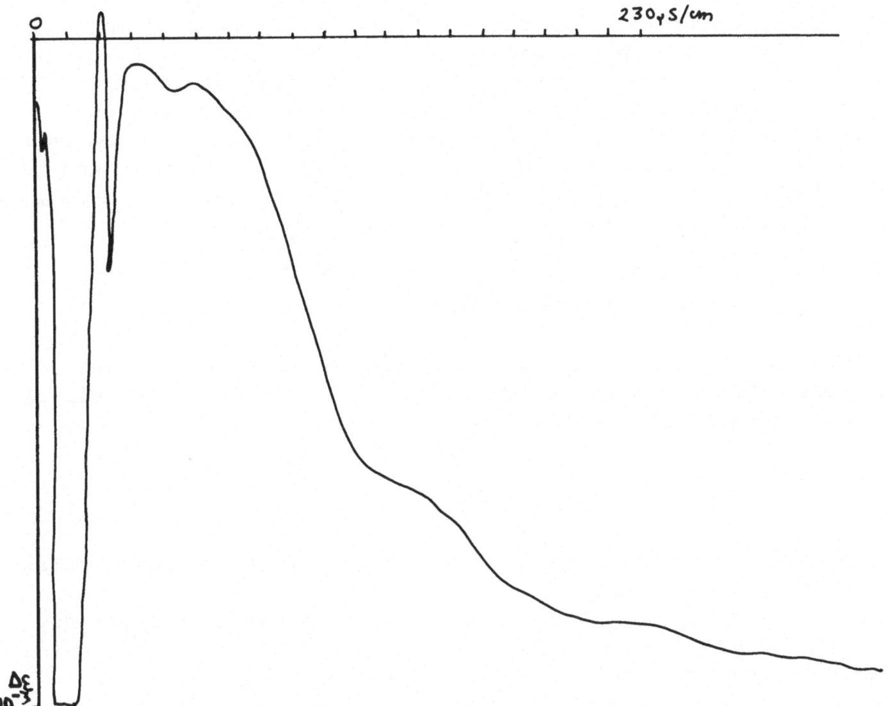


Fig. 3. Dielectric change in chloroaluminium chlorophtalocyanine.

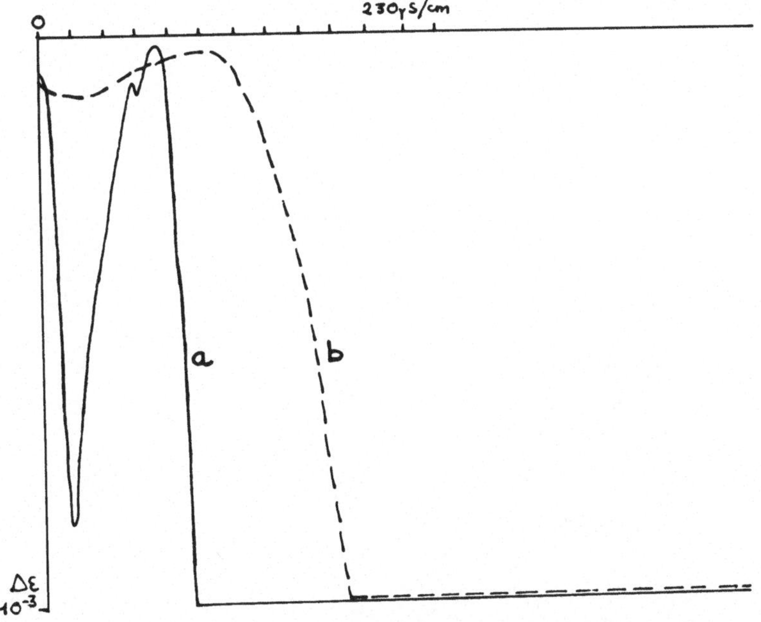


Fig. 4. Dielectric change in dicyanine A. a: full laser power (0.5 Joule); b: reduced laser power (0.05 Joule).

All these artefacts being ruled out, the observed phenomena must be associated with such processes as electron-hole pair photoionization. However, it is not yet possible to give any detailed explanation. One must consider anyway that this phenomenon must be a very general one (since it is observed on so highly chemically different dyes as phtalocyanines and dicyanines), and further investigation, which is now being performed in our laboratory, should turn it into a practical tool for the study of photosynthesis.

## ACKNOWLEDGEMENTS

I have to congratulate Dr. J. P. Leicknam for providing the initial idea of such dielectric studies, and Dr. E. Roux for helpful discussions and for the supply of chloroplasts.

## REFERENCES

1 Reucroft P J and Simpson W H, Discussions of the Faraday Society, 51, (1971) 202-211.
2 Terenin A, Putzeiko E and Akimov I, Discussions of the Faraday Society, 27, (1959) 83.
3 Barber J, The Intact Chloroplast, 1 (1976) 71-85,
4 Bogomolni R A and Klein M P, Nature, 258 (1975) 88-89.
5 Anitoff O E and Leicknam J P, C.R. Acad. Sc. Paris, t 286, (1978) serie B, 203-205.

# RELATION BETWEEN ELECTRIC FIELD-INDUCED OPTICAL RECTIFICATION AND ELECTRO-OPTIC KERR EFFECT IN MACROMOLECULAR SOLUTIONS

Bolesława Kasprowicz-Kielich and Stanisław Kielich

Nonlinear Optics Division, Institute of Physics,
A. Mickiewicz University, 60-780 Poznań, Poland

The conditions for a relationship between the new effect of Electric Field-Induced Optical Rectification (EFIOR) and the Electro-Optical Kerr Effect (EOKE) are analyzed. Measurements of the two effects yield the same information only in the case of molecular substances, acted on by a static electric field of not excessive strength. We show that in solutions of macromolecules exhibiting considerable reorientation effects of permanent and induced electric dipoles, the above relation between the two methods does not hold since, under external conditions, optical saturation of macromolecular alignment takes place in EFIOR whereas it is electric saturation in EOKE. We prove however that the measurement of laser light - induced electric anisotropy leads to identical results as EFIOR in the molecular as well as macromolecular case.

Strong light can cause nonlinear changes in the electric permittivity of media (1) measurable as optically induced electric anisotropy (2) or as electric field-induced optical rectification (EFIOR) (3,4). Recently, Ward and Guha (3) have shown the nonlinear susceptibility tensor components $\chi_{ijkl}(0,0,\omega,-\omega)$, determined by the EFIOR method for nitrobenzene, to be the same as the respective components $\chi_{lkji}(-\omega,\omega,0,0)$, determined from electro-optical Kerr effect (EOKE) measurements, in accordance with permutation symmetry. Now this relation holds prevalently in molecular substances, where processes of electric dipole reorientation are not too intense. Here, it is our aim to show that the relation in question is not in general valid in macromolecular solutions, in which both electric reorientation (5) and optical reorientation of the macromolecules (6) can attain a state of complete ordering, i.e. electric or optical saturation.

We assume the solution to be sufficiently dilute for interactions between the macromolecules to be negligible. The molecules have a permanent electric dipole moment $\mu_i$ as well as anisotropic electric ($\alpha_{ij}$) and optical ($a_{ij}$) polarizabilities. If the static electric field E(0) is not too strong, and omitting nonlinear polarizabilities, we have for the i-component of the total dipole moment of the macromolecule:

$$m_i = \mu_i + \alpha_{ij} F_j (0) \tag{1}$$

In the same approximation, its potential energy is:

$$u (E_o, E_\omega) = - \mu_i F_i (0) - \frac{1}{2} a_{ij} \langle F_i (\omega) F_j (\omega) \rangle_t \tag{2}$$

The brackets $\langle \ \rangle_t$ indicate time-averaging over the oscillation period of the electric field of the highly intense laser light wave. Denoting by $\rho$ the number density of macromolecules, the electric dipole polarisation is, by definition,

$$P_i (E_o, E_\omega) = \rho \langle m_i \rangle_{E_o, E_\omega} \tag{3}$$

For axially symmetric macromolecules, eq. 3 becomes:

$$P_y (E_o, E_\omega) = \rho (\alpha + \frac{\mu^2}{3kT}) F_y (0) + + \frac{\rho}{3} (\gamma + \frac{\mu^2}{kT}) \langle 3 \cos^2\theta - 1 \rangle_{E_\omega} F_y (0) , \tag{4}$$

where $\alpha = (\alpha_{33} + 2\alpha_{11})/3$ is the mean value and $\gamma = \alpha_{33} - \alpha_{11}$ the anisotropy of the polarizability of the macromolecules, with their axis of symmetry subtending an angle $\theta$ with the direction of the static field $E_y(0)$, acting along the Y-axis.

Let the laser beam propagate along the Z-axis and the angle between the fields $E(\omega)$ and $E_y(0)$ be $\Omega$. From eqs. 2 and 4, the change in polarization component is now:

$$\Delta P_y (E_o, E_\omega) = \frac{2}{3} \rho (\frac{\varepsilon + 2}{3})^2 (\pm|\gamma| + \frac{\mu^2}{kT}) \Phi (\pm q_\Omega^\omega) E_y(0), \tag{5}$$

where (5,6)

$$\Phi (\pm q_\Omega^\omega) = \frac{1}{2} \langle 3 \cos^2\theta - 1 \rangle_{E_\omega} = \frac{3}{2} L_2 (\pm q_\Omega^\omega) - \frac{1}{2} \tag{6}$$

is the function of macromolecular reorientation in the optical field; the parameter $q_\Omega^\omega$ of the reorientation being

$$q_\Omega^\omega = \frac{|a_{33}^\omega - a_{11}^\omega|}{4kT} (\frac{n_\omega^2 + 2}{3})^2 (3 \cos^2\Omega - 1) \langle E_\omega^2 \rangle_t . \tag{7}$$

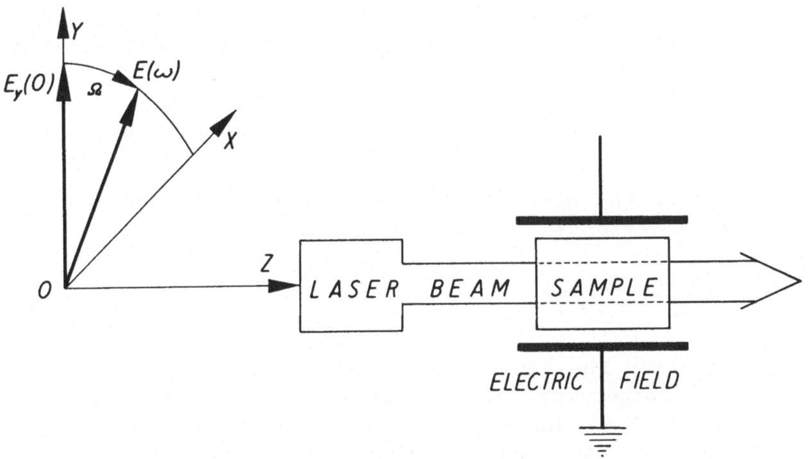


Fig. 1. The laser beam propagates along the Z-axis, its electric vector $\vec{E}(\omega)$ subtending an angle $\Omega$ with the Y-axis. The static electric field $E_Y(0)$ acts along Y.

In eqs. 5 and 6 the positive sign is for macromolecules having positive optical anisotropy ($a_{33} > a_{11}$) i.e. for cigar-shaped particles, whereas the minus sign refers to disc-shaped macromolecules with negative anisotropy ($a_{33} < a_{11}$).

If optical reorientation of the macromolecules is weak, eq. 5 gives:

$$\Delta P_y(\Omega) = \frac{\rho(3\cos^2\Omega - 1)}{45\,kT}\left(|\gamma| \pm \frac{\mu^2}{kT}\right)\left|a_{33}^{\omega} - a_{11}^{\omega}\right| \times$$

$$\left(\frac{\epsilon + 2}{3}\right)^2 \left(\frac{n_\omega^2 + 2}{3}\right) \langle E_\omega^2 \rangle_t \qquad (8)$$

If the optical reorientation is extremely strong, leading to complete alignment of the macromolecules, eq. 5 gives for cigar- and disc-shaped particles, respectively:

$$\Delta P_y(E_o, \infty) = \frac{2}{3}\rho\left(\frac{\varepsilon + 2}{3}\right)^2\left(|\gamma| + \frac{\mu^2}{kT}\right) E_y(0), \qquad (9)$$

$$\Delta P_y(E_o, \infty) = \frac{1}{3}\rho\left(\frac{\varepsilon + 2}{3}\right)^2\left(|\gamma| - \frac{\mu^2}{kT}\right) E_y(0). \qquad (10)$$

Hence we note that formulae 8 to 10 can be useful for determinations of the anisotropy in polarizability, and the permanent dipole moment, of macromolecules.

For comparison we adduce the formula of O'Konski et al. (5) for the electro-optical birefringence of a macromolecular solution.

$$n_{\parallel} - n_{\perp} = \pm 2\pi\rho n \left(\frac{n^2+2}{3n}\right)^2 \left| a_{33}^{\omega} - a_{11}^{\omega} \right| \Phi(p, \pm q), \quad (11)$$

where $\Phi(p, \pm q)$ is the electric reorientation function of macromolecules with the Langevin-Benoit parameters:

$$p = \frac{\mu}{kT}\left(\frac{\varepsilon+2}{3}\right)E_o; \qquad q = \frac{|\gamma|}{2kT}\left(\frac{\varepsilon+2}{3}\right)^2 E_o^2. \quad (12)$$

One sees that in the general case of strong optical (eq.5) or electric (eq.11) macromolecular reorientation, no direct relationship between the methods of EOKE and EFIOR occurs. In the case of weak reorientation only, we obtain the following relationship, resulting from formulae 8 and 11:

$$6\pi \Delta P_y(\Omega) E_y(0) = n(n_{\parallel} - n_{\perp})(3\cos^2\Omega - 1)\langle E_{\omega}^2\rangle_t . \quad (13)$$

Consequently, in this approximation, the two methods of EOKE and EFIOR provide the same information regarding the macromolecular parameters.

Since by definition

$$\Delta\varepsilon_y(\Omega) E_y(0) = 4\pi \Delta P_y(\Omega) \quad (14)$$

with regard to eq. 13 the change in electric permittivity due to intense light is related as follows (1) with the Kerr constant K:

$$\Delta\varepsilon_y(\Omega) = \frac{2}{3} n^2 (3\cos^2\Omega - 1) K \langle E_{\omega}^2\rangle_t . \quad (15)$$

This relation provides another experimental procedure for gaining information concerning the electro-optical properties of macromolecules. This has been attempted for nitrobenzene (7).

If an AC electric field of frequency below the Debye dispersion is applied instead of the static field, no direct analytical relation exists between the three above-discussed methods even in the approximation of weak reorientation (4,8,9).

In statistically inhomogeneous fluids, yet other fluctuational processes take place (e.g. fluctuations in density, concentration, molecular fields, etc.) which affect the above discussed EFIOR and EOKE effects to varying degrees (10). In particular, in a quadratic approximation, these fluctuational processes are of an isotropic nature. They are independent of the angle $\Omega$ and involve an additional variation in electric permittivity:

$$\Delta \mathcal{E} = \frac{2\pi}{9VkT} \left(\frac{\mathcal{E}+2}{3}\right)^2 \left(\frac{n^2+2}{3}\right)^2 \left\langle \left(\Delta A + \frac{1}{kT}\Delta M^2\right)\Delta A^{\omega} \right\rangle x$$

$$\left\langle E_{\omega}^2 \right\rangle_t . \tag{16}$$

Here, $\Delta A$ and $\Delta A^{\omega}$ are the statistical fluctuations in electric polarizability of the whole sample of volume V, whereas $\Delta M^2$ is the fluctuation of the squared electric moment of the sample in the absence of external fields. The effects due to statistical fluctuations (eq. 16), can be detected only by measuring the absolute variation $\Delta \mathcal{E}$ and do not intervene in measurements of optically induced electric anisotropy (2) or Kerr effect.

The preceding theory can be extended to comprise various radial and angular macromolecular correlations as well as statistical-fluctuational processes (10). The theory of the EFIOR and EOKE methods can also be developed for liquid crystals, intensely studied of late by new measuring techniques involving lasers (11-18).

## REFERENCES


1 Piekara A and Kielich S, J. Chem. Phys., 29, (1958) 1297. Kielich S and Piekara A, Acta Physica Polonica, 18, (1959) 439.

2 Kielich S, Physica, 34, (1967) 586.

3 Ward J F and Guha J K, Appl. Phys. Lett., 30 (1977) 276.

4 Buchert J, Kasprowicz-Kielich B and Kielich S, Adv. Mol. Relax. Interaction Processes, 11 (1977) 115.

5 O'Konski C T, Yoshioka K and Orttung W H, J. Phys. Chem., 63, (1959) 1558.

6 Kielich S, J. Colloid Interface Sci., 33, (1970) 142.

7 Drobnik A, Piekara A and Kaczmarek F, Second Conference on Coherence and Quantum Optics, Rochester 1966, Abstract p. 156.

8 Kasprowicz-Kielich B and Kielich S, Adv. Mol. Relax. Proc., 7, (1975) 275.

9 Alexiewicz W, Buchert J and Kielich S, Acta Phys. Polonica, A52, (1977) 445.

10 Kielich S, in "Dielectric and Related Molecular Processes", Ed. M. Davies (Wright, London, 1972) Vol.1 p.192.

11 Prost J and Lalanne J R, Phys. Rev. A8,(1973) 2090.

12 Wong G K L and Shen Y R, Phys. Rev. A10, (1974) 1277.

13 Coles H J and Jennings B R, Mol. Phys. 31, (1976) 571; 1225.

14 Jennings B R and Coles H J, Proc. Roy. Soc. London, A348, (1976) 525.

15 Hanson E G, Shen Y R and Wong G K L, Phys. Rev. A14, (1976) 1281.

16 Lalanne J R, Martin B, Pouligny B and Kielich S, Optic Comm. 19, (1976) 440; Mol. Cryst., Liquid Cryst. 42, (1977) 153.

17 Schadt M, J. Chem. Phys., 67 (1977) 210.

18 Filippini J C and Poggi Y, Phys. Letters, 65A, (1978) 30.

# Subject Index

'And in such indexes, although small pricks
To their subsequent volumes, there is seen
The baby figure of the giant mass
Of things to come at large.'
William Shakespeare (1564-1616) Troilus and Cressida

Underlining indicates reference to a complete article